Immunochemistry of the Extracellular Matrix

Volume I
Methods

Editor
Heinz Furthmayr, M.D.
Associate Professor
Department of Pathology
Yale University
New Haven, Connecticut

CRC Press
Taylor & Francis Group
Boca Raton London New York

CRC Press is an imprint of the
Taylor & Francis Group, an **informa** business

First published 1982 by CRC Press
Taylor & Francis Group
6000 Broken Sound Parkway NW, Suite 300
Boca Raton, FL 33487-2742

Reissued 2018 by CRC Press

Library of Congress Cataloging in Publication Data

Main entry under title:

Immunochemistry of the extracellular matrix.

Bibliography: p.
Includes index.
1. Collagen. 2. Immunochemistry.
3. Ground substance (Anatomy) 4. Extra-
cellular space. I. Furthmayr, Heinz.
II. Title: Extracellular matrix.
QP552.C6I45 616.7'9 81-18062
ISBN 0-8493-6196-6 (v. 1) AACR2
ISBN 0-8493-6197-4 (v. 2)

A Library of Congress record exists under LC control number: 81018062

Publisher's Note
The publisher has gone to great lengths to ensure the quality of this reprint but points out that some imperfections in the original copies may be apparent.

Disclaimer
The publisher has made every effort to trace copyright holders and welcomes correspondence from those they have been unable to contact.

ISBN 13: 978-1-315-89431-7 (hbk)
ISBN 13: 978-1-351-07341-7 (ebk)

PREFACE

Fibrous connective tissue is the single most prevalent tissue in the body and in a variety of forms, such as bone, tendon, cartilage, and fascia, it gives structural rigidity. Collagen, as the major protein constituent of connective tissue is found in addition throughout the body forming structures such as basement membranes, and it is part of the extracellular matrix to which cells adhere to form specific organs. Five genetically distinct types of collagen with distinct physicochemical and structural properties have been described thus far. These collagens are synthesized by various mesenchymal and epithelial cells and apparently are distributed in tissues and organs in specific ways. In a variety of pathological conditions expression, distribution and control of production and/or degradation are affected.

The main purpose of the two volumes on *Immunochemistry of the Extracellular Matrix* is to describe state of the art methods, which have been proven to provide antibody reagents of defined specificity to collagens as well as other glycoproteins found in association within connective tissue. The isolation and purification of collagens, procollagens, and related connective tissue proteins are described for several tissues, tissue culture cells and species. Immunization with these collagens in laboratory animals yields antibodies with different characteristic specificities: to the procollagen extension fragments, the non-helical segments of the α-chains, to helical and denatured determinants. Methods are described to obtain, isolate, and characterize these antibody specificities by selection of the appropriate animal for immunization, immunoadsorbtion procedures, and sensitive serological assays. Experimental data are given and previous results are discussed in terms of producing antibody reagents of desired specificity. Also included are immunohistological methods at the level of the light microscope and the electron microscope aimed at elucidating the localization of these collagens in vivo in a variety of tissues.

Connective tissue biochemistry, biology, and pathology are rapidly expanding fields today and the information assembled in these volumes should encourage and facilitate the use of immunochemical tools for studies in these important areas of research. Finally, the authors tried to define several challenging areas of inquiry in connective tissue biology in which specific antibodies can contribute to our understanding of cellular processes.

The task of assembling these two volumes was greatly facilitated by the enthusiasm of the contributing authors, all experts in the field, who regarded this project as a worthwhile and timely effort and I am grateful for their support, effort, and valuable contributions.

In writing my own contributions I came to realize more than ever, how much I owe to my teachers Rupert Timpl and Klaus Kühn, and to the many friends I have found amongst the group of connective tissue biochemists. Amongst the individuals at Yale I particularly wish to thank Joseph Roll and Joseph Madri, who not only helped to bring this project to fruition, but who were also responsible for initiating connective tissue research in the Department of Pathology together with me.

I welcome comments and criticisms from readers.

November 1981 | Heinz Furthmayr

THE EDITOR

Heinz Furthmayr, M.D. University of Vienna, Austria, is Associate Professor of Pathology at Yale University. He received his training in immunochemistry in the Department of Immunology at the University of Vienna, in protein chemistry at the Max Planck Institute for Biochemistry in Munich, Germany, and in membrane biochemistry at the Department of Pathology at Yale University. He has been a Visiting Fellow in the Department of Experimental Biology and Genetics at the CSSR Academy of Sciences in Prague and at the Hôpital St. Louis in Paris. In 1978 he received a Research Career Award from the American Cancer Society. His publications include over 70 scientific articles, several reviews and chapters on connective tissue proteins and plasma membrane glycoproteins. He is a member of the American Society of Immunologists and the American Society of Pathologists and serves on Advisory Committees of the U.S. Public Health Service.

CONTRIBUTORS

VOLUME I

William T. Butler, Ph.D.
Senior Scientist
Institute of Dental Research
University of Alabama in Birmingham
Birmingham, Alabama

Ronald G. Crystal
Chief, Pulmonary Branch
National, Heart, Lung, and Blood Institute
National Institutes of Health
Bethesda, Maryland

Samantha F. Curran, Ph.D.
Research Associate
Department of Biochemistry
CMDNJ Rutgers Medical School
Piscataway, New Jersey

Saryu N. Dixit, Ph.D.
Associate Professor of Biochemistry
College of Medicine
University of Tennessee Center for Health Services
Memphis, Tennessee

Jean-Michel Foidart, M.D.
Research Associate
Department of Obstetrics
University of Liege
Liege, Belgium

Heinz Furthmayr, M.D.
Associate Professor
Department of Pathology
Yale University School of Medicine
New Haven, Connecticut

R. W. Glanville, Ph.D.
Max-Planck-Institute for Biochemistry
Department of Connective Tissue Research
Munich, West Germany

Mary J. C. Hendrix
Postdoctoral Fellow
Department of Anatomy
Harvard Medical School
Boston, Massachusetts

A. Tyl Hewitt, Ph.D.
NIH Expert
Laboratory of Developmental Biology and Anomalies
National Institute of Dental Research
National Institutes of Health
Bethesda, Maryland

Klaus Kühn, Ph.D.
Fogerty Scholar
National Institute of Dental Research
National Institutes of Health
Bethesda, Maryland

Thomas F. Linsenmayer, Ph.D.
Associate Professor
Department of Anatomy
Harvard Medical School
Boston, Massachusetts

Joseph A. Madri, Ph.D., M.D.
Assistant Professor
Department of Pathology
Yale University School of Medicine
New Haven, Connecticut

George R. Martin, Ph.D.
Chief, Laboratory of Developmental Biology and Anomalies
National Institute of Dental Research
National Institutes of Health
Bethesda, Maryland

Bjorn R. Olsen, M.D., Ph.D.
Professor of Biochemistry
College of Medicine and Dentistry
Rutgers Medical School
Piscataway, New Jersey

Donna M. Pesciotta
Research Associate
Department of Anatomy
Albert Einstein College of Medicine
Bronx, New York

Stephen I. Rennard
Senior Staff Fellow
Pulmonary Branch
National Heart, Lung, and Blood Institute
National Institutes of Health
Bethesda, Maryland

Charles A. Reese, Ph.D.
Medical Student
University of South Alabama College of Medicine
Mobile, Alabama

Leila Risteli, Ph.D.
Research Assistant
Department of Medical Biochemistry
University of Oulu
Oulu, Finland

Rupert Timpl, Ph.D.
Head of Research Group
Max-Planck-Institute for Biochemistry
Munich, West Germany

Robert L. Trelstad
Professor and Chairman
Department of Pathology
Rutgers Medical School
Piscataway, New Jersey

Hugh H. Varner, Ph.D.
Postdoctoral Fellow
Laboratory of Developmental Biology and Anomalies
National Institute of Dental Research
National Institutes of Health
Bethesda, Maryland

Kenneth M. Yamada
Senior Surgeon
Laboratory of Molecular Biology
National Cancer Institute
National Institutes of Health
Bethesda, Maryland

TABLE OF CONTENTS

Volume I

Chapter 1

CHEMICAL PROPERTIES OF COLLAGEN

Klaus Kühn

TABLE OF CONTENTS

I. INTRODUCTION

Connective tissue appears in many different forms, its properties always being adapted to its physiological function. Tendon, for example, which has to transmit tensile forces from muscle to skeleton, are built of parallel fiber bundles. Articular cartilage, however, which forms the smooth surface of the joint has to resist enormous pressure and is built of a three-dimensional network of thin fibrils, embedded in water-binding proteoglycans. These two examples show that the appropriate biochemical properties of a connective tissue are provided by a specific macromolecular organization of the extracellular matrix. The organism comprises several genetically distinct collagen types, and each connective tissue contains a specific spectrum of these collagens (for a recent review see Reference [1]). Thus, tendon consists mainly of Type I collagen and cartilage of Type II collagen, whereas skin or aortic walls contain a mixture of Types I and III collagen, but in different ratios. Distinctly different in structure and function from interstitial connective tissue are the basement membranes which also contain collagen. They separate the interstitial space from the parenchymal tissues, form a support for endothelial and epithelial cells, and serve as a filter barrier in tissues, such as the glomeruli of the kidney and in the capillary walls.[2]

It has been suggested that the organism regulates, at least in part, the formation of the macromolecular structure of the extracellular matrix and its physiological function, using a specific tissue distribution of the different collagen types. Therefore every effort has been made to investigate the common and different structural features of the collagen types which may be important for their specific role in connective tissue. In this respect, those characteristic amino acid sequences of the polypeptide chains, responsible for the formation of higher structures, such as triple-helical molecules and fibrils, or those sequence features which stabilize these structures by intermolecular cross-links, are of particular interest.[3] In addition, the formation of the higher structures appears to be influenced by interaction of the different collagen types with other connective tissue constituents, e.g., the proteoglycans.

The biological activities of mesenchymal cells such as gene expression, proliferation, maintenance of cell phenotype, and differentiation seem to be regulated by interactions with extracellular matrix components. There is evidence for a direct interaction between collagen and cells,[4,5] but also for an indirect contact that is mediated by the glycoprotein fibronectin.[6]

In this chapter, the chemical properties of the different collagen types which determine the formation of the higher structures important for the physiological function of connective tissue will be described. No attempts were made to cover the entire field of chemistry of collagen and its application to biological problems. The reader is referred to recent reviews that deal with collagen chemistry,[1,3,6–8] immunochemistry,[9] and biosynthesis of collagen.[10,11]

II. COLLAGEN TYPES

The main collagenous constituents of interstitial connective tissue are Types I, II, and III collagen[8] (see Table 1). Type I collagen which predominates in bone, tendon, skin, and vessel walls is synthesized by osteoblast, fibroblast, and smooth muscle cells. A small amount of a collagen molecule containing three $\alpha 1(I)$-chains has been detected in a variety of tissues.[12–16] Whereas fibroblasts from tendon and cornea and osteoblasts synthesize only Type I collagen, fibroblasts from skin and smooth muscle cells produce both Type I and Type III collagens. Type II collagen,

Table 1
GENETICALLY DISTINCT VERTEBRATE COLLAGENS

Type	Molecular chain[a] composition	Native polymer	Tissue distribution
I	$[\alpha1\ (I)]_2\alpha2\ (I)$	Fibril	Skin, bone, tendon, ligament, dentin, blood vessels, widespread in interstitial connective tissue
II	$[\alpha1\ (II)]_3$	Fibril	Cartilage, nucleus pulposus, vitrous body
III	$[\alpha1\ (III)]_3$	Fibril	Same as Type I except for bone and tendon; prominent in blood vessels, "reticulin fibers"?
IV	α1 (IV)[b] α2 (IV) α3 (IV)	Unknown	Basement membranes
V	$[\alpha1\ (V)]_2\alpha2\ (V)$[c]	Unknown	Widespread, muscles, basement membrane, exoskeleton of mesenchymal cells?

[a] Molecular weight of α-chains 95,000 daltons, except for α1 (IV) (mol wt. 160,000 to 190,000 daltons).
[b] Chain composition not known; there are indirect indications for three identical α-chains in each triple-helical molecule.
[c] Type V has been termed AB collagen; chain B and A are designated here as α1 (V) and α2 (V), respectively (according to Bornstein and Sage);[1] chain composition $[\alpha1\ (V)]_2\alpha2\ (V)$ are not generally accepted; $[\alpha1\ (V)]_3$ and $[\alpha2\ (V)]_3$ are also discussed.

the collagenous constituent of cartilage, is produced by chondrocytes. All three collagens have been isolated as native molecules, whose main compositions are known. The amino acid sequence of the α1(I)-, α2(I)-,[1,3] and α1(III)-chains[17–22] have been completed and that of the α1(II)-chain[23–25] in part elucidated.

Type V collagen has been localized in the form of denatured α1(V)- and α2(V)-chains in several tissues, such as fetal membranes,[26,27] skin,[27] synovial membrane,[28] and striated[29] and smooth muscles.[30] It is mainly synthesized by smooth muscle cells[30] and myoblasts,[29] but has also been found in small amounts in the biosynthetic products of several other cells.[31]

Type IV collagen, synthesized by endothelial and epithelial cells, is not as well characterized as interstitial collagen. It has been isolated predominantly as pepsin-derived molecular fragments. There is evidence of three different chains, α1(IV), α2(IV), and α3(IV), which differ however in their structure from the α-chains of the interstitial collagens.[32]

III. POSTTRANSLATIONAL MODIFICATIONS OF COLLAGEN

Collagen is synthesized in the cell as single polypeptide chains in a precursor form carrying at the N terminal and at the C terminal additional procollagen peptides. Its physiological, functional form, however, is the insoluble and mechanically stable fibril in the extracellular space. Along the biosynthetic pathway from the newly synthesized product to the fibril, the proα-chains are subjected to a series of post-translational modifications, including hydroxylation of certain proline and lysine residues by specific hydroxylases and attachment of carbohydrate moieties to some of the hydroxylysine residues by galactosyl and glucosyl transferases (for a review see References [10] and [11]). Formation of the triple helix which occurs after these alterations must be preceded by correct alignment of the three constituent peptide chains which is assumed to be facilitated by the C terminal procollagen peptides. The completed triple-helical procollagen molecule is secreted into the extracellular

Table 2
POSTTRANSLATIONAL PATHWAY OF COLLAGEN α-CHAINS

Modification	Enzyme	Location
Hydroxylation of Pro and Lys	Peptidyl proline hydroxylase Peptidyl lysine hydroxylase	Cisternae of endoplasmatic reticulin
Glycosylation of Hyl	Galactosyl transferase Glucosyl transferase	Cisternae of endoplasmatic reticulin
Chain selection and assembly	?	Cisternae of endoplasmatic reticulin
Triple-helix formation	Self-assembly	Cisternae of endoplasmatic reticulin
Translocation	—	Golgy complex
Conversion from procollagen to collagen	N and C terminal procollagen peptidase	Extracellular space
Fibril formation	Self-assembly	Extracellular space
Cross-link formation	Lysyloxydase	Extracellular space

space. The procollagen peptides are then enzymatically cleaved off, and the molecule thus acquires the capability to aggregate into fibrils. The mechanical stability of the fibrils is provided by intermolecular cross-links. Certain lysine or hydroxylysine residues located in the nonhelical regions at both ends of the molecule are oxidatively desaminated by a lysine oxidase.[33] The resulting carbonyl compounds are subsequently capable of interacting with distinct hydroxylysine residues present in the triple-helical cross-link region of adjacent molecules within the fibrillar array, thus establishing covalent intermolecular bonds of the Schiff's base type.[33]

The posttranslational pathway of Type IV collagen has not been investigated to such an extent as that of the interstitial collagens. Since it also contains hydroxyproline and hydroxylysine and galactosyl and glucosyl residues attached to hydroxylysine, these modification steps must be similar to those described for interstitial collagen. It is, however, not clear whether the α(IV)-chains are synthesized in a precursor form which has to be converted. It is also not known how the molecules aggregate in the extracellular space into macromolecular structures. The intercellular matrix appears to be stabilized by lysine-derived intermolecular cross-links, similar to interstitial connective tissue and in addition by disulfide bridges. It is not known whether these intermolecular disulfide bonds are formed intra- or extracellularly.

IV. INTERSTITIAL COLLAGENS

A. Amino Acid Compositions

The amino acid composition of the α1(I), α2(II), α1(II), and α1(III) is characterized by a content of about 33% glycine and 20% of the amino acids proline and hydroxyproline. Each chain has its own specific features. The α1(III)-chain, for example, shows, in comparison to the other α-chains, a higher content of hydroxyproline and a smaller amount of phenylalanine and alanine. A peculiarity is the presence of two cysteine residues per chain. The cysteine residues form interchain disulfide bridges so that denaturation of the triple-helical Type III collagen molecules results in trimeric γ components which can be converted by reduction to the usual α-chains. Type II collagen is characterized by its high content of hydroxylysine and by 4% carbohydrates which are attached to the hydroxyline residues in the form of a glucosyl galactosyl disaccharide or as galactosyl monosaccharide. The α2-chain contains the highest amount of hydrophobic amino acid residues.

B. Amino Acid Sequence of the α-Chains

The α-chains of the interstitial collagens Types I, II, and III are homologous. Each chain contains two structurally and functionally different sequence regions, the central triple-helical part of 1014 to 1023 residues and the nontriple-helical regions at both ends of the chain, whose length can vary in the range of 9 to 50 residues.[3] The nonhelical end regions are important for the formation of intermolecular bonds, whereas the triple-helical region controls are responsible on the one hand for the formation and stabilization of the triple helix, and on the other hand, are responsible for the self-assembly of the molecules into fibrillar structures.

1. Structures Important for Triple-Helix Formation

In the triple-helical structure, every third position along the peptide chains come to lie in the center of the helix, where there is only space for a glycine residue.[40] It is therefore essential for the formation of a continuous triple helix that in the central part of the α-chain every third position is occupied by glycine, and indeed, following this rule, the central part of the α1(I)- and the α2(I)-chain consists of 338 tripeptide units and that of α1(III) consists of 341 tripeptide units of the general formula Gly-X-Y.[3] Important for the stability of the triple helix is the presence of the amino acids proline and hydroxyproline. There are three types of tripeptides containing proline and hydroxyproline which contribute to the stability of the helix in the following order: Gly-Pro-Hyp > Gly-Pro-Y > Gly-X-Hyp.[41] Exceptionally important for the stability of the triple-helical structure is the hydroxyl group of the hydroxyproline residues in position Y, since underhydroxylated collagen has a more than 15°C lower denaturation temperature than normal hydroxylated collagen.[42,43]

2. Structures Important for Fibril Formation

Responsible for the ordered aggregation of the collagen molecules to fibrils are the polar-charged amino acids Asp, Glu, Arg, and Lys and the hydrophobic residues Val, Leu, Ile, Met, and Phe. These functional residues are not uniformly distributed along the peptide chains, but are arranged in polar clusters and in hydrophobic areas. The polar clusters have been first observed in the electron microscope with the help of segment-long-spacing crystallites (see Figure 1). The characteristic cross-striation pattern, which arises after staining with phosphotungstic acid and uranyl acetate, reflects the sequence of the polar-charged residues along the α-chains.[44] The band patterns of the crystallites formed by Types I, II, III, and V collagen indicate a homology in the sequence of the charged residues between the different collagen types, although some differences in intensity and position of individual striations can be observed. Comparison of the amino acid sequence of the α1(I), α2(I), α1(III) and in part of the α1(II)-chain confirmed this first electron microscopical estimation of the collagen sequence and indicated in addition a general sequence homology of the α-chains. Between α1(I) and α1(III), for example, 50% of the residues remained conserved during evolution, and the polar residues were found to be extremely stable.[48]

Closer analysis of the amino acid sequence of the α-chains with statistical methods revealed periodic arrangements of amino acids which lead to several homologous repeating units along the polypeptide chain.[48–50] The most prominent repeating unit in all three α-chains is $D = 234$ residues and it is formed by the polar-charged and the hydrophobic amino acids. This repeat determines the axial aggregation of the collagen molecules into fibrils.[49–54] A three-dimensional analysis of the sequence of Type I and recently of Type III collagen indicated that the highest polar and hydrophobic contact between two neighboring molecules occurs when they are shifted

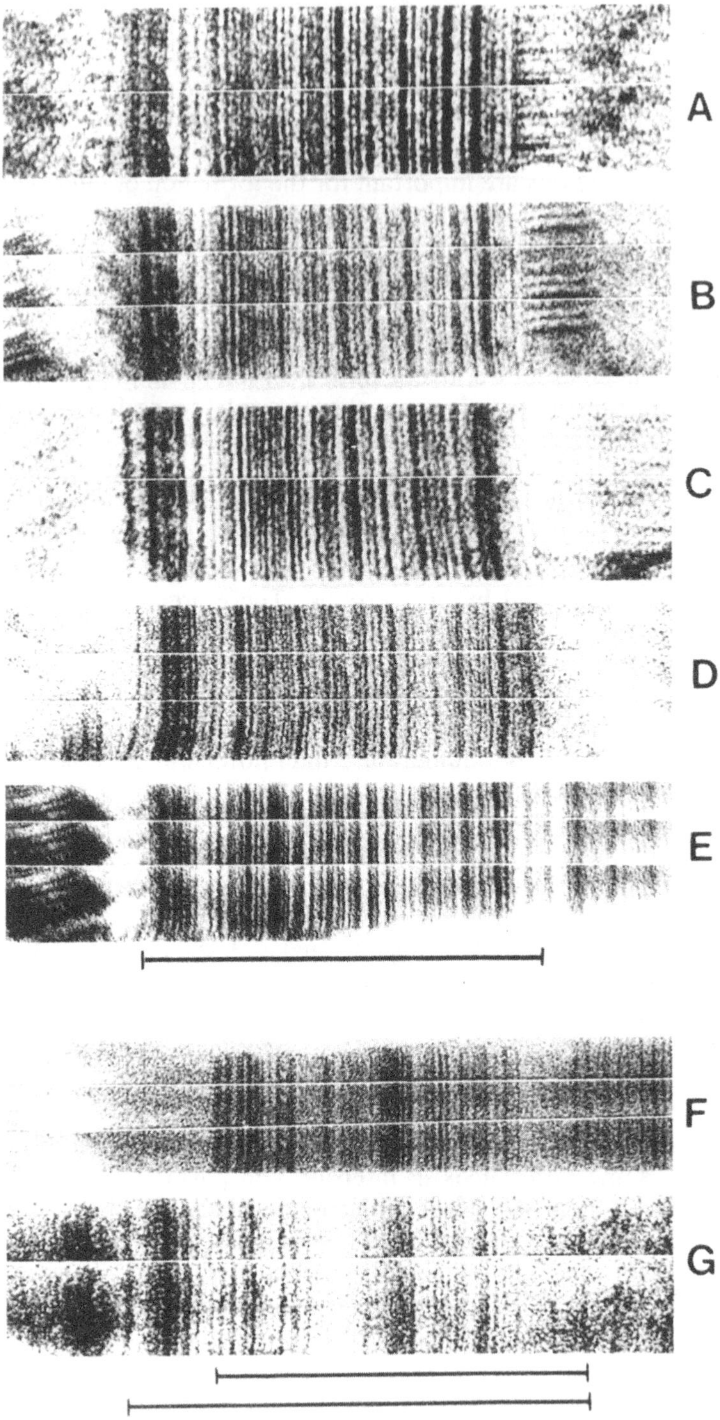

FIGURE 1. Segment-long-spacing crystallites precipitated from collagen solutions in 1 *M* acetic acid by ATP. The molecules are arranged parallel to each other with their ends in register; stained with phosphotungstic acid and uranyl acetate. (A) $[\alpha 1(I)]_3$ formed in vitro;[45] (B) $[\alpha 2(I)]_3$ formed in vitro;[45] (C) $[\alpha 1(II)]_3$ Type II collagen;[46] (D) $[\alpha 1(III)]_3$ Type III collagen;[47] (E) $[\alpha 1(V)]_2\alpha 2(V)$ Type V collagen;[39] (F) pepsin-derived triple-helical fragment of Type IV collagen isolated from human placenta;[38] (G) trypsin derived triple-helical fragment of Type IV collagen isolated from mouse EHS sarcoma.[142] Length: A to E = 290 nm; F = 260 nm; G = 330 nm.

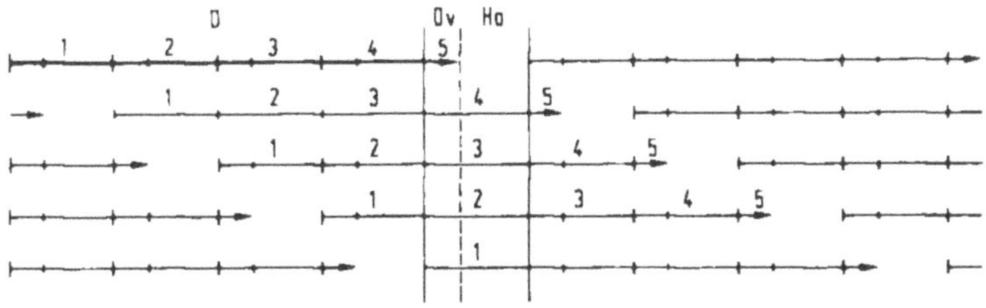

FIGURE 2. Schematic representation of the axial arrangement of the interstitial collagen molecules in the 670 Å cross-striated fibril. The molecules are staggered against each other by the distance D of 234 amino acid residues or 670 Å.[3]

against each other by a distance D = 234 residues. This distance, calculated from the sequence, is consistent with the periodicity of 600 to 700 Å along the collagen fibrils found in X-ray and electron microscopic investigations.[55] The axial array of the molecules in the fibrils derived from these investigations is schematically depicted in Figure 2. As a consequence of this common internal sequence periodicity, the molecules of all three interstitial collagen types aggregate in the same axial D staggered array to the typical cross-striated native fibrils.

Beside the axial arrangement of the molecules, their lateral array is also important, as this controls the diameter of the collagen fibers. Unfortunately, one has not been able to find such a clear relationship between amino acid sequence and lateral array as was possible for sequence and axial stagger. Also X-ray and electron microscopical investigations of fibrils did not lead to a clear idea about the lateral arrangement of the molecules. Two concepts are presently discussed. According to the microfibril concept, small subunits of four or five strands of molecules are first formed and subsequently aggregate parallelly into fibrils.[52–55] This model is favored by electron microscopical observations of subfibrillar filaments with a diameter of 40 to 50 Å.[55] However, a hexagonal packing of the molecules[56] seems to be in better agreement with the x-ray reflection patterns. Although the axial arrangement of the molecules of the collagen types appears to be identical, it is not known whether there exist differences in the lateral array from type to type which may be controlled by type-specific sequence features.

The macromolecular organization of the individual collagens has been investigated in tissues which contain predominantly one collagen, such as tendon and cornea which contain Type I or hyaline cartilage which contains Type II collagen. These studies revealed that Type II collagen in general forms fibrils with a smaller diameter than Type I collagen and that the often-observed tendency of Type I fibrils to be arranged in parallel fiber bundles cannot be found in Type II collagen.[57,58] The macromolecular structure of Type III collagen has not yet been directly investigated, since there is no tissue which contains predominantly Type III collagen. Indirect immunofluorescence using specific antibodies against Type III collagen led to the impression that mainly the fine network of the argyophilic reticulin is stained.[59] At present it is unknown whether Type I and Type III collagens which are synthesized by the same cell[60] and which are secreted into the same extracellular space form separate fibrils or aggregate in a common macromolecular structure. The mechanisms which regulate the formation of the suprafibrillar structure of collagen remain unclear. It is not known, for example, how the fibril diameter is regulated and how the arrangement of the fibrils is directed into parallel fiber bundles or into a three-dimensional fiber network. It has been suggested that these processes are strongly

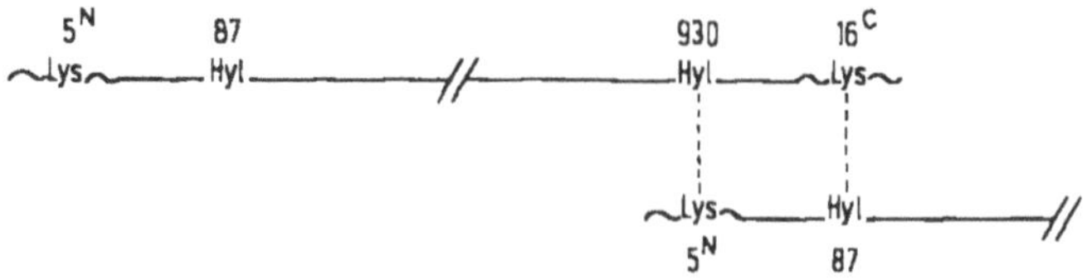

FIGURE 3. Schema of the cross-link sites of the α1(I) chain.[3] Lys 5^N and 17^c are the active sites of the N and C terminal nonhelical cross-link regions which have to be oxidized by lysyloxidase to aldehyde derivatives before cross-link formation can take place. Hyl 87 and 930 are the active sites of the triple-helical cross-link regions which serve as reaction partners for the lysine-derived aldehydes. Cross-links between two 4D staggered neighboring molecules from Lys 5^N to Hyl 980 or from Lys 16^c to Hyl 87 have been established by isolating the respective CNBr or trypsin-derived cross-linked peptides.

```
                         5N                  10N                  15N
α1 (I)   pGlu-Leu-Ser-Tyr-Gly-Tyr-Asp-Glu-Lys-Ser-Thr-Gly-Ile-Ser-Val-Pro
α2 (I)                      pGlu-Phe-Asp-Ala-Hyl-Gly-Gly-Gly-Pro
α1 (III)     pGlu-Tyr-Glu-Ala-Tyr-Asp-Val-Lys-Ser-Gly-Val-Ala-Gly-Gly-Gly-Ile-Ala
```

FIGURE 4. Amino acid sequence of the nonhelical cross-link region of the α1(I), α2(I), and α1(III) chains at the N terminal end of calf skin collagens. Note that in Type I collagen, in the N terminal region, the distance between the oxidizable lysine residue and the triple helix is variable, 7 residues in α1(I), 4 residues in α2(I). Literature: α1(I);[65] α2(I);[66] α1(III).[67]

influenced by a specific interaction with matrix proteoglycans or glycoproteins such as fibronectin.

3. Sequence Regions Important for Intermolecular Cross-links

Intermolecular cross-links are essential for the biomechanical stability of collagen fibrils. Genetic diseases where the formation of the cross-links is disturbed or blocked lead to serious disfunction of connective tissue.[61,62] The same is true if cross-link formation is inhibited by β-aminopropionitrile, a specific inhibitor of the enzyme lysyloxidase.[63] The α1(I)-chain provides two different kinds of sequence regions involved in intermolecular bonds, the nonhelical and the triple-helical cross-link sites (see Figure 3). The nonhelical cross-link regions are located at the N and C terminal ends of the α1(I)-chain.[3] They do not contain a glycine in every third position and therefore form flexible extensions which bear as an active site one lysine or hydroxylysine residue (see Figures 4 and 5). The cross-link formation is initiated by oxidation of these regions to lysyl aldehydes by the enzyme lysyl oxidase. The two triple-helical cross-link regions (position 82 to 93 and position 925 to 936) are homologous (see Figure 6). Each of them contains, as reaction partner for the lysyl-derived aldehydes of the nonhelical region, in position 87 and 930, a hydroxylysine residue. The chemistry of the cross-link compounds formed, such as dehydrohydroxylysinonorleucine, is summarized.[64]

Several cross-linked peptides have been isolated after cleavage of insoluble collagen with CNBr or trypsin. Their characterization established cross-links between the N terminal nonhelical region and hydroxylysine residue in position 930[73] or between the C terminal end and hydroxylysine in position 87.[74,75] These intermolecular bonds connect neighboring molecules which are staggered against each other by a distance of 4D (see Figure 3). According to the amino acid sequence, the formation of cross-links between 1D staggered molecules appears to be impossible.[3]

In contrast to the triple-helical body of the collagen molecules which are fairly resistant against proteolytic attack, the nontriple helical end regions can be cleaved

```
                        5c                        10c                     15c
α1 (I)    Ser-Gly-Gly-Phe-Asp-Phe-Ser-Phe-Leu-Pro-GlN-Pro-Pro-GlN-Glu-
α2 (I)    Gly-Gly-Gly-Tyr-Glu-Val-Gly-Phe-Asp-Ala-Glu  -    -    -    -

                        20c                      25c       ↓
α1 (I)    Lys-Ala-His-Asp-Gly-Gly-Arg-Tyr-Tyr-Arg-Ala-Asp-Asp-Ala-AsN
α2 (I)     -    -    -    -    -    -    -   Tyr-Tyr-Arg-Ala-Asp-GlN-Pro-Ser
```

FIGURE 5. Amino acid sequence of the nonhelical cross-link region at the C terminal end of the α1(I) and α2(I) chain of chick collagen.[78] Note that the α2(I) chain does not contain a lysine residue in the C terminal nonhelical region. In comparison to α1(I), a region of 11 residues containing the lysine residue is deleted. Arrow indicates the putative cleavage site of the C terminal procollagen peptidase. (From Forrest, F. and Boedker, H., *Biochemistry*, 20, 996, 1981.)

```
              82                  87                        93
α1 (I)     -Gly-Leu-Hyp-Gly-Met-Hyl-Gly-His-Arg-Gly-Phe-Ser-
α2 (I)     -Gly-Leu-Hyp-Gly-Phe-Hyl-Gly-Ile-Arg-Gly-His-AsN-
α1 (II)    -Gly-Leu-Hyp-Gly-Val-Hyl-Gly-His-Arg-Gly-Thr-Hyp-
α1 (III)   -Gly-Phe-Hyp-Gly-Met-Hyl-Gly-His-Arg-Gly-Phe-Asp-

             925                 930                       936
α1 (I)     -Gly-Asp-Arg-Gly-Ile-Hyl-Gly-His-Arg-Gly-Phe-Ser-
α2 (I)     -Gly-His-Glu-Gly-Leu-Pro-Gly-Leu-Hyl-Gly-His-AsN-
α1 (III)   -Gly-Ala-Met-Gly-Ile-Hyl-Gly-His-Arg-Gly-Phe-Ser-
```

FIGURE 6. Amino acid sequence in the vicinity of the cross-link sites within the triple helix of calf skin collagen. Hydroxylysine 87 and 930 react with the lysine-derived aldehydes in the C terminal and N terminal nontriple-helical regions, respectively. Note the high degree of homology between the two regions within one chain and between corresponding regions in different chains. In the α2(I) chain, Hyl in position 930 is substituted by a proline residue. Hyl which may react with the lysine derived aldehyde is located in position 933. Literature: α1(I);[68-70] α2(I);[71,157] α1(II);[72] α1(III).[17,21,22]

off by proteases.[76] This is particularly true for the C terminal region which is so labile that it is difficult to isolate soluble collagen molecules from tissue with intact C terminal ends.[77] The α2-chain, e.g., for which the sequence of the entire triple-helical region is known, has not been isolated with a complete C terminal nonhelical end. However, the sequence of the C terminal end of α2 has been recently elucidated at the level of the complementary DNA.[78] From these results, we learned that the nonhelical region at the C terminus of α2 does not contain a lysine or hydroxylysine residue (see Figure 5). Thus, the α2-chain can participate in a cross-link reaction only with its N terminal end. The α1(III)-chain contains, similar to α1(I) and α2(I), an oxidizable lysine residue at the N terminal nontriple-helical region[17] and two triple-helical cross-link regions with hydroxylysine in positions 87 and 930.[21,22] The nonhelical part at the C terminus of α1(III) which may also contain a lysine residue has not yet been isolated. Recently cross-linked peptides of Type III collagen have been isolated which suggest an intermolecular bond between the N terminal nonhelical cross-link site with hydroxylysine in position 930.[79] The sequence of the α1(II)-chain is only partly known. From several experimental results,[80] however, and from the presence of the typical triple-helical cross-link region around the hydroxylysine residue in position 87, it is possible that Type II collagen forms similar cross-links to those found in Types I and III collagens.

Due to the protease sensitivity of the nonhelical regions, it is possible to dissolve insoluble cross-linked collagen by limited proteolysis without destroying the triple-helical part of the collagen molecules. The general observation, however, that collagen, particularly from tissue of older individuals, can be dissolved only in part by proteolytic attack, may be explained by the presence of unknown cross-links which cannot be cleaved by proteolytic attack.

4. Sequence Region Important for Collagenase Attack

Insoluble cross-linked collagen is largely resistant to proteolytic enzymes. This is due to the triple-helical structure which protects the peptide bonds against proteolytic attack. The triple-helical structure of collagen molecules incorporated into a cross-linked fibril is additionally stabilized, so that its denaturation temperature is increased from 37°C in a dissolved state to more than 60°C. The organism has enzymes which specifically cleave peptide bonds involved in triple helical structures.[81,81a] These vertebrate collagenases first discovered by Gross and co-workers split the intact molecules into two pieces, with lengths of 3/4 and 1/4 of a molecule.[82] Many collagenases from different sources have been isolated and characterized. Most of them cleave the collagen molecule of Types I, II, and III at one specific cleavage site which, in the α1(I)-chain is the peptide bond between Gly and Ile in position 775 and 776.[83,84] In α2(I), α1(II), and α1(III), the residue Ile in position 776 is substituted by a Leu residue (see Figure 7). Treatment of denatured α1(I) or the respective cyanogen bromide peptide α1(I)CB6[83] with highly purified collagenase resulted in the cleavage of the same Gly-Ile bond. It has been observed, however, that the triple-helical molecules are considerably faster cleaved than the denatured chains.[85] This suggests that the triple-helical structure enhances the binding of the enzyme to the substrate.

The rate of collagenase cleavage of Types I, II, and III collagen was found to be different. Thus, rheumatoid synovial collagenase splits Type III and I five times faster than Type II.[84,85a] Comparison of the amino acid sequences of the three collagen types around the cleavage site does not provide an explanation for this fact. It has been observed that the triple helix of type III molecules,[86] but also of Type I collagen,[87] can also be cleaved in the same area by trypsin. For example, trypsin splits the Arg-Gly bond in triple helical Type III collagen which is located five positions C terminal to the collagenase cleavage site,[86] (see Figure 7). Because of the low content of proline and hydroxyproline in this region, one has suggested a weaker triple-helical structure which can be more easily disturbed and so digested. In this connection, it is striking that highly cross-linked collagen, where the triple helix appears extraordinarily stable, cannot,[88] or can be only slowly[89] attacked by vertebrate collagenase. Breakdown of highly cross-linked collagen may therefore require the cooperation of several enzymes, such as proteases which depolymerize insoluble collagen by splitting the cross-link-bearing nonhelical end regions, vertebrate collagenase which cleaves the molecules in two pieces, and finally proteolytic enzymes which digest the polypeptide chains after denaturation of the triple-helical molecule fragment.[90]

5. Sequence Regions Important for Cell Interaction

Connective tissue cells are surrounded by an extracellular matrix. There are many experimental data suggesting that the biological activity of the cells, such as differentiation during embryonic development, proliferation, and biosynthetic activity, or stability of their phenotype, are controlled by specific interaction with extracellular matrix components. The interaction of collagen with cells has been measured using several methods, for instance, by cell attachments to collagen substrata,[91] binding of dissolved radioactively labeled collagen on cells,[4] or by testing the chemotactic attraction of cells to collagen.[92] According to these experiments, collagen can, on

```
                    765              770              775↓             780              785
α1 (I)       Gly-Pro-Ala-Gly-Ala-Hyp-Gly-Thr-Pro-Gly-Pro-Gln-Gly-Ile-Ala-Gly-Gln-Arg-Gly-Val-Val-Gly-Leu-Hyp-Gly-Gln-Arg
α2 (I)       Gly-Thr-Ala-Gly-Pro-Hyp-Gly-Thr-Thr-Gly-Pro-Gln-Gly-Leu-Leu-Gly-Ala-Hyp-Gly-Phe-Leu-Gly-Leu-Hyp-Gly-Ser-Arg
α1 (II)                                                  Gly-Ile-Ala-Gly-Gln-Arg-Gly-Phe-Leu-Gly-Leu-Hyp-Gly-Ser-Arg
α1 (III)     Gly-Pro-Hyp-Gly-Ala-Hyp-Gly-Pro-Leu-Gly-Ile-Ala-Gly-Leu-Thr-Gly-Ala-Arg-Gly-Leu-Ala-Gly-Pro-Hyp-Gly-Met-Hyp
```

FIGURE 7. Amino acid sequence surrounding the cleavage site (marked by an arrow) for vertebrate collagenase in three different chains of calf skin collagen. Note the low content of imino acids C terminal to the cleavage site Gly-Ile. This sequence region is also important for the interaction of collagen with fibronectin. Literature: α1(I);[3] α2(I);[3] α1(II);[84] α1(III).[20]

the one hand, interact directly with the cell surface, but on the other hand uses the glycoprotein fibronectin as mediator for interaction with cells.

The specific binding of fibronectin to collagen, which can be used to isolate fibronectin from tissue extract or serum by chromatography on collagen affinity columns,[93,94] has been extensively investigated. Using cyanogen bromide peptides from the α-chains of the different collagen types, it was found that in all collagens the same region, between residue 751 and 791, is responsible for collagen binding.[6,95,96] Surprisingly, this region overlaps with the cleavage site of vertebrate collagenase, and it has been shown that the integrity of the collagen-sensitive bond between residues 775 and 776 and of a chymotryptic sensitive bond between positions 779 and 780 are required for fibronectin binding[6] (see Figure 7). Fibronectin itself contains two different binding sites, one which interacts with collagen and another which binds to the cell surface.[97,98] The latter can be inhibited by gangliosides. According to cell attachment experiments, the first step is formation of a collagen fibronectin complex which then interacts via fibronectin with fibroblasts.[6] Important for the discussion on the physiological role of fibronectin is the observation that it has a higher affinity to denatured than to native collagen.[93,99] Heparin, however, appears to stabilize the binding of native collagen to fibronectin.[100]

Recent studies suggest that hepatocytes[101] and hamster kidney cells[102] do not require fibronectin to bind to collagen. The latter requires fibronectin only for attachment to denatured collagen, whereas native collagen seems to interact directly with the cells. Hepatocytes attach equally well to all collagen types, whereby the native collagens were more efficient in promoting cell attachment than denatured collagen. Since all cyanogen bromide peptides of Type I collagen showed similar binding activities, it has been suggested that cell binding sites in collagen have a fairly simple structure and occur in multiple copies along the α-chains. This is in agreement with observations that collagen-like synthetic peptides also show a low but significant cell attachment effect.[102] The results of these cell attachment experiments suggest that collagen-cell interaction can occur in two different manners, directly and via fibronectin. Which of the two is preferentially used seems to depend upon the experimental conditions applied,[103] and the physiologically more important mechanism has still to be identified.

Beside cell attachment, collagen has also the ability to act as a chemotactic agent on fibroblasts.[92] It was found that all three interstitial collagens and their constituent α-chains can serve as chemoattractants. This is also true for the bacterial collagenase-derived peptide mixture and for collagen-like synthetic peptides. These results suggest that collagen and its degradation products provide chemotactic stimuli at sites of tissue injury and inflammation. The fibroblast membrane receptors involved in the chemotactic response to collagen appear to respond to typical collagen amino acid sequences common to Types I, II, and III collagens, with hydroxyproline as an important constituent.

6. *Sequence Regions Important for Collagen-Platelet Interaction*

When subendothelial components of a vascular wall, after injury to the endothelium, are exposed to the blood stream, platelets rapidly adhere, undergo a release reaction, and subsequently aggregate to form a primary hemostatic plug. It has been shown that collagen is the most thrombogenic subendothelial component.[104] The interaction between platelets and the different collagen types has been investigated in artificial systems. Whereas in vivo native grown cross-linked fibrils are exposed to platelets, in the in vitro experiments dissolved and purified collagen types were reacted with platelets in the form of reaggregated fibrils, triple-helical molecules,

	478 486
α1 (III)	Gly-Lys-Hyp-Gly-Glu-Hyp-Gly-Pro-Lys
α1 (I)	Gly-Lys-Hyp-Gly-Glu-Gln-Gly-Val-Hyp

FIGURE 8. Smallest sequence stretch of α1(III) necessary for adhesion of platelets to Type III calf collagen. Fragments of α1(I), including the corresponding region are not active in platelet adhesion. (From Fauvel, F., Legrand, Y. J., Kühn, K., Bentz, H., Fietzek, P. P., and Caen, J. P., *Throm. Res.*, 16, 269, 1979. With permission.)

denatured α-chains, and CNBr-derived α-chain fragments. It is now well established that aggregation of platelets can only be caused by reformed fibrillar structure of native triple-helical molecules.[105,106] The activity of the fibrillar structures seems to be independent of the array of the molecules. They can be arranged either in a D-staggered order or in an irregular manner.[107] Blockage of the charged carboxylic or amino groups did not cause significant alteration in the platelet aggregation activity.[107]

The adhesion of platelets does not need collagen in fibrillar form, but requires intact triple-helical structures.[108] If denatured collagen chains or CNBr peptides are used, the collagen activity is drastically reduced.[108] However, there are some CNBr peptides, for example, α1(I)CB6 and α1(III)CB4, which exhibit in contrast to the other CNBr peptides, a small but significant adhesion activity.[108–110] According to these observations, the nonhelical C terminal end of the α1(I)-chain and a sequence region from position 478 to 486 in Type III collagen appears to be responsible for this rest activity[111] (see Figure 8). Pepsin-treated Type I collagen, where the C terminal end has been removed, and untreated Type I collagen with an intact C terminus, both in fibrillar form, show the same activity in aggregating platelets.[107] Therefore, the C terminal sequence region which seems to be important for adhesion does not influence platelet aggregation. It has been also suggested that the first step of collagen platelet interaction involves the carbohydrate moiety of collagen.[112] The fact, however, that collagen types with a high content of carbohydrates, such as Type II and Type IV collagen, are less active than Type I and Type III collagen with a lower carbohydrate content speaks against an important role for these structures.[113]

C. The Structure of the Procollagen Peptides

The original form in which collagen is synthesized is a preproα-chain which contains the typical hydrophobic leading sequence of extracellular proteins.[114] This sequence could be only detected by cell-free translation of the respective mRNA. The proα-chains can be isolated from the medium of tissue culture of fibroblasts or chondrocytes, but cannot be observed in tissue extracts. The pro forms of α1(I), α2, α1(II), and α1(III) appear to be homologous and contain at both the N and C terminus globular peptides with molecular weights of 15,000 and 25,000, respectively.

The amino acid sequence of the presumably entire proα1(I)-chain is now known.[78,115,116] The N terminal procollagen peptide has been isolated from skins of dermatosparactic calves or sheep. Dermatosparaxis,[117,118] an inherited connective tissue disease, is characterized by the absence of an intact procollagen peptidase which specifically cleaves off the N terminal propeptide.[119] Consequently, the p_Nα1(I)-chain that still contains the N terminal extension peptide, but where the C terminal end is cleaved off, is enriched in skin and could be isolated in sufficient amounts for sequence analysis.[115,116,120] The C terminal procollagen peptide has been isolated from tissue culture in too small amounts for a complete sequence analysis.[11] Using recombinant DNA technology, however, it was recently possible

```
                                      ↓
pNα1 (I)      -Gly-Gly-Asn-Phe-Ala-Pro-Gln-Leu-Ser-Tyr-Gly-Tyr-
pNα1 (III)    -Gln-Asn-Asn-Tyr-Ala-Pro-Gln-Tyr-Glu-Ala-Tyr-Asp-
```

FIGURE 9. Amino acid sequence surrounding the cleavage site of the N terminal procollagen peptidases in the $p_N\alpha1$(I) and the $p_N\alpha1$(III) chains. The procollagen peptidases split the proline-glutamine bond. For the type specificity of the Type I cleaving procollagen peptidase, the phenylalanine residues three positions to the left of the cleavage site seem to be important[127] and are substituted in Type III by a tyrosine. Literature: $p_N\alpha1$(I) bovine;[124] $p_N\alpha1$(III) bovine.[128]

to sequence the C terminal procollagen peptide of the α1(I)- and the α2-chain at the level of the c-DNA.[78,121]

The N terminal procollagen peptide of proα1(I) is 139 residues long and contains two domains, an N terminal globular part of 86 residues and a triple-helical collagenous part of 45 residues which is connected to the N terminal nonhelical region of the α-chain by a stretch of eight amino acid residues.[115] The conformation of the globular part was found to be stable even under extreme conditions of temperature, pH, and presence of denaturants.[122] The reason for this appears to be the presence of five disulfide bridges and a high content of hydrophobic amino acid residues. The triple-helical part is characterized by an unusually high content of proline and hydroxyproline which occupies 35% of the position X and Y of the tripeptide unit Gly-X-Y. Consistent with this amino acid composition, the procollagen peptides are able to reform triple-helical structures in solution with the remarkable thermal stability of 35°C to 45°C.[122] The precursor segment of proα2 also contains a triple-helical segment homologous to the corresponding region in proα1(I), so that the precursor molecule of Type I collagen, consisting of two proα1(I)- and one proα2-chain, contains in the N terminal procollagen extension, an intact triple helix. The globular part of the N terminal proα2 peptide, however, is absent.[123]

The nonhelical part which connects the triple-helical domain of the N terminal propeptide and the triple-helical region of the α1-chain contains one proline-glutamine bond which is specifically split by the procollagen peptidase during conversion of procollagen to collagen[124] (see Figure 9). The enzyme has an approximately 1000-fold greater affinity for native collagen than for the denatured chains.[125] In addition it has been shown that the highly purified enzyme does not cleave denatured proα-chains.[126] This indicates the region comprising the cleavage site has a distinct three-dimensional structure important for the action of the enzyme. Further support comes from the fact that collagen contains several proline-glutamine bonds which cannot be cleaved by the enzyme both in the native and denatured state. Using a variety of synthetic peptides as inhibitors for the Type I collagen-specific N terminal procollagen peptidase, which were similar to the sequence surrounding the cleavage site, it has been shown that phenyl alanine, three positions N terminal to the cleavage site, seems to be particularly important for the procollagen peptidase[127] (see Figure 9). The cleavage site in the corresponding region of Type III collagen is also Pro-Gln. The phenylalanine residue, however, important for the activity of the type I specific enzyme is substituted by a tyrosine[128] which may explain why the N terminal extension of Type III collagen cannot be removed by the type I p_N collagen cleaving enzyme.[129]

P_N collagen from Type III collagen could be isolated by neutral salt extraction of embryonic calf skin.[37] Its N terminal procollagen peptide is similar to the respective peptide of Type I collagen.[128] It is 130 residues in length and contains also a globular and a triple helical domain.[130] $P_N\alpha1$(I) and $p_N\alpha1$(III) are in particular homologous in the globular part. The Type III propeptide contains at the C terminal end of the triple-helical domain two additional cysteine residues not present in Type I collagen which are involved in an interchain disulfide bridge.[130] These bridges stabilize the triple-helical structure, so that its denaturation temperature is increased to more than 53°C and its peptide bonds are protected against the attack of bacterial collagenase at 37°C, a temperature where the triple-helical body of the collagen molecules will be immediately digested.[131]

According to the sequence of the cDNA, the C terminal procollagen peptide of α1(I) and of α2(I) are 245 and 244 amino acids in length, respectively.[78] There is no triple-helical domain as is observed in the N terminal propeptide. In α1(I) the C terminal peptide contains eight cysteine residues from which two are thought to be involved in interchain and six are thought to be involved in intrachain disulfide bridges. The α2(I) propeptide contains seven cysteine residues; only one of them seems to be involved in the interchain disulfide link. Since the C terminus of the α1(I)-chain from rat appears to be Tyr-Tyr, Ala, and, Arg,[132] the cleavage site of the C terminal procollagen peptidase should be an Ala-Asp bond (see Figure 5).

Several biological functions have been postulated for the collagen precursor. Procollagen has been thought of as a transport form which prevents intracellular fibrogenesis and facilitates transcellular and transmembrane movement of collagen. There is some indirect evidence that the C terminal extension is involved in alignment and selection of the α-chains before triple-helix formation. Thus, the interchain disulfide bonds located in Type I collagen in the C terminal peptide is formed before the triple-helix formation occurs,[133] and it can be also observed if peptidyl hydroxylation and therefore triple-helix formation is inhibited.[134] Recently it was shown that embryonic chick tendon fibroblasts synthesized in the presence of low concentration of puromycine shortened proα-chains. These largely lacked the C terminal extension, and although relatively normal in hydroxylation, they were not triple-helical and were not secreted.[135]

Recently, an important physiological function of the N terminal pro extension was discovered. Starting with the observation that fibroblasts from dermatosparactic sheep skin synthesize quantitatively more collagen than control cells, the influence of the N terminal procollagen α1(I) peptide on the rate of collagen synthesis of fibroblast was tested. The experiments led to the conclusion that the N terminal propeptide of the α1(I)-chain is able to specifically inhibit the collagen synthesis of fibroblasts, whereas the synthesis of Type II collagen in chondrocytes is not affected.[136]

In cell-free systems, however, the procollagen peptide of proα1(I) chain inhibits not only the mRNA of the proα1(I), but also of the proα1(II)-chain.[137] It has been therefore suggested that the cell specificity of the inhibition capacity is given by specific receptor sites on the cell surface responsible for the uptake of the peptide in the cells, whereas the actual inhibition takes place at the level of translation. Experiments with fragments of the extension peptide revealed that only the globular domain was necessary for the inhibitory activity.[136]

V. BASEMENT MEMBRANE COLLAGENS

The special function of basement membrane as semipermeable filter and support for endo- and epithelial cells requires a macromolecular organization different from that of interstitial connective tissue. Basement membranes are described as continuous homogeneous sheets of electron-dense material with an amorphous matrix, sometimes containing 4-nm-thick filaments randomly arranged.[138] From this, one should expect that the molecular structure of basement membrane collagen also differs from that of interstitial collagen. For a long time, it was thought that the collagens from both sources are homologously built.[139] Only recently has new information on the characteristic differences between interstitial and basement membrane collagens been gained. A number of these characteristics, including insolubility and proteolytic susceptibility, are responsible for the difficulties encountered when investigating collagenous basement membrane constituents. As sources for pure basement membranes, the glomeruli of kidney and the anterior lens capsule have been used. Recently, a mouse tumor (EHS sarcoma) was discovered which synthesizes basement membrane in large amounts. Also human placenta which contains, beside other tissues, a large amount of basement membrane is used for the isolation of basement membrane collagen.

Basement membranes can be solubilized in the presence of strong denaturants, such as guanidine hydrochloride, urea, or sodium dodecyl sulfate, after reduction of the intermolecular disulfide bridges. However, such procedures led to highly polymeric compounds which cannot be investigated without further fragmentation. Another way to prepare soluble basement membrane collagen is treatment with proteolytic enzymes, mainly pepsin, under conditions where the noncollagenous structures are destroyed, but the triple-helical domains of collagen remain, at least in part, intact. Experimental procedures used for the isolation and characterization of the pepsin-derived fragment are described in this volume.[139a] A more detailed description of the structure of basement membrane collagen gained from recent work in several laboratories is presented in Volume II.[32] In the following, only a general view of the structure of basement membrane collagen in comparison to that of interstitial collagen is given.

A. Type IV Collagen

Depending on the experimental conditions, pepsin treatment of basement membrane from anterior lens capsule, mouse tumor, and placenta resulted in a variety of triple-helical fragments of different sizes containing α-chain fragments with molecular weights between 140,000 and 15,000 daltons. Characterization of these polypeptides by amino acid composition, CNBr peptide pattern, and immunological methods gave evidence for three different α-chains, designated as α1(IV), α2(IV), and α3(IV).[32] In Table 3, the amino acid compositions of the most characteristic pepsin-derived fragments of α1(IV) and α2(IV) with a molecular weights of 95,000 and 75,000 daltons, respectively, are listed. The typical composition of basement membrane collagen differs from interstitial collagen by a higher degree of hydroxylation of the proline and lysine residues. Most of the hydroxylysines are substituted by galactosyl glucosyl groups. Type IV collagen is further characterized by a low content of alanine and arginine and a high amount of large hydrophobic amino acids.

First sequence studies of the various CNBr peptides of the α1(IV)- and the α2(IV)-chain revealed the striking result that the triple-helical part of both chains seems to be frequently interrupted by nonhelical regions.[141,142] The smallest interruption found is caused by a deletion of only one glycine, but also longer nonhelical sequence stretches were found. These regions may be responsible for the occurrence of so many triple-helical fragments after pepsin treatment. Another difference to interstitial collagen is the distribution of the polar-charged amino acids at has been deduced from the cross striation pattern of the respective SLS crystallites.[38,143–145] According to the denaturation temperature of the triple helical fragments, Type IV collagen has a thermal stability similar to or slightly higher than interstitial collagen.[143] In addition, a triple helical fragment was isolated with an unusually high melting point of 70°C and a high cysteine content of about 10%, which has been

Table 3
AMINO ACID COMPOSITION OF COLLAGEN α-CHAINS; COMPOSITION CALCULATED AS AMINO ACID RESIDUES PER 1000

	α1 (I)	α2 (I)	α1 (II)	α1 (III)	α1 (IV)	α2 (IV)	α1 (V)	α2 (V)
Hyp	86	85	93	120	146	127	108	107
Asp	45	47	43	47	48	52	51	55
Thr	16	17	22	16	18	28	22	27
Ser	34	24	26	48	34	26	26	34
Glu	77	71	87	71	77	62	99	90
Pro	135	120	129	106	79	69	119	92
Gly	327	328	333	341	318	313	220	318
Ala	120	101	102	95	31	49	46	59
Val	18	34	17	16	30	24	19	30
Cys	—	—	—	2	—	—	—	—
Met	7	4	11	8	15	12	7	10
Ile	9	17	9	14	34	42	20	18
Leu	21	34	26	18	54	59	40	37
Tyr	4	3	1	4	5	6	1	—
Phe	12	16	14	9	24	37	12	12
Hyl	5	11	23	7	61	42	36	25
Lys	32	21	15	28	5	6	7	18
His	3	8	2	7	6	8	19	10
Arg	50	57	51	45	20	45	48	57

Note: Literature: α1 (I), bovine skin;[34]
α2 (I), bovine skin;[35]
α1 (II), bovine cartilage;[36]
α1 (III), bovine skin;[37]
α1 (IV), α2 (IV), human placenta;[38]
α1 (V), α2 (V), human placenta.[39]

designated 7-S collagen according to its sedimentation constant.[146] The high thermal stability can be explained by disulfide bridges between the three chains of a triple helix.

The protein chemical investigations of Type IV collagen have not led to a generally accepted structural model. Thus it was not clear whether 7-S collagen represents a new type of basement membrane collagen or whether it is a part of Type IV collagen. The length of the triple helical part of the α(IV) chains was unknown and there was not conception as to how the molecules aggregate to form the macromolecular structure of basement membrane matrix. Many of these open questions could recently be answered using rotary shadowing, an electron microscopical technique, suitable for the visualization of individual molecules or molecular fragments of collagen. For these investigations, basement membrane collagen was dissolved by limited proteolysis or by acidic extraction, the molecular fragments obtained were characterized in the electron microscope and finally the fragments were used to reconstruct the entire type IV collagen molecule.[147,148,149]

Short pepsin treatment of basement membranes at 4°C yielded polymeric Type IV collagen with a molecular weight of about two million daltons. The electronmicroscopy of this polymer showed four 390-nm long molecules, held together via their 30-nm long identical end regions (Figure 10).[147] More intensive treatment of this polymer with pepsin at 37°C destroyed the long arms of the spider-like structure, leaving intact only the 30-nm long crosslink areas as well as the short arms 28 nm

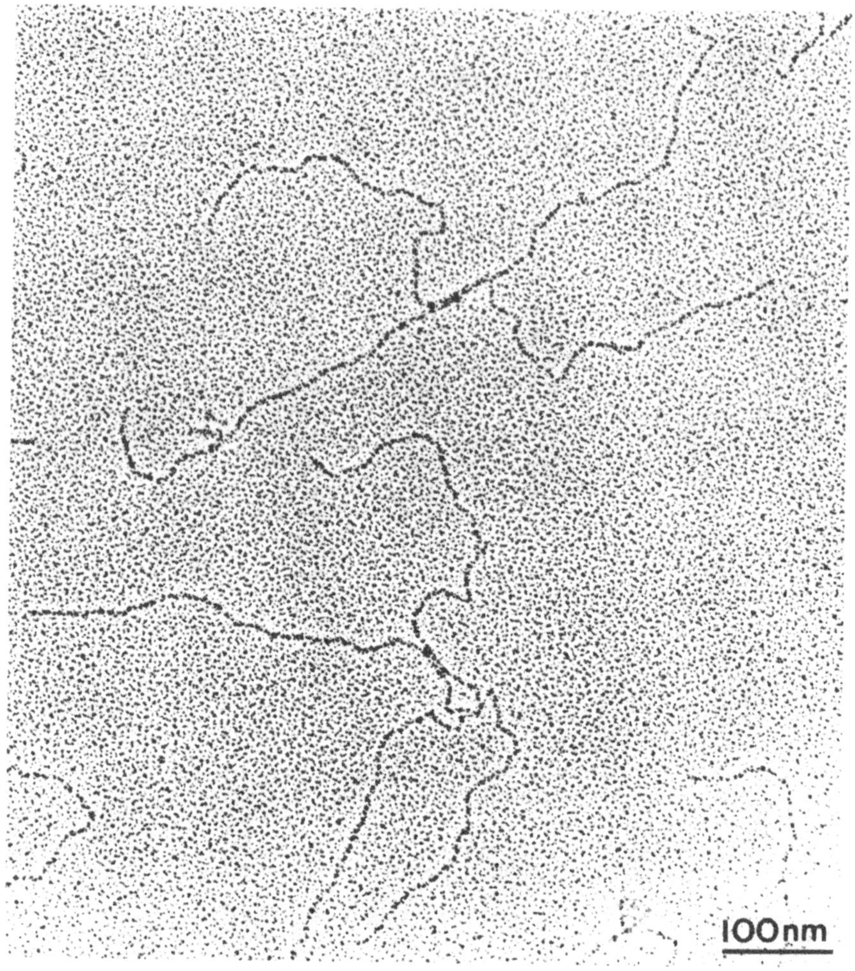

FIGURE 10. Electronmicrographs (rotary shadowing) of polymeric Type IV collagen obtained by limited pepsin digestion from human placenta. It consists of four 390-nm long strands connected to one another via their 30-nm long end regions.[147]

in length (Figure 11). These pictures were identical with the micrographs of 7-S collagen molecules, which indicated that 7-S collagen is the terminal crosslink site of the Type IV collagen molecules.[147] A second crosslink area located at the opposite end of the molecules was observed in polymers present in the acetic acid extract of mouse tumor matrix (Figure 12). It contained dimeric molecular fragments connected via a globular domain. Frequently one observed molecules containing the 7-S domain at the one end and the globular domain at the other.[148,149] These electron microscopical investigations finally led to a model of a macromolecular structure of Type IV collagen (Fig. 13). It is a network of individual 390-nm long molecules which are aggregated and crosslinked via identical ends.[148,149] Electronmicrographs of intact Type IV molecules isolated from the medium of basement membrane constituents synthesizing mouse teratocarcinoma cells exhibited in accordance with the model a 390-nm long thread with a globular part at one end.[149] According to biosynthetic studies the entire α1(IV) and α2(IV) chains seem to have a molecular weight of 180,000 and 160,000 daltons, respectively.[150–152] Also these values are in

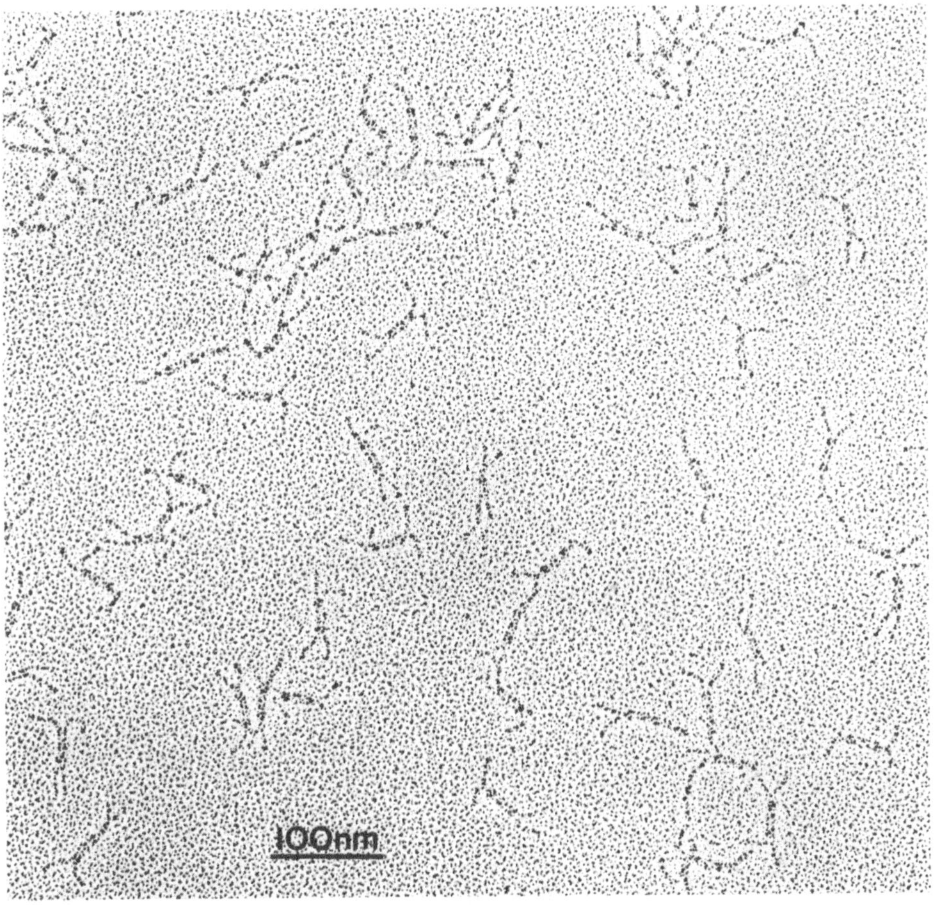

FIGURE 11. Electronmicrographs of 7-S collagen, obtained by treatment of polymeric Type IV collagen with pepsin at 20°. They represent the terminal crosslink domain of polymeric Type IV collagen. (From Kühn, K., Wiedemann, H., Timpl, R., Risteli, J., Dieringer, H., Voss, T., and Glanville, R. W., *FEBS Lett.*, 125, 123, 1981. With permission.)

excellent agreement with the length of the molecules found in the electron microscope. The chain composition of Type IV molecules is, however, not yet clear.

The fact that Type IV collagen does not form fibrillar structures, but is organized in a network of individual molecules, explains the amorphous appearance of basement membranes in the electron microscope. It is in agreement with the finding, that beside the typical low angle reflex of 0.29 nm, caused by the triple-helical structure, no strong wide-angle reflection of 63 nm, which is characteristic for the native fibril of interstitial collagen, can be observed.[153]

B. Type V Collagen

Two α-chains of Type V collagen, α1(V) and α2(V), formerly named A, B, have been isolated from chorionic amniotic membranes[26] and a variety of other tissues.[27] In contrast to Type IV collagen, each chain type has been observed only in one form with a molecular weight of 95,000 daltons, similar to α-chains from interstitial collagen. The amino acid composition of the α1(V)- and the α2(V)-chain (see Table 3) resembles basement membrane collagen. The resistance against pepsin, however,

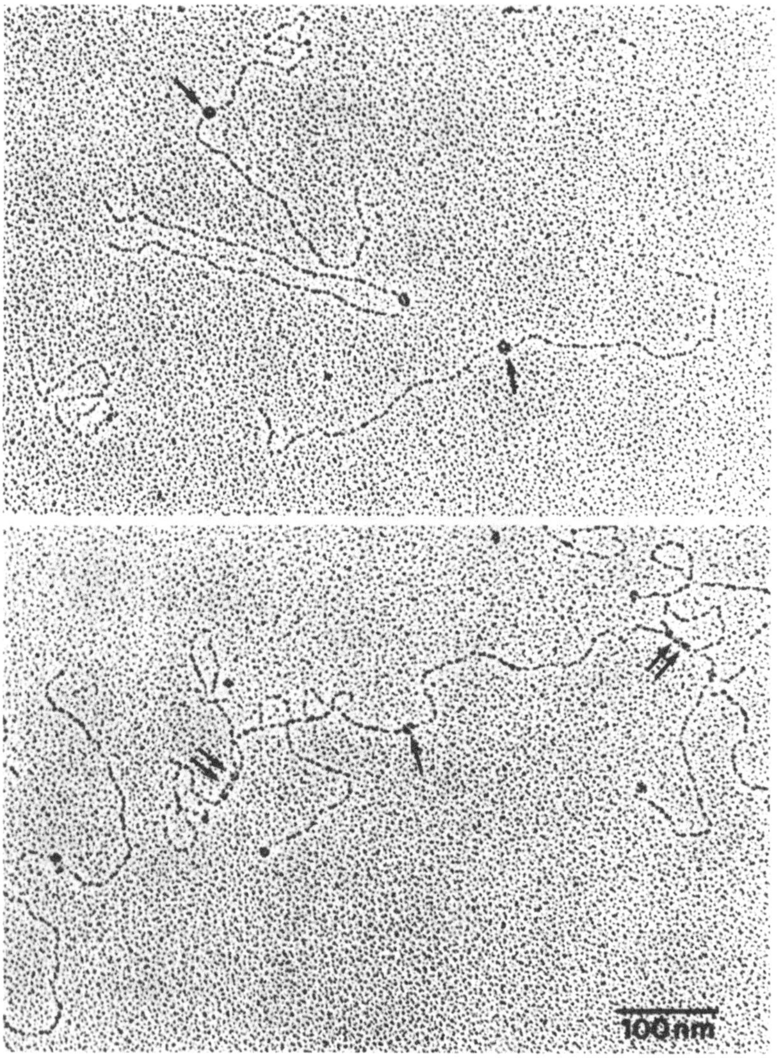

FIGURE 12. Electronmicrograph of Type IV collagen polymers present in an acetic acid extract of mouse tumor matrix. There are dimers of two identical strands 320 nm in length connected by a globule (arrow) and polymeric molecules connected at one end by the globular domain (arrow) and at the other end by the 7-S collagen domain (double arrow). (From Timpl, R., Wiedemann, H., Van Delden, V., Furthmayr, H., and Kühn, K., *Eur. J. Biochem.*, 1981, in press. With permission.)

is more similar to interstitial collagen. The distribution of the polar-charged amino acid residues along the α-chains of Type V collagen also seems to be homologous to the interstitial collagens and different to Type IV collagen as deduced from the cross-striation pattern of the SLS crystallites (see Figure 1).

From the amniotic chorionic membranes, Type V was isolated in the form of a native molecule with a chain composition of $[\alpha 1(V)]_2\alpha 2(V)]$. According to segment-long-spacing crystallites, denaturation and renaturation behavior of this Type V collagen molecule resembles Type I collagen.[39] In different tissues, the ratio of α1(V):α2(V) was not always found to be 2:1, which suggests the existence of molecules with different chain compositions.[27] In many tissues, but not in all, α1(V)

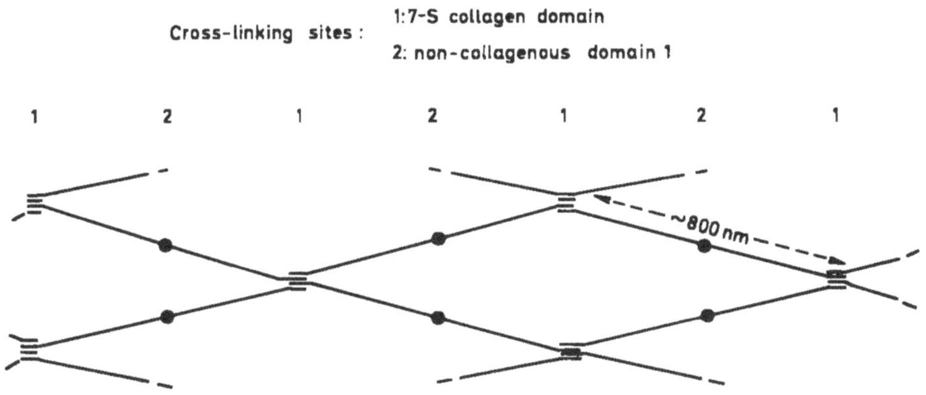

FIGURE 13. Two dimensional model of the macromolecular structure of Type IV collagen in basement membranes. The distance between two identical crosslink sites (about 800 nm) is similar to the length of two stretched molecules.[148]

and α2(V) are accompanied by a third chain, α3(V) (formerly named C).[154] It is not yet clear whether α3(V) actually belongs to Type V collagen.

Information on nontriple helical domains similar to the procollagen peptides in interstitial collagens is not available. Recently, it has been shown that Type V molecules are able to aggregate in vitro into fibrils of the native type, showing the typical cross-striation period of 67 nm.[155] For the in vivo situation, it has been suggested that Type V collagen is located in the amorphous material surrounding smooth muscle cells and that fibroblast and other mesenchymal cells are also probably surrounded by this collagen which may function as exoskeleton for these cells.[156]

C. Poorly Characterized Collagens

In addition to the interstitial and the basement membrane collagens, several other collagens have been isolated from different tissues after limited pepsin digestion. The structures of the entire molecules are not yet known.

Beside Type II collagen, hyaline cartilage contains small proportions of three additional polypeptide chains of the size of an α chain. They were first found in human cartilage[157] but also in cartilage from pig,[158] bovine,[159] and chick.[160] Two of these chains designated 1α and 2α are homologous but genetically different to the α1(V) and α2(V) chains. The third chain 3α resembles the α1(II) chain and could be a post-translational modification of this chain. In addition, a high soluble triple helical disulfide bonded fragment with an apparent molecular weight of 110,000 daltons has been isolated from cartilage. After reduction it forms small fragments with a molecular weight of about 38,000 daltons.[158,159,160] The high soluble collagenous molecule from chick hyaline cartilage has recently been characterized in the electron microscope using the rotary shadowing technique.[161] It is a 134-nm long triple-helix which exhibits a flexible region 96 nm from one end. It consists of two different chains at a ratio of 2:1. The entire molecule from which this fragment is derived remains to be isolated and characterized.

Another high soluble, high molecular weight and cysteine rich fragment, designated by some authors as intima collagen, has been isolated from basement membrane rich tissues such as human aorta[162] and human and bovine placenta.[146,163–165] Two different peptide chains were obtained after reduction of the disulfide bridges.

Their amino acid compositions resemble Type IV collagen. They contain less than one third glycine, indicating the presence of nonhelical regions in this protein. The native fragment has recently been characterized in the electron microscope.[32] The intact parent molecule of this fragment is also not known.

VI. CONCLUDING REMARKS

The purpose of this article was to describe the principles involved in the formation of the collagenous extracellular matrix and its interactions with other connective tissue components and cells. The relationship between the amino acid sequence and higher structures and the features of those regions responsible for specific interactions with cells and enzymes important for collagen metabolism have been presented. Best described are the properties of the interstitial collagens. Investigations of basement membrane collagens have revealed that the classical structure of interstitial collagens, represented as a rod-like molecule which is able to aggregate into cross-striated fibrils, can vary according to physiological requirements. The collagen specific structure, the triple-helix, is frequently interrupted in Type IV collagen by short nontriple helical regions,[141-142] bestowing the molecule with a higher flexibility.[166] The most striking difference between interstitial and basement membrane collagen is, however, that the latter do not form fibrillar structures; they build a flexible network of individual molecules, optimally adapted to the special function of basement membranes.[148]

ACKNOWLEDGMENTS

Work in the laboratory of the author was aided by grants from the Deutscje Forschungsgemeinschaft, SFB 51 and from the Arbeitsgemeinschaft Industrieller Forschungsvereinigungen (AIF).

REFERENCES

1. **Bornstein, P. and Sage, H.,** Structure of distinct collagen types, *Annu. Rev. Biochem.,* 48, 957, 1980.
2. **Kefalides, N. A., Alper, R., and Clark, C. C.,** Biochemistry and metabolism of basement membranes, *Int. Rev. Cytol.,* 61, 167, 1979.
3. **Fietzek, P. P. and Kühn, K.,** The primary structure of collagen, *Int. Rev. Connect. Tissue Res.,* 7, 1, 1976.
4. **Goldberg, B.,** Binding of soluble type I collagen molecules to the fibroblast plasma membrane, *Cell,* 16, 265, 1979.
5. **Grinnell, F. and Minter, D.,** Attachment and spreading of baby hamster kidney cells to collagen substrate: effects of cold insoluble globulin, *Proc. Natl. Acad. Sci. U.S.A.,* 75, 4408, 1978.
6. **Kleinmann, H. K., Hewitz, H. T., Murray, J. C., Liotta, L. A., Rennard, S. I., Pennypacker, J. P., McGoodwin, E. B., Martin, G. R., and Fishman, P. H.,** Cellular and metabolic specificity in the interaction of adhesion proteins with collagen and with cells, *J. Supramol. Struct.,* 11, 69, 1979.
7. **Bornstein, P. and Traub, W.,** Structurally distinct collagens, in *The Proteins,* Vol. 4, 3rd ed., Neurath, H., and Hill, R. L., Eds., Academic Press, New York, 1979, 411.
8. **Miller, E. J.,** Biochemical characterization and biological significance of the genetically distinct collagens, *Mol. Cell. Biochem.,* 13, 168, 1976.
9. **Furthmayr, H. and Timpl, R.,** Immunochemistry of collagens and procollagens, *Int. Rev. Connect. Tissue Res.,* 7, 61, 1976.

10. **Fessler, J. H. and Fessler, L. I.,** Biosynthesis of procollagen, *Annu. Rev. Biochem.*, 47, 129, 1978.
11. **Prockop, D. J., Kivirikko, K. I., Tudermann, L., and Guzman, N. A.,** The biosynthesis of collagen and its disorders, *N. Engl. J. Med.*, 301, 13, 1979.
12. **Müller, P. K., Meigel, W. N., Pontz, B. G., and Raisch, K.,** Influence of α,α'-dipyridyl on the bionsynthesis of collagen in organ culture, *Hoppe-Seylers Z. Physiol. Chem.*, 355, 985, 1974.
13. **Benya, P. D., Padilla, S. R., and Nimni, M. E.,** The progeny of rabbit articular chondrocytes synthesize collagen types I and III and type I trimer but not type II: verification by cyanogen bromide peptide analysis, *Biochemistry*, 16, 865, 1977.
14. **Jimenez, S. A., Bashey, R. I., Bendit, M., and Yankowski, R.,** Identification of collagen α1(I) trimer in embryonic chick tendons and calvaria, *Biochem. Biophys. Res. Commun.*, 78, 1354, 1977.
15. **Little, C. D., Church, R. L., Miller, R. A., and Ruddle, F. H.,** Procollagen and collagen produced by teratocarcinoma-derived cell-line TS04: evidence for a new molecular form of collagen, *Cell*, 10, 278, 1977.
16. **Mayne, R., Vail, M. S., and Miller, E. J.,** Analysis of changes in collagen biosynthesis that occur when chick chondrocytes are grown in 5-bromo-2'deoxyuridine, *Proc. Natl. Acad. Sci. U.S.A.*, 72, 4511, 1975.
17. **Fietzek, P. P., Allmann, H., Rauterberg, J., Henkel, W., Wachter, E., and Kühn, K.,** The covalent structure of calf skin type III collagen. I. The amino acid sequence of the amino terminal region of the α1(III) chain (position 1-222), *Hoppe-Seylers Z. Physiol. Chem.*, 360, 809, 1979.
18. **Dewes, H., Fietzek, P. P., and Kühn, K.,** The covalent structure of calf skin type III Collagen. II. The amino acid sequence of the cyanogen bromide peptide α1(III)CB1,8,10,2 (position 223-402), *Hoppe-Seylers Z. Physiol. Chem.*, 360, 821, 1979.
19. **Bentz, H., Fietzek, P. P., and Kühn, K.,** The covalent structure of calf skin type III collagen. III. The amino acid sequence of the cyanogen bromide peptide α1(III)CB4 (position 403-551), *Hoppe-Seylers Z. Physiol. Chem.*, 360, 833, 1979.
20. **Lang, H., Glanville, R. W., Fietzek, P. P., and Kühn, K.,** The covalent structure of calf skin type III collagen. IV. The amino acid sequence of the cyanogen bromide peptide α1(III)CB5 (position 552-788), *Hoppe-Seylers Z. Physiol. Chem.*, 360, 841, 1979.
21. **Dewes, H., Fietzek, P. P., and Kühn, K.,** The covalent structure of calf skin type III collagen. V. The amino acid sequence of the cyanogen bromide peptide α1(III)CB9A (position 789-927), *Hoppe-Seylers Z. Physiol. Chem.*, 360, 851, 1979.
22. **Allmann, H., Fietzek, P. P., Glanville, R. W., and Kühn, K.** The covalent structure of calf skin type III collagen. VI. The amino acid sequence of the carboxyterminal cyanogen bromide peptide α1(III)CB9B (position 928-1028), *Hoppe-Seylers Z. Physiol. Chem.*, 360, 861, 1979.
23. **Butler, W. T., Finch, J. E., Jr., and Miller, E. J.,** Covalent structure of cartilage collagen. Amino acid sequence of residues 363-551 of bovine α1(II) chains, *Biochemistry*, 16, 4981, 1977.
24. **Butler, W. T., Miller, E. J., and Finch, J. E., Jr.,** The covalent structure of cartilage collagen. Amino acid sequence of the NH_2-terminal helical portion of the α1(II) chain, *Biochemistry*, 14, 3000, 1976.
25. **Butler, W. T., Miller, E. J., Finch, J. E., Jr., and Inagami, T.,** Homologous region of collagen α1(I) and α1(II) chains: apparent clustering of variable and invariant amino acid residues, *Biochem. Biophys. Res. Commun.*, 57, 190, 1974.
26. **Burgeson, R. E., El Adli, F. A., Kaitila, I. I., and Hollister, D. W.,** Fetal membrane collagens: identification of two new collagen alpha chains, *Proc. Natl. Acad. Sci. U.S.A.*, 73, 2579, 1976.
27. **Chung, E., Rhodes, R. K., and Miller, E. J.,** Isolation of three collagenous components of probable basement membrane origin from several tissues, *Biochem. Biophys. Res. Commun.*, 71, 1167, 1976.
28. **Brown, R. A., Shuttleworth, C. A., and Weiss, J. B.,** Three new α chains of collagen from non-basement membrane source, *Biochem. Biophys. Res. Commun.*, 80, 866, 1978.
29. **Duance, V. C., Restall, D. J., Beard, H., Bourne, F. J., and Bailey, A. J.,** The location of three collagen types in skeletal muscle, *FEBS Lett.*, 72, 248, 1977.
30. **Mayne, R., Vall, M. S., and Miller, E. J.,** Characterization of the collagen chains synthesized by cultured smooth muscle cells derived from Rhesus monkey thoracic aorta, *Biochemistry*, 17, 446, 1978.
31. **Pöschl, A. and von der Mark, K.,** Synthesis of type V collagen by chick corneal fibroblasts in vivo and in vitro, *FEBS Lett.*, 115, 100, 1980.
32. **Martin, G. R. and Timpl, R.,** Structural components of the basement membranes, in *Immunochemistry of the Extracellular Matrix*, Vol. 2, Furthmayr, H., Ed., CRC Press, Boca Raton, Fla., 1981, in press.
33. **Tanzer, M. L., Crosslinking,** in *Biochemistry of Collagen*, Ramachandran, G. N. and Reddi, A. H., Eds., Plenum, New York, 1976, 137.

34. **Rauterberg, J. and Kühn, K.,** Acid soluble calf skin collagen. Characterization of the peptides obtained by cyanogen bromide cleavage of its α1 chain, *Eur. J. Biochem.*, 19, 398, 1971.
35. **Fietzek, P. P., Münch, M., Breikreuz, D., and Kühn, K.,** Isolation and characterization of the cyanogen bromide peptides from the α2 chain of calf skin collagen, *FEBS Lett.*, 9, 229, 1970.
36. **Miller, E. J. and Lunde, L. G.,** Isolation and characterization of the cyanogen bromide peptides from the α1(II) chain of bovine and human cartilage collagen, *Biochemistry*, 12, 3153, 1973.
37. **Timpl, R., Glanville, R. W., Nowack, H., Wiedemann, H., Fietzek, P. P., and Kühn, K.,** Isolation, chemical and electron microscopical characterization of neutral salt soluble type III collagen and procollagen from fetal bovine skin, *Hoppe-Seylers Z. Physiol. Chem.*, 356, 1783, 1975.
38. **Glanville, R. W., Rauter, A., and Fietzek, P. P.,** Isolation and characterization of a native placental basement membrane collagen and its component α chains, *Eur. J. Biochem.*, 95, 383, 1979.
39. **Bentz, H., Bächinger, H. P., Glanville, R., and Kühn, K.,** Physical evidence for the assembly of A and B chains of human placental collagen in a single triple helix, *Eur. J. Biochem.*, 92, 563, 1978.
40. **Ramachandran, G. N. and Ramakkrishnan, C.,** Molecular structure, in *Biochemistry of Collagen*, Ramachandran, G. N. and Reddi, A. H., Eds., Plenum Press, New York, 1976, 45.
41. **Segal, D. M.,** Polymerism of tripeptides as collagen models. VII. Synthesis and solution properties of four collagen-like poly hexapeptides, *J. Mol. Biol.*, 43, 497, 1969.
42. **Sakakibara, S., Inonye, K., Shudo, K., Kishida, Y., Kobayashi, Y., and Prockop, D. J.,** Synthesis of $(Pro\text{-}Hyp\text{-}Gly)_n$ of defined molecular weights. Evidence for the stabilization of collagen triple helix by hydroxyproline, *Biochim. Biophys. Acta*, 303, 198, 1973.
43. **Berg, R. A. and Prockop, D. J.,** The thermal transition of a non-hydroxylated form of collagen. Evidence for a role for hydroxyproline in stabilizing the triple helix of collagen, *Biochem. Biophys. Res. Commun.*, 52, 115, 1973.
44. **von der Mark, K., Wendt, P., Rexrodt, F., and Kühn, K.,** Direct evidence for a correlation between amino acid sequence and cross striation pattern of collagen, *FEBS Lett.*, 11, 105, 1970.
45. **Tkocz, C. and Kühn, K.,** The formation of triple helical collagen molecules from α1 or α2 polypeptide chains, *Eur. J. Biochem.*, 7, 454, 1969.
46. **Stark, M., Miller, E. J., and Kühn, K.,** Comparative electron microscope studies on the collagens extracted from cartilage, bone and skin, *Eur. J. Biochem.*, 27, 192, 1972.
47. **Wiedemann, H., Chung, E., Fujii, T., Miller, E. J., and Kühn, K.,** Comparative electron microscope studies on type III and type I collagen, *Eur. J. Biochem.*, 51, 363, 1975.
48. **Hofmann, H., Fietzek, P. P., and Kühn, K.,** Comparative analysis of the sequences of the three collagen chains α1(I), α2 and α1(III). Functional and genetic aspects, *J. Mol. Biol.*, 141, 293, 1980.
49. **Hulmes, D. J. S., Miller, A., Parry, D. A. D., Piez, K., and Woodhead-Galloway, J.,** Analysis of primary structure of collagen for the origins of molecular packing, *J. Mol. Biol.*, 79, 137, 1973.
50. **Cunningham, L. W., Davis, H. A., and Hammonds, R. G.,** An analysis of the association of collagen based on structural models, *Biopolymers*, 15, 483, 1976.
51. **Trus, B. L. and Piez, K. A.,** Molecular packing of collagen: three dimensional analysis of electrostatic interaction, *J. Mol. Biol.*, 108, 705, 1976.
52. **Piez, K. A. and Trus, B. L.,** Sequence regularities and packing of collagen molecules, *J. Mol. Biol.*, 122, 419, 1978.
53. **Traub, W.,** Molecular assembly in collagen, *FEBS Lett.*, 92, 114, 1978.
54. **Hofmann, H., Fietzek, P. P., and Kühn, K.,** The role of polar and hydrophobic interactions for the molecular packing of type I collagen: a three-dimensional evaluation of the amino acid sequence, *J. Mol. Biol.*, 125, 137, 1978.
55. **Miller, A.,** Molecular packing in collagen fibrils, in *Biochemistry of Collagen*, Ramachandran, G. N. and Reddi, A. H., Eds., Plenum, New York, 1976, 85.
56. **Hulmes, D. J. and Miller, A.,** Quasi-hexagonal molecular packing in collagen fibrils, *Nature*, 282, 879, 1979.
57. **Kühn, K. and von der Mark, K.,** The influence of proteoglycans on the macromolecular structure of collagen, in *Collagen — Platelet Interaction*, Gastpar, H., Kühn, K., and Marx, R., Eds., Schattauer Verlag, Stuttgart, 1978, 123.
58. **Hay, E. D., Hasty, D. L., and Kiehnau, K. L.,** Fine structure of collagens and their relation to glucosamino glycans, in *Collagen — Platelet Interaction*, Gastpar, H., Kühn, K., and Marx, R., Eds., Schattauer Verlag, Stuttgart, 1978, 129.
59. **Nowack, H., Gay, S., Wick, G., Becker, U., and Timpl, R.,** Preparation and use in immunology of antibodies specific for type I and type III collagen and procollagen, *J. Immunol. Methods*, 12, 117, 1976.
60. **Gay, S., Martin, G. R., Müller, P. K., Timpl, R., and Kühn, K.,** Simultaneous synthesis of type I and III collagen by fibroblasts in culture, *Proc. Natl. Acad. Sci. U.S.A.*, 73, 4037, 1976.

61. **Pinnell, S. R., Krane, S. M., Kenzora, J. E., and Glimcher, E. J.**, A heritable disorder of connective tissue, hydroxylysine-deficient collagen disease, *N. Engl. J. Med.*, 286, 1013, 1972.
62. **Byers, P. H., Siegel, R. C., Holbrook, K. A., Narayanan, A. S., Bornstein, P., and Hale, J. G.**, Crosslinked cutis laxa. Defective crosslink formation in collagen due to decreased lysyl oxidase activity, *N. Engl. J. Med.*, 303, 61, 1980.
63. **Tanzer, M. L.**, Experimental lathyrism, *Int. Rev. Connect. Tiss. Res.*, 3, 91, 1965.
64. **Tanzer, M. L.**, Cross-linking, in *Biochemistry of Collagen*, Ramachandran, G. N. and Reddi, A. H., Eds., Plenum Press, New York, 1976, 137.
65. **Rauterberg, J., Timpl, R., and Furthmayr, H.**, Structural characterization of N terminal antigenic determinants in calf and human collagen, *Eur. J. Biochem.*, 27, 231, 1972.
66. **Fietzek, P. P., Breitkreutz, D., and Kühn, K.**, Amino acid sequence of the amino terminal region of calf skin collagen, *Biochim. Biophys. Acta*, 365, 305, 1974.
67. **Glanville, R. W. and Fietzek, P. P.**, Amino acid sequence of the N terminal non-triple helical crosslink region of type III collagen, *FEBS Lett.*, 71, 99, 1976.
68. **Fietzek, P. P. and Kühn, K.**, The covalent structure of collagen: amino acid sequence of the cyanogen bromide peptides α1-CB2, α1-CB4 and α1-CB5 from calf skin collagen, *Eur. J. Biochem.*, 52, 77, 1975.
69. **Wendt, P., von der Mark, K., Rexrodt, F., and Kühn, K.**, The covalent structure of collagen. The amino acid sequence of the 112 residues amino terminal part of peptide α1-CB6 from calf skin collagen, *Eur. J. Biochem.*, 30, 169, 1972.
70. **Fietzek, P. P., Rexrodt, F. W., Wendt, P., Stark, M., and Kühn, K.**, The covalent structure of collagen. The amino acid sequence of peptide α1-CB6-C2, *Eur. J. Biochem.*, 30, 163, 1972.
71. **Fietzek, P. P. and Rexrodt, F. W.**, The covalent structure of collagen. The amino acid sequence of α2-CB4 from calf skin collagen, *Eur. J. Biochem.*, 59, 105, 1975.
72. **Butler, W. T., Miller, E. J., and Finch, J. E., Jr.**, The covalent structure of cartilage collagen. Amino acid sequence of the NH_2 terminal helical portion of α1(II) chain, *Biochemistry*, 13, 3000, 1976.
73. **Kang, A. H.**, Studies on the location of intermolecular crosslinks in collagen. Isolation of a CNBr peptide containing δ-hydroxylysinenorleucine, *Biochemistry*, 11, 1828, 1972.
74. **Henkel, W., Rauterberg, J., and Stirtz, T.**, Isolation of a crosslinked cyanogen bromide peptide from insoluble rabbit collagen. Tissue differences in hydroxylation and glycosylation of the crosslinks, *Eur. J. Biochem.*, 69, 223, 1976.
75. **Becker, U., Furthmayr, H., and Timpl, R.**, Tryptic peptides from the crosslink regions of insoluble calf skin collagen, *Hoppe-Seylers Z. Physiol. Chem.*, 356, 21, 1975.
76. **Zimmermann, B. K., Pikkarainen, J., Fietzek, P. P., and Kühn, K.**, Cross-linkages in collagen. Demonstration of three different intermolecular bonds, *Eur. J. Biochem.*, 16, 217, 1970.
77. **Zimmermann, B. K., Timpl, R., and Kühn, K.**, Intermolecular crosslinks of collagen. Participation of the carboxyl terminal non-helical region of the α1 chain, *Eur. J. Biochem.*, 35, 216, 1973.
78. **Forrest, F. and Boedker, H.**, Sequence determination and analysis of the 3′ region of chicken pro-α1(I) and pro-α2(I) collagen messenger ribonucleic acid including the carboxyterminal propeptide sequences, *Biochemistry*, 20, 996, 1981.
79. **Henkel, W., Rauterberg, J., and Glanville, R. W.**, Isolation of crosslinked peptides from insoluble human leiomyoma. The involvement of the N terminal non-helical region of type III collagen in intermolecular crosslinking, *Eur. J. Biochem.*, 96, 249, 1979.
80. **Miller, E. J. and Robertson, P. B.**, The stability of collagen crosslinks when derived from hydroxylysyl residues, *Biochem. Biophys. Res. Commun.*, 54, 432, 1973.
81. **Gross, J.**, Aspects of the animal collagenases, in *Biochemistry of Collagen*, Ramachandran, G. N. and Reddi, A. H., Eds., Plenum Press, New York, 1976, 275.
81a. **Woolley, D. E. and Evanson, J. M., Eds.**, *Collagenase in Normal and Pathological Connective Tissue*, John Wiley and Sons, New York, 1980.
82. **Gross, J. and Nagar, Y.**, Specific degradation of the collagen molecule by tadpole collagenolytic enzyme, *Proc. Natl. Acad. Sci. U.S.A.*, 54, 1197, 1965.
83. **Gross, J., Harper, E., Harris, E. D., Jr., McCroskery, P., Highberger, J. H., Corbett, C., and Kang, A. H.**, Animal collagenases, specificity of action and structure of the substrate cleavage site, *Biochem. Biophys. Res. Commun.*, 61, 605, 1974.
84. **Miller, E. J., Harris, E. D., Chung, E., Finch, J. E., Jr., McCroskery, P. A., and Butler, W. T.**, Cleavage of type II and III collagens with mammalian collagenase: site of cleavage and primary structure at NH_2 terminal position of the smaller fragments released from both collagens, *Biochemistry*, 15, 787, 1976.
85. **McCroskery, P. A., Wood, S., and Harris, E. D.**, Gelatin: a poor substrate for a mammalian collagenase, *Science*, 182, 70, 1973.

85a. **Woolley, D. E., Lindberg, K. A., Glanville, R. W., and Evanson, J. M.,** Action of rheumatoid synovial collagenase on cartilage collagen different susceptibilities of cartilage and tendon collagen to collagenase attack, *Eur. J. Biochem.*, 50, 437, 1975.
86. **Miller, E. J., Finch, J. E., Chung, E., and Butler, W. T.,** Specific cleavage site of the native type III collagen with trypsin. Similarity of the cleavage products to collagenase cleavage site. *Arch. Biochem. Biophys.*, 173, 631, 1976.
87. **Olsen, B. R.,** Electron microscopic studies on collagen III. Tryptic digestion of tropocollagen macromolecules, *Z. Zellforsch.*, 61, 913, 1964.
88. **Leibovich, S. J. and Weiss, J. B.,** Failure of human rheumatoid synovial collagenase to degrade either normal or rheumatoid arthritic polymeric collagen, *Biochim. Biophys. Acta*, 251, 109, 1971.
89. **Woolley, D. E., Glanville, R. W., Crossley, M. J., and Evanson, J. M.,** Purification of rheumatoid collagenase and its action on soluble and insoluble collagen, *Eur. J. Biochem.*, 54, 611, 1975.
90. **Weiss, J. B.,** Enzymic degradation of collagen, *Int. Rev. Connect. Tissue Res.*, 7, 101, 1976.
91. **Kleinman, H. K., McGoodwin, E. B., Martin, G. R., Klebe, R. J., Fietzek, P. P., and Woolley, D. E.,** Localisation of the binding site for cell attachment in the α1(I) chain of collagen, *J. Biol. Chem.*, 253, 5642, 1978.
92. **Postlethwaite, A. E., Seyer, J. M., and Kang, A. H.,** Chemotactic attraction of human fibroblasts to type I, II and III collagens and collagen derived peptides, *Proc. Natl. Acad. Sci. U.S.A.*, 75, 871, 1978.
93. **Hopper, K. E., Adelmann, B. C., Gentner, G., and Gay, S.,** Recognition by guinea-pig peritoneal exudate cells of conformationally different states of collagen molecules, *Immunology*, 30, 249, 1976.
94. **Engvall, E. and Ruoslahti, E.,** Binding of soluble form of fibroblast surface protein fibronectin to collagen, *Int. J. Cancer*, 20, 1, 1977.
95. **Kleinmann, H. K., McGoodwin, E. B., and Klebe, R. J.,** Localization of cell attachment region in types I and II collagen, *Biochem. Biophys. Res. Commun.*, 72, 426, 1976.
96. **Dessau, W., Adelmann, B. C., Timpl, R., and Martin, G. R.,** Identification of the sites of collagen α chains that bind serum antigelatin factor (cold insoluble globulin), *Biochem. J.*, 169, 55, 1978.
97. **Yamada, K. M. and Kennedy, D. W.,** Fibroblast cellular and plasma fibronectins are similar but not identical, *J. Cell. Biol.*, 80, 492, 1979.
98. **Balian, G., Click, F. M., Crouch, E., Davidson, J. M., and Bornstein, P.,** Isolation of a collagen-binding fragment from fibronectin and cold insoluble gelatin, *J. Biol. Chem.*, 254, 1429, 1978.
99. **Jilek, F. and Hörmann, H.,** Cold-insoluble globulin (fibronectin), IV affinity to soluble collagen of various types, *Hoppe-Seylers Z. Physiol. Chem.*, 359, 247, 1978.
100. **Jilek, F. and Hörmann, H.,** Fibronectin (cold insoluble globulin). Influence of heparin and hyaluronic acid on the binding of native collagen, *Hoppe-Seylers Z. Physiol. Chem.*, 360, 597, 1979.
101. **Rubin, K., Kjellen, L., and Öbrink, B.,** Intercellular adhesion between juvenile liver cells. A method to measure the formation of stable lateral contracts between cells attached to a collagen gel, *Exp. Cell Res.*, 109, 413, 1977.
102. **Rubin, K., Höök, M., Öbrink, B., and Timpl, R.,** Attachment of rat hepatocytes to collagen. Binding structure in collagen molecules, in preparation.
103. **Kleinman, H. K., McGoodwin, E. B., Rennard, S. I., and Martin, G. R.,** Preparation of collagen substrates for cell attachment: effect of collagen concentration and phosphate buffer, *Anal. Biochem.*, 94, 308, 1979.
104. **Beachey, E. H., Chiang, T. M., and Kang, E. H.,** Collagen-platelet interaction, *Int. Rev. Connect. Tissue Res.*, 8, 1, 1979.
105. **Muggli, R. and Baumgartner, H. R.,** Collagen induced platelet aggregation: requirement for tropocollagen multimers, *Thromb. Res.*, 3, 715, 1973.
106. **Jaffe, R. and Deykin, D.,** Evidence for a structural requirement for the aggregation of platelets by collagen, *J. Clin. Invest.*, 53, 875, 1974.
107. **Balleisen, L., Marx, R., and Kühn, K.,** Platelet-collagen interaction. The influence of native and modified collagen (type I) on the aggregation of human platelets, *Haemostasis*, 5, 155, 1976.
108. **Balleisen, L., Marx, R., and Kühn, K.,** Uber die stimulierende Wirkung von Kollagen und Kollagenderivaten auf Ausbreitung und Folienadhasion von Thrombozyten in defibrinogenisiertem Menschen-Citratplasma und tierischem Citratplasma, *Blut*, 31, 95, 1975.
109. **Fauvel, F., Legrand, Y. J., and Caen, J. P.,** Platelet adhesion to type I collagen and α1(I) trimers: involvement of C-terminal α1(I)CB6A peptide, *Thromb. Res.*, 12, 273, 1978.
110. **Fauvel, F., Legrand, Y. J., Bentz, H., Fietzek, P. P., Kühn, K., and Caen, J. P.,** Platelet-collagen interaction: adhesion of human blood platelets to purified CB4 peptide from type III collagen, *Thromb. Res.*, 12, 841, 1978.

111. **Fauvel, F., Legrand, Y. J., Kühn, K., Bentz, H., Fietzek, P. P., and Caen, J. P.,** Platelet adhesion to type III collagen; involvement of a sequence of nine amino acids from α1(III)CB4 peptide, *Thromb. Res.*, 16, 269, 1979.
112. **Jamieson, G. A., Urban, C. L., and Barber, A. J.,** Enzymatic basis for platelet collagen adhesion as the primary step in haemostasis, *Nat. New Biol.*, 234, 5, 1971.
113. **Balleisen, L., Gay, S., Marx, R., and Kühn, K.,** Comparative investigation on the influence of human bovine collagen type I, II and III on the aggregation of human platelets, *Klin. Wochenschr.*, 52, 903, 1975.
114. **Olsen, B. R. and Berg, R. A.,** Posttranslational processing and secretion of procollagen in fibroblasts, *Proc. Br. Soc. Exp. Biol.*, 33, 57, 1979.
115. **Hörlein, D., Fietzek, P. P., Wachter, E., Lapiere, C. M., and Kühn, K.,** Amino acid sequence of the aminoterminal segment of dermatosparactic calf skin procollagen type I, *Eur. J. Biochem.*, 99, 31, 1979.
116. **Rhode, H., Wachter, E., Richter, W. J., Brückner, P., Helles, O., and Timpl, R.,** Amino acid sequence of the N-terminal non-collagenous segment of dermatosparactic sheep procollagen type I, *Biochem. J.*, 179, 631, 1979.
117. **Ansay, M., Gillet, A., and Hanset, R.,** La dermatosparaxie hereditaire des bozides: biochemie descriptive de la peau, *Ann. Med. Vet.*, 112, 449, 1968.
118. **Helle, O. and Nes, N. N.,** A hereditary skin defect in sheep, *Acta Vet. Scand.*, 13, 443, 1972.
119. **Lapiere, C. M., Lenaers, A., and Kohn, L. D.,** Procollagen peptidase, an enzyme excising the coordinate peptides of procollagen, *Proc. Natl. Acad. Sci. U.S.A.*, 68, 3054, 1971.
120. **Lenaers, A., Ansay, M., Nusgens, B. V., and Lapiere, C. M.,** Collagen made of extended α chains, procollagen, in genetically defective dermatosparactic calves, *Eur. J. Biochem.*, 23, 533, 1971.
121. **Shocralter, A. M., Pesciotta, D. M., Eikenberry, E. F., Yamamoto, T., Pastan, I., DeCrombrugghe, B., Fietzek, P. P., and Olsen, B. R.,** Nucleotide sequence of a collagen cDNA fragment coding for the carboxyl end of proα1(I) chains, *FEBS Lett.*, 111, 61, 1980.
122. **Engel, J., Brückner, P., Becker, U., Timpl, R., and Rutschmann, B.,** Physical properties of the amino terminal precursor specific portion of type I procollagen, *Biochemistry*, 16, 4026, 1977.
123. **Becker, U., Helle, O., and Timpl, R.,** Characterization of the amino terminal segment in procollagen pα2 chain from dermatosparactic sheep, *FEBS Lett.*, 72, 197, 1977.
124. **Hörlein, D., Fietzek, P. P., and Kühn, K.,** Pro-Gln: the procollagen peptidase cleavage site in the α1(I) chain of dermatosparactic calf skin procollagen, *FEBS Lett.*, 89, 279, 1978.
125. **Kohn, L. D., Isersky, C., Zupnick, J., Lenaers, A., Lee, G., and Lapiere, C. M.,** Calf tendon procollagen peptidase: its purification and endopeptidase mode of action, *Proc. Natl. Acad. Sci. U.S.A.*, 71, 40, 1974.
126. **Tuderman, L., Kivirikko, K. I., and Prockop, D. J.,** Partial purification and characterization of a neutral protease which cleaves the N terminal propeptide from procollagen, *Biochemistry*, 17, 2948, 1978.
127. **Morikawa, T., Tuderman, L., and Prockop, D. J.,** Inhibitors of procollagen N-protease. Synthetic peptides with sequences similar to the cleavage site in the proα1(I) chain, *Biochemistry*, 19, 2646, 1980.
128. **Brandt, A., Hörlein, D., Glanville, R. W., and Kühn, K.,** Amino acid sequence of the amino terminal segment of calf skin procollagen type III, in preparation.
129. **Fessler, L. I. and Fessler, J. H.,** Characterization of type III collagen from chick embryo blood vessels, *J. Biol. Chem.*, 254, 233, 1979.
130. **Nowack, H., and Olsen, B. R., and Timpl, R.,** Characterization of the amino terminal segment in type III procollagen, *Eur. J. Biochem.*, 70, 205, 1976.
131. **Bruckner, P., Bächinger, H. P., Timpl, R., and Engel, J.** Three conformationally distinct domains in the amino terminal segment of type III procollagen and its rapid triple helix coil transition, *Eur. J. Biochem.*, 90, 595, 1978.
132. **Becker, U., Fietzek, P. P., Furthmayr, H., and Timpl, R.,** Non-helical sequences of rabbit collagen. Correlation with antigenic determinants detected by rabbit antibodies in homologous regions of rat and calf collagen, *Eur. J. Biochem.*, 54, 359, 1975.
133. **Uitto, J. and Prockop, D. J.,** Rate of helix formation by intracellular procollagen and protocollagen: evidence for a role for disulfide bonds, *Biochem. Biophys. Res. Commun.*, 55, 904, 1973.
134. **Fessler, L. I. and Fessler, J. H.,** Protein assembly of procollagen and effect of hydroxylation, *J. Biol. Chem.*, 249, 7637, 1974.
135. **Rosenblom, J., Endo, R., and Harsch, M.,** Termination of procollagen chain synthesis by puromycin. Evidence that assembly and sectetion required a COOH terminal extension, *J. Biol. Chem.*, 251, 2070, 1976.

136. **Wiestner, M., Krieg, T., Hörlein, D., Glanville, R. W., Fietzek, P. P., and Müler, P. K.,** Inhibiting effect of procollagen peptides on collagen biosynthesis in fibroblast cultures, *J. Biol. Chem.*, 10, 7016, 1979.
137. **Paglia, L., Wilczek, J., de Leon, L. D., Martin, G. R., Hörlein, D., and Müller, P.,** Inhibition of procollagen cell free synthesis by amino terminal extension peptides, *Biochemistry*, 18, 5030, 1979.
138. **Kefalides, N. A., Alper, R., and Clark, C.,** Biochemistry and metabolism of basement membranes, *Int. Rev. Cytol.*, 61, 167, 1979.
139. **Kefalides, N. A.,** Isolation of collagen from basement membrane containing three identical α chains, *Biochem. Biophys. Res. Commun.*, 45, 226, 1971.
139a. **Dixit, S. N. and Glanville, R. W.,** Isolation and characterization of pepsin solubilised basement membrane collagens (Type IV) from human placenta, bovine kidney cortices and bovine lens capsule, in *Immunochemistry of the Extracellular Matrix*, Vol. 1, Furthmayr, H., Ed., CRC Press, Boca Raton, Fla., 1982, in press.
140. **Dixit, S. and Glanville, R. W.,** Isolation and characterization of pepsin solubilized basement membrane collagen (type IV) from human placenta, bovine, kidney, cartices and bovine lens capsule, in *Immunochemistry of the Extracellular Matrix*, Vol. 1, Furthmayr, H., Ed., CRC Press, Inc., Boca Raton, Fla., 1982, in press.
141. **Schuppan, D., Timpl, R., and Glanville, R. W.,** Discontinuities in the triple helical sequence Gly-X-Y of basement membrane (Type IV) collagen, *FEBS Lett.*, 115, 297, 1980.
142. **Glanville, R. W. and Rauter, A.,** Pepsin fragments of human placental basement-membrane collagen showing interrupted triple-helical amino acid sequences, *Hoppe-Seylers Z. Physiol. Chem.*, 362, 943, 1981.
143. **Timpl, R., Martin, G. R., Bruckner, P., Wick, G. and Wiedemann, H.,** Nature of the collagenous protein in a tumor basement membrane, *Eur. J. Biochem.*, 84, 43, 1978.
144. **Schwartz, D. and Veis, A.,** Characterization of basement membrane collagen of bovine anterior lens capsule via segment-long-spacing crystallites and the specific cleavage of the collagen by pepsin, *FEBS Lett.*, 85, 326, 1978.
145. **Schwartz, D. and Veis, A.,** Structure of bovine anterior lens capsule basement membrane collagen molecules from electron microscope, *Biopolymers*, 18, 2363, 1979.
146. **Risteli, J., Bächinger, H. P., Engel, J., Furthmayr, H., And Timpl, R.,** 7-S collagen: characterization of an unusual basement membrane structure, *Eur. J. Biochem.*, 108, 239, 1980.
147. **Kühn, K., Wiedemann, H., Timpl, R., Risteli, J., Dieringer, H., Voss, T., and Glanville, R. W.,** Macromolecular structure of basement membrane collagens. Identification of 7S collagen as a crosslinking domain of Type IV collagen, *FEBS Lett.*, 125, 123, 1981.
148. **Timpl, R., Wiedemann, H., Van Delden, V., Furthmayr, H., and Kühn, K.,** A network model foi the organization of type IV collagen molecules in basement membranes, *Eur. J. Biochem.*, 1981, in press.
149. **Timpl, R., Oberbäumer, I., Furthmayr, H. and Kühn, K.,** Macromolecular organization of Type IV collagen, in *New Trends in Basement Membrane Research*, Kühn, K., Timpl, R. and Schöne, H., Eds., Raven Press, New York, 1981, in press.
150. **Crouch, E. and Bornstein, P.,** Characterization of a Type IV procollagen synthesized by human amniotic fluid cells in culture, *J. Biol. Chem.*, 254, 4197, 1979.
151. **Tryggvason, K., Gehron-Robey, P., and Martin, G. R.,** Biosynthesis of type IV procollagens, *Biochemistry*, 19, 1284, 1980.
152. **Alitalo, K., Vaheri, A., Krieg, T., and Timpl, R.,** Biosynthesis of two subunits of Type IV procollagen and of other basement membrane proteins by a human tumor cell line, *Eur. J. Biochem.*, 109, 247, 1980.
153. **Roveri, N., Ripamonti, A., Bigi, A., Volpin, D., and Giro, M. G.,** X-ray diffraction study of bovine lens capsule collagen, *Biochim. Biophys. Acta*, 576, 404, 1973.
154. **Sage, H. and Bornstein, P.,** Characterization of novel collagen chain in human placenta and its relation to AB-collagen, *Biochemistry*, 18, 3815, 1979.
155. **Chiang, T. M., Mainardi, C. L., Seyer, M. J., and Kang, A. H.,** Collagen-platelet interaction; Type V (A-B) collagen induces platelet aggregation, *J. Lab. Clin. Med.*, 95, 99, 1980.
156. **Eyre, D. R.,** Collagen: molecular diversity in the body's protein scaffold, *Science*, 207, 1315, 1980.
157. **Burgeson, R. E. and Hollister, D.,** Collagen heterogeneity in human cartilage: identification of several new collagen chains, *Biochem. Biophys. Res. Commun.*, 71, 1124, 1979.
158. **Shimokomaki, M., Duance, V. C., and Bailey, A. J.,** Identification of a new disulfide bonded collagen from cartilage, *FEBS Lett.*, 121, 51, 1980.
159. **Ayad, S., Abedin, M. Z., Grundy, S. M., and Weiss, J.,** Isolation and characterization of an unusual collagen from hyaline cartilage and invertebral disc, *FEBS Lett.*, 123, 195, 1981.

160. **Reese, C. A. and Mayne, R.,** Minor collagens of chicken hyaline cartilage, *Biochemistry*, 1981, in press.
161. **Reese, C. A., Wiedemann, H., Kühn, K., and Mayne, R.,** Characterization of a high soluble collagenous molecule isolated from chicken hyaline cartilage, *Biochemistry*, 1981, in press.
162. **Chung, E., Rhodes, R. K., and Miller, E. J.,** Isolation of three collagenous components of probable basement membrane origin from several tissues, *Biochem. Biophys. Res. Commun.*, 71, 1167, 1976.
163. **Furuto, D. K. and Miller, E. J.,** Isolation of a unique collagenous fraction from limited pepsin digest of human placental tissue, *J. Biol. Chem.*, 255, 290, 1980.
164. **Furuto, D. and Miller, E. J.,** Characterization of a unique collagenous fraction from limited pepsin digests of human placental tissue: molecular organization of the native aggregate, *Biochemistry*, 20, 1635, 1981.
165. **Jander, R., Rauterberg, J., Voss, B., V., and Basewitz, D. B.,** A cysteine-rich collagenous protein from bovine placenta, *Eur. J. Biochem.*, 1, 14, 1981.
166. **Glanville, R. W., Voss, T., and Kühn, K.,** A comparison of the flexibility of molecules of basement membrane and interstitial collagens, in *New Trends in Basement Membrane Research*, Kühn, K., Timpl, R., and Schöne, H., Eds., Raven Press, New York, 1981, in press.

Chapter 2

NATIVE COLLAGEN FRACTIONATION

Robert L. Trelstad

TABLE OF CONTENTS

I. COLLAGEN SOLUBILIZATION

The initial step in collagen purification is to disrupt the tissue in the cold (4°C), render the collagens soluble, and separate them from the noncollagenous materials with which they are associated, either as a result of the tissue disruption or because of natural in vivo interaction. Homogenization of tissues prior to extraction is essential to obtain satisfactory sampling, and a suitable yield and can be done on the fresh tissues or after freezing and/or lyophilization. With the recognition that endogenous tissue proteases may cleave the collagens either within the body of the triple helix or at the nonhelical extensions, it has become important to minimize such proteolysis by initial treatment of the tissue with protease inhibitors. The protocols described by Gay and Miller[1] and Mayne and Zettergren[2] employ a neutral solution of 1 *M* NaCl, 50 m*M* Tris-HCl, pH 7.5, containing 1 m*M* phenylmethanesulfonyl fluoride, 10 to 20 m*M* EDTA (ethylenediaminetetraacetic acid), 5 to 10 m*M* N-ethylmaleimide, 50 m*M* E-aminocaproic acid, and 1 m*M* *p*-chloromercuriphenyl sulfonic acid. The use of inhibitors in relatively high ionic strength buffers promotes dissociation of endogenous proteases which might be bound within the tissue. Following homogenization and washing, the starting materials are then directly placed into the various solvents for collagen extraction.

In general, native collagen molecules at 4°C are soluble below pH 5 only if the ionic strength of the solvent is low, whereas above pH 5, a moderate ionic strength is necessary. Based on this "salting in" and "salting out" behavior of the collagens, a rational solubilization and fractionation can be effected (see Figure 1). The conditions generally used for the solubilization of the collagens in a native state are

1. Low ionic strength solvents at acid pH
2. Moderate ionic strength solvents at neutral pH
3. Limited proteolysis at acid pH

Solutions at acid pH cause collagen fibrils to swell by disruption of intermolecular salt linkages and also by breaking Schiff base covalent cross-linkages. The acids most commonly used as solvents are acetic acid, citric acid, formic acid, and hydrochloric acid. The organic acids are used at concentrations from 0.1 to 0.5 *M*, and HCl is generally used at a concentration close to 0.01 *M*.

At neutral pH, the collagens are not soluble at low ionic strength, and the buffers should contain additional salts at concentrations from 0.15 to 1 *M*. The most commonly employed buffers are tris-HCl and phosphate, and the most commonly added salt is NaCl. The dissolution of tissue collagens into polydisperse macroaggregates can be achieved by treatment of the tissue with EDTA or glucose in a neutral solvent[3,4] and also by cyclic exposures of such treated tissues to an acid pH, a process which has been called lyotrophic relaxation.[5]

Solubilization of tissue collagens by limited proteolysis has generally been accomplished with pepsin at temperatures from 4 to 18°C.[6,7] The tissues are suspended in approximately three to five volumes (in respect to the wet weight of the tissue in grams) of 0.5 *M* acetic acid containing pepsin at a concentration ranging from 0.1 to 50 mg of pepsin per gram wet weight of the tissue. An alternative manner of estimating the pepsin concentration is to use 1 mg of pepsin per milliliter of solvent. Prior extraction of the tissues with neutral solutions containing protease inhibitors, as described above, and/or reagents, such as 1 *M* NaCl, 2 *M* $MgCl_2$, 4 *M* guanidinium hydrochloride, or 5 *M* urea, with 10 m*M* 2-mercaptoethanol[2,7] are useful to remove tissue components, such as the glycosaminoglycans and noncollagenous proteins,

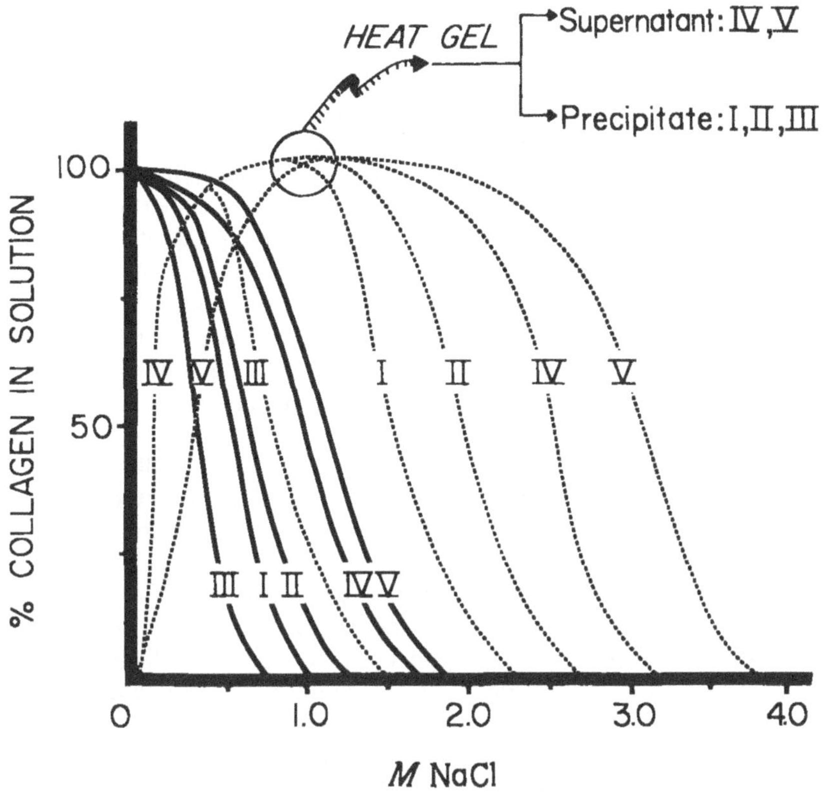

FIGURE 1. Graphic presentation of the "salting in" and "salting out" behavior of the various collagen types at acid pH (solid lines) and at neutral pH (dotted lines). At acid pH, all collagens are soluble in the absence of salt, and its addition leads to progressive precipitation. At neutral pH, the collagens are not soluble at very low ionic strengths, and salt is necessary for solubility. The salt fractionation techniques most commonly employed involve precipitation along the "right" side of the illustrated curves. Recent studies[1] with collagen Types IV and V in neutral solution indicate that fractional precipitation in the presence of 2 *M* urea at low ionic strengths, i.e., the "left" side of the neutral pH curves, effectively separates these two collagens. Attention to the behavior of the other collagen types at low ionic strength in neutral solution may provide additional fractionation procedures.

The curves in this figure are estimates based on data in the literature. They should be used as a guide for fractionation, but the specific concentrations of NaCl appropriate for any preparation should be individually determined.

If a neutral solution of collagen is heated to 37°C for 4 to 16 hr, a gel will form from Types I, II, and III, but not from IV and V. Heat gelation fractionation coupled with salt fractionation provides an effective purification of collagen Types IV and V.

which interfer with access of the pepsin to cleavage sites in the collagens. Adjustment of the pH of the tissue homogenate to 2.0 to 2.5 with HCl assures optimal digestion with the pepsin. The buffering capacity of the tissue homogenate is considerable, and the acidity obtainable with 0.5 *M* acetic acid is insufficient to maximize the peptic activity. Following digestion the centrifuged supernatant should be adjusted to a pH above 7 to inactivate the pepsin.

The rationale for using pepsin to solubilize the collagens is based on the relative resistance of the helical portion of the molecule to digestion with this enzyme. Other enzymes such as trypsin can attack the triple-helical portion of the molecule as, for example, in Type III collagen,[8] and Type IV collagen can be partially cleaved by pepsin.[9] Although the tryptic sensitivity of Type III collagen has not been employed in a bulk preparative manner, the coupled sensitivity of the Type IV collagens to pepsin and the resistance of some fragments to collagenase have been useful in solubilization.[9] Further developments using selective sensitivities of the collagens to different proteases can be anticipated.

Solubilization of tissue collagens and the yields obtained depend on the age, species, and prior history of the animal. Starvation diminishes the yields of neutral and acid-soluble collagen from tissues as does increasing age.[10] By administration of the lathyrogen, beta aminopropionitrile (BAPN), it is possible to inhibit the covalent cross-linking which is responsible, in part, for the insolubility of the collagen.[11] Treatment with BAPN is usually carried out for 2 to 3 weeks prior to sacrifice of the animal. The drug can be administered via the drinking water at a concentration of from 0.01 to 0.1% w/v. BAPN can also be conveniently administered to chick embryos by injection of 0.1 mℓ of saline containing the drug into the blunt end (air sac) of the egg. The weight of BAPN which we have found will provide extremely lathyritic tissues in 3 days is equal to 10% of the dry weight of the embryo in milligrams on the day of the injection. A table of dry weights of chick embryos at various days of incubation is available.[12]

The native collagen fractionation schemes described below which separate the various interstitial collagen types (I, II, and III) are relatively independent of the methods of solubilization. That is, Type I rat tail tendon collagen solubilized by acetic acid extraction generally behaves the same during subsequent purification, as does neutral-soluble, lathyritic, or pepsin-solubilized rat tail tendon collagen. Caution should be exercised, however, in generalizing in that collagens solubilized by proteolysis, especially Types IV and V, have differing solubility properties, depending on the proteolytic conditions.[13]

In the course of fractionating the collagens, the best results are obtained by cyclic precipitation and resolubilization. We have found that the best results are obtained if three or more precipitations are employed, each under different conditions. It should be noted that the conditions which promote solubilization of the tissue-bound collagens also pertain to partially purified pellets obtained after centrifugation. That is, solubilization with solvents of pH less than 5.0 should be done in the relative absence of salt, whereas solvents above pH 5.0 require moderate ionic strengths. For example, the centrifuged pellet from a neutral solution precipitated by high (20%) NaCl will contain a good deal of salt. Resolubilization of this pellet in 0.1 *M* acetic acid may be hindered by such salt which should be removed by dialysis.

II. BULK FRACTIONATION BY TISSUE SELECTION

The distribution of the various collagens is different from tissue to tissue, and an important method for obtaining one particular collagen type is to begin with a tissue enriched in that collagen. The tissues often used as starting sources for particular collagen types are listed in Table 1.

III. BULK FRACTIONATION BY SOLUBILITY

The solubilities of the various collagens are sufficiently different from each other that effective procedures have been devised which accomplish relatively rapid fractionation from bulk solutions. The most important solution conditions which dictate the solubilities of the collagens are the pH and ionic strength of the buffer and the collagen concentration. The relative solubilities of each of the collagen types under

Table 1
TISSUES ENRICHED IN THE LISTED COLLAGEN TYPES

Collagen type	Tissue source	Ref.
Type I	Tendon	34
	Skin	35
	Bone	36
Type I trimer	Gingiva	37
	Skin	38
	Dentine	39
	Tendon	40
Type II	Cartilage	41
	Nucleus pulposus	42
	Vitreous body	43
Type III	Uterine leiomyoma	41
	Skin	16
	Aorta	45
	Parenchymal organs	44, 46
Type IV	Kidney	20
	Mouse tumor	28
	Lens capsule	2, 13
	Gizzard	1
	Placenta	47
	Aorta	45
Type V	Placenta membranes	48
	Chick embryo	21

carefully controlled conditions have not been thoroughly studied, but a number of purification schemes have been derived from these solubility properties which form the basis for most fractionation schemes currently in use.[14–16] The relative solubilities of the various collagens at acid and neutral pH are graphically illustrated in Figure 1. Although the profiles in Figure 1 are estimates, they indicate that despite the similarities of the precipitation profiles for each of the collagens, their differences are adequate to achieve separation if the fractionation is carefully conducted.

The conditions at acid pH usually employed in bulk salt fractionation consist of collagen at concentrations below 1.5 mg/mℓ in solutions of acetic acid (0.1 to 0.5 *M*); the conditions at neutral pH use similar collagen concentrations and are usually conducted in solutions of 0.15 *M* phosphate buffer or 0.05 to 0.1 *M* Tris-HCl at pH 7.5. The precipitants which have been used most commonly at both acid and neutral pH are NaCl, ethanol, and NH_4SO_4.[15] The manner in which the precipitants are added to the collagen solution is important. The most effective approach when using a salt is to employ a concentrated solution of that salt which is slowly added to the collagen solution with stirring so that the salt concentration never exceeds the final concentration desired. This, of course, can also be achieved by dialysis. If solid salt is added, the environment around the salt crystals during their solubilization will transiently exceed that which is achieved at equilibrium and lead to altered results. The duration of the precipitation period is also important in that longer time periods (24 hr) produce different solubility profiles from those obtained rapidly (30 min).[15]

The most commonly employed low ionic strength condition for collagen precipitation is achieved by dialysis against 0.01 *M* Na_2HPO_4. This condition quantitatively precipitates all of the collagens and is also useful in separating native from denatured preparations in that denatured collagens are very soluble at this low ionic strength. It is not uncommon to dialyze collagen solutions against either distilled or tap water to induce their precipitation. This is an effective low ionic strength condition and is suitable as long as the pH of the solution is above 5. Selective precipitation at low ionic strength in neutral buffers occurs from mixtures of Types IV and V collagens in the presence of 2 *M* urea.[1] Further study of the precipitation of collagen Types I, II, and III at low ionic strengths from neutral solutions in the presence or absence of urea may provide additional fractionation procedures.

Another fractionation approach which has not been explored is to change the pH of a solution of collagen while keeping the salt concentration constant. The precipitation of Type V collagen upon dialysis from neutral solutions to the pH 4.8 sodium citrate buffer employed in carboxymethyl cellulose chromatography, however, suggests that such fractionation procedures should be useful.[17]

In addition to salt concentration and pH, the concentration of the collagen and the relative purity of the solution dictate the behavior of the preparation during fractional salt precipitation. Higher collagen concentrations and/or the presence of several different collagen types tend to render the collagens more precipitable. Part of this effect is entrapment of more soluble species by those that are less soluble. In addition, when the basement membrane collagens become more isolated from the interstitial collagens, losses during purification increase owing to their increased adsorption to glassware and other surfaces.

The influence of salt concentration on the precipitation of the collagens from solution can also be used to selectively solubilize collagens which are already in a precipitated state. For example, if a mixture of collagen Types I and III in precipitate form is extracted with a neutral solution containing 1.5 *M* NaCl, the majority of the Type I collagen will be solubilized, while the Type III will remain insoluble.[18]

IV. COLUMN FRACTIONATION BY SOLUBILITY

The preferential solubilization of precipitated collagens by solutions of varying ionic strengths has been adapted by Ehrlich[19] for rapid separation of collagen mixtures. Collagens in solution are applied to the top of a column of Sephadex® G-25 which has been equilibrated with 30% NaCl at neutral pH. As the collagens enter the resin in solution, they precipitate because of the high ionic strength within the column. The column is then eluted by a stepwise series of neutral solutions containing lower ionic strength buffers which solubilize the collagens and allow them to reenter the mobile buffer phase. The column is finally washed with acetic acid and then a denaturing solvent such as SDS in order to strip all bound materials before reequilibration. This procedure is also effective for the more soluble collagen types if done under acid conditions.

V. BULK FRACTIONATION BY HEAT GELATION

Neutral solutions of the interstitial collagens gel when heated to 37°C, whereas collagen Types IV and V do not.[20,21] When heterogeneous mixtures of collagens in the early stages of purification are subjected to heat gelation fractionation, a near complete separation between collagen Types I, II, and III from IV and V is achieved.[20,21] Neutral solutions at an ionic strength of from 0.08 to 0.15 are incubated at 37°C for from 4 to 16 hr and then centrifuged to separate the gelled from nongelled portions. The majority of the Type IV and V in this one step remain in the nongelling supernatant and can be further purified by salt fractionation. Analysis of the native state of the nongelling fractions by protease resistance, precipitability at low ionic strength, and by direct measurement of the circular dichroism spectrum and melting temperature indicate that the collagens are in a native state.[21] Schwartz et al.[13] have reported that the Type IV collagens isolated from the bovine lens capsule by peptic digestion are progressively degraded by prolonged exposure to pepsin in a manner which alters their precipitability and thermal gelation behavior.

It has become common to isolate microtubular proteins by cyclic heat gelation and cooling.[22] Although collagen gels will dissociate upon cooling, this technique has not been extensively employed to study the different collagens primarily because those fractions of most current interest and for which the technique has been applied, i.e., Type IV and V, are in the nongelling supernatants. In addition to its potential use in the separation of collagens of different types, the cyclic process of heat gelation and cooling may also yield fractions of differing biochemical character within a particular collagen type and provide some clues about the nature of the gelling process itself. The existence of different fractions within a relatively homogeneous preparation based on heat gelation has been noted.[23]

VI. BULK FRACTIONATION BY DENATURATION/ RENATURATION

Covalent cross-links between alpha chains maintain their association in a registered fashion during denaturation. Molecules containing such cross-links generally renature more rapidly upon cooling. Accordingly mixtures of molecules which contain molecular species which have different rates of renaturation can be separated in bulk by cyclic denaturation/renaturation. ChandraRajan[24] isolated the collagens in human fetal skin by pepsinization and then separated the Type I from the Type III collagen by denaturation/renaturation. The collagen preparation was denatured in 2 *M* guanidine-HCl at pH 7.5 at 45°C for 30 min, and then the mixture of collagens was allowed to renature during dialysis against deionized water at 4°C. Because of the disulfide cross-links in the pepsin-resistant portion of the Type III collagen, this species renatured rapidly and precipitated in the low ionic strength deionized water, whereas Type I did not. By keeping the concentration of the denatured solution below 0.25 mg/mℓ, entrapment of denatured Type I within the Type III aggregate was minimized.

The melting temperatures of collagens Types I, II, III, IV, and V in the forms obtained by any of the isolation techniques listed above are similar and in the range of 33 (at acid pH) to 41°C (at neutral pH). The major difference which can be exploited for fractionation purposes is therefore the kinetic differences in the renaturation process such as employed by Bentz et al.[25] in studies of Type V chains. Recently, Timpl and coworkers[9] have isolated a basement membrane collagen fragment which has a melting temperature of about 70°C. Should other collagens and/ or fragments derived from them show similar properties, selective denaturation will also be a useful approach to the isolation of different collagen types. Moreover the increased thermal stability of the collagens in an intact tissue (60°C)[26] may differ for the various collagens such that pretreatment of whole tissues at various temperatures may promote extraction of certain collagen types.

VII. COLUMN FRACTIONATION BY DENATURATION/ RENATURATION

Denatured collagen fractions recovered from either molecular sieve or ion exchange chromatography will renature upon dialysis at low temperature and can be precipitated at low ionic strengths. Mixtures of Type I and III collagens, which have been denatured in 1 *M* $CaCl_2$ and chromatographed on an agarose A-5M resin, can be resolved into several fractions dictated by size. That fraction which elutes as a three-chained "gamma" fraction (mol wt approximately 300,000 daltons) will be considerably enriched in Type III collagen, since the nonreducing conditions of the chromatography do not allow dissociation of the chains. These disulfide cross-linked molecules renature rapidly when dialyzed against cold distilled water and then against 0.01 *M* Na_2HPO_4. In principal this approach is little different from that described above for bulk fractionation except that prior selection of fractions to be renatured can be achieved by the chromatrography. The renatured molecules obtained in this manner are not fully native as determined by their recovery of optical rotation,[18,25] but they do form SLS crystallites, are relatively resistant to peptic digestion, and will form striated fibrils upon thermal gelation at 26 to 30°C.

A separation of Type III collagen by transient formation of a thiol-mediated cross-link to a solid state has been described by Sykes.[27] Collagens from lyophilized tissue powder were solubilized (4 mg/mℓ), reduced, and denatured in 0.1 *M* Tris-HCl, pH 8.0, containing 1.0 *M* urea, 0.3 *M* NaCl, 1 m*M* EDTA, and 5 m*M* dithiothreitol (DTT) for 30 min at 60°C. The reduced solution was passed through Sephadex® G-25 to remove the DTT and then mixed with thiol-activated Sepharose® 4B which had been equilibrated with the above buffer minus DTT. The slurry of resin and collagen was then poured into a jacketed column maintained at 42°C and eluted with buffer minus DTT, buffer plus 1 *M* NaCl, and finally with buffer plus DTT. The Type III collagen which was covalently bound to the column was readily eluted following the addition of DTT.

VIII. COLUMN FRACTIONATION BY CHARGE OR ADSORPTION

Chromatography of native collagens on ion exchange resins has been used to separate procollagens, especially pro-Type I from pro-Type III[18] and pepsin-solubilized Types IV and V[25,28,29]. Collagen solutions at concentrations of approximately 1 mg/mℓ or less are applied to the columns in the buffer with which the column is equilibrated. Polyanionic materials such as the glycosaminoglycans can be removed by passage of neutral solutions of collagen in the buffer noted below over a column of DEAE-cellulose which is equilibrated with this buffer containing NaCl to a final concentration of at least 0.20 *M*.

Conditions for anion exchange chromatography on DEAE-cellulose designed to allow protein binding and subsequent elution with competing anions are neutral pH (5 m*M* Tris-HCl, pH 7.5 to 8.6), little or no salt (0 to 20 m*M* NaCl), and deionized urea (2 *M*). The column is jacketed and chilled (4 to 8°C), and the bound collagens are eluted with a linear gradient of NaCl to a final concentration of 0.3 to 0.5 *M* NaCl.[18,25]

Conditions for cation exchange chromatography are an acid pH (40 m*M* sodium acetate, pH 4.8), little or no salt, and deionized urea (2 *M*). The bound proteins are eluted by a linear gradient of NaCl to a final concentration of 0.4 to 0.6 *M*.[29]

Native procollagen will adsorb to glass beads (CPG-10, 120/200 mesh, Electro

Nucleonics Inc., Fairfield, N.J.) equilibrated with Dulbecco's phosphate buffered saline without $CaCl_2$ containing 25 m*M* EDTA.[30] Serum-free culture media is applied to such a column, and noncollagenous proteins are removed by elution with 1 *M* potassium phosphate, pH 8.2, containing 25 m*M* EDTA. The procollagens can then be eluted with 1 *M* Tris-HCl, pH 8.2, containing 25 m*M* EDTA.

IX. BULK AND COLUMN FRACTIONATION BY LIGAND INTERACTION

Specific collagen ligands in a free or solid state have not been exploited for either bulk precipitation or column separations. Since native collagen binds to and can be precipitated by a number of different materials, including proteoglycans,[31] antibodies,[32] and nonspecific IgG coated latex particles,[33] it is likely that procedures for the selective precipitation of native collagens in chemical amounts will become as useful as they are for the precipitation of radiolabeled native collagens.

REFERENCES

1. **Mayne, R. and Zettergren, J. G.**, Type IV collagen from chicken muscular tissues. Isolation and characterization of the pepsin-resistant fragments, *Biochemistry*, 19, 4065, 1980.
2. **Gay, S. and Miller, E. J.**, Characterization of lens capsule collagen: evidence for the presence of two unique chains in molecules derived from major basement membrane structures, *Arch. Biochem. Biophys.*, 1980, 370, 1979.
3. **Schofield, J. D., Freeman, I. L., and Jackson, D. S.**, The isolation, and amino acid composition and carbohydrate composition, of polymeric collagens prepared from various human tissues, *Biochem. J.*, 124, 467, 1971.
4. **Hayashi, T. and Nagai, Y.**, Increased amount of soluble collagen by new extraction methods and its relation to the nature of associating forces among collagen molecules, *Connect. Tissue Res.*, 1, 39, 1972.
5. **Veis, A., Bhatnagar, R. S., Shuttleworth, C. A., and Mussell, S.**, The solubilization of mature, polymeric collagen fibrils by lyotropic relaxation, *Biochim. Biophys. Acta*, 200, 97, 1970.
6. **Rubin, A. L., Drake, M. P., Davison, P. F., Pfahl, D., Speakman, P. T., and Schmitt, F. O.**, Effects of pepsin treatment on the interaction properties of tropocollagen macromolecules, *Biochemistry*, 4, 181, 1965.
7. **Miller, E. J.**, Structural studies on cartilage collagen employing limited cleavage and solubilization with pepsin, *Biochemistry*, 11, 4903, 1972.
8. **Miller, E. J., Finch, J. E., Jr., Chung, E., Butler, W. T., and Robertson, P. B.**, Specific cleavage of the native type III collagen molecule with trypsin, *Arch. Biochem. Biophys.*, 173, 631, 1976.
9. **Timpl, R., Risteli, J., and Bachinger, H. P.**, Identification of a new basement membrane collagen by the aid of a large fragment resistant to bacterial collagenase, *FEBS Lett.*, 101, 265, 1979.
10. **Gross, J.**, The influence of growth rate on neutral salt extracts of guinea pig dermis, *J. Exp. Med.*, 107, 265, 1958.
11. **Levene, C. I. and Gross, J.**, Alterations in state of molecular aggregation of collagen induced in chick embryos by β-aminopropionitrile (Lathyrus Factor), *J. Exp. Med.*, 110, 771, 1959.
12. **Neuman, R. E.**, Hydroxyproline content of the developing chick embryo, *Proc. Soc. Exp. Biol. Med.*, 75, 37, 1950.
13. **Schwartz, D., Chin-Quee, T., and Veis, A.**, Characterization of bovine anterior lens-capsule basement membrane collagen. I. Pepsin susceptibility, salt precipitation and thermal gelation: a property of non-collagen component integrity, *Eur. J. Biochem.*, 103, 21, 1980.
14. **Trelstad, R. L., Kang, A. H., Toole, B. P., and Gross, J.**, Collagen heterogeneity: high resolution separation of native $[\alpha 1(I)]_2\alpha 2$ and $[\alpha 1(II)]_3$ and their component α chains, *J. Biol. Chem.*, 247, 6469, 1972.

15. **Trelstad, R. L., Catanese, V. M., and Rubin, D. R.,** Collagen fractionation: separation of native types I, II and III by differential precipitation, *Anal. Biochem.*, 71, 114, 1976.
16. **Epstein, E. H., Jr.,** $[\alpha 1(III)]_3$ human skin collagen. Release by pepsin digestion and preponderance in fetal life, *J. Biol. Chem.*, 249, 3225, 1974.
17. **Hong. B-S., Davison, P. F., and Cannon, D. J.,** Isolation and characterization of a distinct type of collagen from bovine fetal membranes and other tissues, *Biochemistry*, 18, 4278, 1979.
18. **Byers, P. H., McKenney, K. H., Lichtenstein, J. R., and Martin, G. R.,** Preparation of Type III procollagen and collagen from rat skin, *Biochemistry*, 13, 5243, 1974.
19. **Ehrlich, H. P.,** Zone precipitation chromatography: its use in the isolation of different collagen types, *Prep. Biochem.*, 9, 407, 1979.
20. **Trelstad, R. L. and Lawley, K. R.,** Isolation and initial characterization of human basement membrane collagens, *Biochem. Biophys. Res. Commun.*, 76, 376, 1977.
21. **Trelstad, R. L., Lawley, K. R., Hayashi, K., Ehrlich, H. P., and Silver, F. S.,** Type V collagen from the chick embryo: biochemical, physicochemical and ultrastructural characteristics, *Collagen Res.*, 1, 39, 1981.
22. **Shelanski, N. L., Gaskin, F., and Cantor, C. R.,** Microtubule assembly in the absence of added nucleotides, *Proc. Natl. Acad. Sci. U.S.A.*, 70, 765, 1973.
23. **Fessler, J. H.,** Some properties of neutral-salt soluble collagen I, *Biochem. J.*, 76, 452, 1960.
24. **ChandraRajan, J.,** Separation of type III collagen from type I collagen and pepsin by differential denaturation and renaturation, *Biochem. Biophys. Res. Commun.*, 83, 180, 1978.
25. **Bentz, H., Bachinger, H. P., Glanville, R., and Kuhn, K.,** Physical evidence for the assembly of A and B chains of human placental collagen in a single triple helix, *Eur. J. Biochem.*, 92, 563, 1978.
26. **Gross, J.,** Thermal denaturation of collagen in the dispersed and solid state, *Science*, 143, 960, 1964.
27. **Sykes, B. C.,** The separation of two soft-tissue collagens by covalent chromatography, *FEBS Lett.*, 61, 180, 1976.
28. **Timpl, R., Martin, G. R., Bruckner, P., Wick, G., and Wiedemann, H.,** Nature of the collagenous protein in a tumor basement membrane, *Eur. J. Biochem.*, 84, 43, 1978.
29. **Kresina, T. F. and Miller, E. J.,** Isolation and characterization of basement membrane collagen from human placental tissue. Evidence for the presence of two genetically distinct collagen chains, *Biochemistry*, 18, 3089, 1979.
30. **Gerard, S. and Mitchell, W. M.,** Isolation and purification of human type I procollagen by adsorption to glass beads, *Anal. Biochem.*, 96, 433, 1979.
31. **Toole, B. P.,** Binding and precipitation of soluble collagens by chick embryo cartilage proteoglycan, *J. Biol. Chem.*, 251, 895, 1976.
32. **Bell, W., Timpl, R., and Furthmayr, H.,** Conformation dependence of antigenic determinants on the collagen molecule, *Immunology*, 24, 13, 1974.
33. **Eeckhout, Y., Cambiaso, C. L., Riccomi, H., Vaes, G., and Masson, P. L.,** Agglutination of IgG-coated particles by collagen, *FEBS Lett.*, 107, 69, 1979.
34. **Bornstein, P.,** Comparative sequence studies of rat skin and tendon collagen. II. The absence of a short sequence at the amino terminus of the skin α1 chain, *Biochemistry*, 8, 63, 1969.
35. **Fietzek, P. P. and Piez, K. A.,** Isolation and characterization of the cyanogen bromide peptides from the α2 chain of rat skin collagen, *Biochemistry*, 8, 2129, 1969.
36. **Miller, E. J., Martin, G. R., Piez, K. A., and Powers, M. J.,** Characterization of chick bone collagen and compositional changes associated with maturation, *J. Biol. Chem.*, 242, 5481, 1967.
37. **Narayanan, A. S. and Page, R. C.,** Biochemical characterization of collagens synthesized by fibroblasts derived from normal and diseased human gingiva, *J. Biol. Chem.*, 251, 5464, 1976.
38. **Uitto, J.,** Collagen polymorphism: isolation and partial characterization of α1(I)-trimer molecules in normal human skin, *Arch. Biochem. Biophys.*, 192, 371, 1979.
39. **Wohllebe, M. and Carmichael, D. J.,** Type-I trimer and type-I collagen in neutral-salt-soluble lathyritic-rat dentine, *Eur. J. Biochem.*, 92, 813, 1978.
40. **Jimenez, S. A., Bashey, R. I., Benditt, M., and Yankowski, R.,** Identification of collagen $[\alpha 1(I)]_3$ trimer in embryonic chick tendons and calvaria, *Biochem. Biophys. Res. Commun.*, 78, 1354, 1977.
41. **Miller, E. J. and Matukas, V. J.,** Chick cartilage collagen: a new type of α1 chain not present in bone or skin of the species, *Proc. Natl. Acad. Sci. U.S.A.*, 64, 1264, 1969.
42. **Eyre, D. R. and Muir, H.,** Collagen polymorphism: two molecular species in pig intervertebral disc, *FEBS Lett.*, 42, 192, 1974.
43. **Swann, D. A. and Sotman, S. S.,** The chemical composition of bovine vitreous-humour collagen fibres, *Biochem. J.*, 185, 545, 1980.

44. **Chung, E. and Miller, E. J.,** Collagen polymorphism: characterization of molecules with the chain composition $[\alpha 1(III)]_3$ in human tissues, *Science*, 183, 1200, 1974.
45. **Trelstad, R. L.,** Human aorta collagens: evidence for three distinct species, *Biochem. Biophys. Res. Commun.*, 57, 717, 1974.
46. **McLees, B. D., Schleiter, G., and Pinnell, S. R.,** Isolation of type III collagen from human adult parenchymal lung tissue, *Biochemistry*, 16, 185, 1977.
47. **Sage, H., Woodbury, R. G., and Bornstein, P.,** Structural studies on human type IV collagen, *J. Biol. Chem.*, 254, 9893, 1979.
48. **Burgeson, R. E., El Adli, F. A., Kaitila, I. I., and Hollister, D. W.,** Fetal membrane collagens: identification of two new collagen alpha chains, *Proc. Natl. Acad. Sci. U.S.A.*, 73, 2579, 1976.

Chapter 3

PREPARATION OF TYPE I AND TYPE III COLLAGENS, THEIR P_N COLLAGENS, AND AMINOTERMINAL EXTENSION PEPTIDES

R. W. Glanville

TABLE OF CONTENTS

I. INTRODUCTION

Native Type I and Type III collagens can be isolated from most animal tissues. The tissue of choice for the preparation of these collagens is usually skin because of the quantities of material available, the ease of preparing skins, and extractability of the collagens. Skins from which native collagens have been prepared include human,[1-3] calf,[4,5] chicken,[6,7] rat,[8] guinea pig,[9] baboon,[10] and rabbit.[11] For immunological studies, the collagen molecules must be native, and it is advantageous to use salt-soluble collagens as opposed to pepsin-solubilized collagens. Salt-soluble collagens have intact telopeptide regions and are better antigens. As the solubility of collagens in neutral salt buffers decreases with age, young animals, or ideally fetal skins, should be used. The proportion of Type III collagen in skin also decreases with age, and therefore the use of fetal skins has a double advantage.[2] Another way of increasing yields of salt-soluble collagen is to use the skins of lathyritic animals, i.e., animals that have been fed the lathrogen β aminopropionitrile which inhibits the formation of cross-links, thus rendering collagens more soluble.

Intact procollagens, with both amino and carboxyterminal extensions intact, cannot be extracted directly from tissues and have to be produced in culture (see Chapter 7). However, it is possible to extract collagen molecules with an intact amino terminal extension only (P_N collagen). P_N Type III collagen has been extracted from fetal calf skin and lathyritic rat skins.[12,13] P_N Type I collagen has also been extracted from rat skins and from dermatosparactic calf and sheep skins.[13-15] Dermatosparaxis is a genetic defect which causes a decrease in Type I procollagen aminopeptidase activity in the skin of the affected animals.[16] Because of this defect, P_N Type I collagen accumulates in the skin of diseased animals and can be extracted to some extent into neutral salt buffers.[14] It has been shown that the antibodies raised against P_N collagens are directed mainly towards the aminoterminal extension peptides and that antibodies specific for P_N collagens can be readily raised by using only a part of the extension as antigen.[17,18] This allows more efficient use of the available tissue, as the insoluble residue left after extraction can be digested with bacterial collagenase and the extension peptides isolated from the digestion mixture.

From the above considerations, three skins appear to have particularly favorable characteristics. Fetal calf skin can be used as a source of Type I, Type III, and P_N Type III collagens, dermatosparactic calf or sheep skin can be used as a source of Type I, and P_N Type I collagens and lathyritic rat skin can be used as a source of Type I, Type III, P_N Type I, and P_N Type III collagens. The methods used to solubilize, isolate, and purify the collagens are very similar for all three skins. In this article, the methods and results described are those obtained from experiments using fetal calf skin and dermatosparactic calf skin extractions.

For the sake of clarity, the nomenclature used for the various peptides and the differences between the extension peptides of P_N Type I and P_N Type III collagens are shown in Figure 1.

II. METHODS

A. Preparation and Extraction of Skins

Fetal calves were removed from the placental tissues at the local slaughterhouse and immediately deep frozen. When sufficient fetuses had been collected (five to ten, depending upon size) they were thawed out in a cold room, and the skins were peeled off the carcases. This was more difficult with fetuses over 60 cm long, and the skin had to be cleanly dissected away. Fetuses with hair were avoided, although

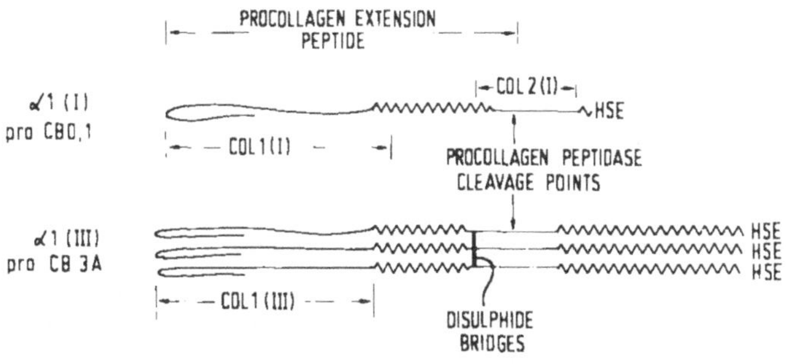

FIGURE 1. A schematic diagram of the aminoterminal cyanogen bromide peptides of Types I and III procollagens. In vivo conversion of procollagen to collagen by procollagen peptidases releases the procollagen extension peptides. Cleavage of the cyanogen bromide peptides with bacterial collagenase (pro-CB3A must be reduced and carboxymethylated) releases the intact globular portion termed Col 1 and a nonhelical peptide Col 2 as major fragments. WWW, triple helical domain; ——, nontriple-helical domain; HSE, homoserine. (From Westner, M., Krieg, T., Hörlein, D., Glanville, R. W., Fietzek, P. P., and Müller, P. K., *J. Biol. Chem.*, 254, 7016, 1979. With permission.)

they can be used. The skins were cut into strips 2 to 3-cm-wide and rinsed under cold running water to remove blood. The pieces of skin were weighed and then frozen before passing twice through an electric meat mincer, the final sieve having holes approximately 1.5 mm in diameter. When necessary crushed ice was added to the skins being minced to maintain their temperature under 10°C. The minced skin was collected in 0.05 *M* Tris-HCl buffer, pH 7.4 (2.5 ℓ/kg skin), containing 0.2 *M* NaCl, 0.01 percent NaN_3, and the protease inhibitors phenylmethanesulphonyl fluoride (3 mg/ℓ), *p*-chloromercuribenzoate (3 mg/ℓ), and 0.01 *M* EDTA. The inhibitors were freshly prepared and added immediately after the skin was minced. The extract was stirred at 5°C for 36 to 48 hr and then centrifuged at 5°C at 14,000 g.

Dermatosparactic calf skins were obtained from the Department of Medical Genetics at the University of Liege in Belgium (courtesy of Dr. C. Lapière). Skins were freed of hair and subcutaneous fat by scraping with a scalpel and then minced and extracted as described above for normal fetal calf skins.

B. Fractional Salt Precipitation of Native Collagens

The extract supernate was brought to 2.5 *M* NaCl by the slow addition of crystalline NaCl, while stirring the solution vigorously. The precipitation and all following steps are carried out at around 5°C. The precipitated collagens were removed from the 2.5 *M* NaCl solution by centrifuging at 40,000 × g. The supernate contains soluble noncollagenous proteins, and in the case of fetal calf skin extracts, some peptide derived from the amino terminal extension of Type III procollagen (see Section II.G). The precipitate (200 g/ℓ) was redissolved in 0.05 *M* Tris-HCl buffer, pH 7.4, containing 0.2 *M* NaCl and 0.01% NaN_3 by stirring overnight. The resulting solution was centrifuged (40,000 × g) to remove insoluble materials which were discarded, and the supernate was brought to 1.7 *M* NaCl by the dropwise addition of a solution of 5 *M* NaCl in 50 m*M* Tris-HCl buffer, pH 7.4. The precipitate which contained mainly Type III collagen with P_N Type III (dermatosparactic skins contain very small amounts of Type III and P_N Type III collagen) was removed by centrifuging at 40,000 × g, redissolved, and the 1.7 *M* precipitation step was repeated. The 1.7

M NaCl supernate was brought to 2.5 *M* NaCl, again by the dropwise addition of a 5 *M* NaCl solution as above. The supernate was discarded, and the precipitate which contains both Type I and P_N Type I collagens in dermatosparactic skin extracts, but only Type I in fetal calf skin extracts, was redissolved, and the 1.7 and 2.5 *M* precipitations were repeated.

C. Chromatographic Purification of Native Collagens

Chromatographic purification was achieved using two columns (2.5 × 20 cm) of DEAE-cellulose (Whatman De52® microgranular) equilibrated under different conditions. The first column was equilibrated at 5 to 10°C with 30 m*M* Tris-HCl, pH 8.6, containing 0.2 *M* NaCl and 2 *M* urea, and was designed to remove acidic impurities. The second column, equilibrated at 5 to 10°C with 30 m*M* Tris-HCl, pH 8.6, containing 0.02 *M* NaCl and 2 *M* urea, effected the separation of normal collagens from P_N collagens.

The 1.7 and 2.5 *M* precipitates (see Section II.B) were dissolved in the equilibration buffer of the first DEAE column and dialyzed overnight once against the same buffer (ten times sample volume). The sample was applied to the column and washed through with equilibration buffer. The material which did not bind to the column was collected and dialyzed once against 30 m*M* tris-HCl containing 2 *M* urea (ten times sample volume) to reduce the salt concentration in the sample from 0.2 to 0.02 *M* NaCl. This was applied to the second DEAE column. Material which did not bind to the column contained normal collagen, while bound material was P_N collagen and was eluted by applying a linear salt gradient from 0.02 to 0.25 *M* NaCl in 2ℓ of equilibration buffer.

D. Preparation of Col 1(I) from Native P_N Type I Collagen

The peptide Col 1(I) can be prepared from extracts of dermatosparactic calf (or sheep skin) by digesting the 2.5 *M* precipitate, prepared as described in Section II.B, with bacterial collagenase. The precipitate was taken up in 0.05 m*M* Tris-HCl, pH 7.8, containing 0.002 *M* $CaCl_2$, denatured at 45°C for 30 min, and incubated with bacterial collagenase (enzyme to substrate, 1 to 50,000) for 4 hr at 37°C. The reaction mixture was centrifuged (27,000 × g) to remove insoluble material, and urea was added to the supernate to give a final concentration of 2 *M* and then dialyzed against a 0.05 *M* Tris-HCl buffer, pH 8.6, containing 2 *M* urea. This buffer was also used to equilibrate a column (2.5 × 25 cm) of DEAE-cellulose (Whatman® DE52) at 40°C. The dialyzed sample was applied and the bound proteins eluted with a salt gradient of 0 to 0.4 *M* NaCl in 2 ℓ buffer. The fraction containing Col 1(I) was desalted over BioGel® P2 (BioRad® 100 to 200 mesh) into 0.01 *M* acetic acid and lyophilized. The final purification was made using a column (2.5 × 140 cm) of agarose 1.5m (BioRad® 200 to 400 mesh), equilibrated at room temperature with 0.05 *M* Tris-HCl buffer, pH 7.5, containing 1 *M* $CaCl_2$.

E. Preparation of Col 1 to 3 and Col 1(III) from Native P_N Type III Collagen

Peptide Col 1(III) cannot be directly prepared from a collagenase digest as described for Col 1(I). This is because of the presence of disulfide bridges in the extension peptides of Type III collagen which protects the short triple helix against degradation.[19] Col 1(III) is best produced by using a two-stage reaction. The 1.7 *M* precipitate from a fetal calf skin extract was first digested with bacterial collagenase at 37°C under conditions described for the production of Col 1(I) in Section II.D above. Under these conditions, the major collagen triple helix was degraded, leaving a peptide Col 1 to 3 which is composed of the two nontriple-helical domains and the short helix intact (see Figure 1). This peptide was purified using DEAE cellulose and agarose 1.5m as described in Section II.D for Col 1(I). In the second stage of the reaction, the peptide Col 1 to 3 was cleaved with bacterial collagenase at 55°C which is above the midpoint melting temperature of the short triple helix.[19] Under these conditions, Col 1(III) was produced which was isolated by chromatography of the digest on agarose 1.5m as described for Col 1(I).

F. Preparation of Col 1 to 3, Col 1(III), Col 1(I) by Direct Bacterial Collagenase Digestion of Skin

This method was particularly useful for dermatosparactic skins, as much of the P_N Type I collagen is not extractable.[14] The skins were prepared as described above for the extractions and the minced tissue (400 × g) was dispersed in 50 m*M* Tris-HCl, pH 7.8, buffer containing 0.002 *M* $CaCl_2$ (the residues from the buffer extracts of dermatosparactic skins can also be digested to gain more Col 1(I), but the residues from the fetal calf skin extracts are not suitable for digestion as practically all P_N Type III collagen has been extracted). The tissue was denatured for 2 hr at 45°C, cooled to 37°C, bacterial collagenase (enzyme to skin, 1:50,000) was added, and the solution was incubated with stirring at 37°C for 24 hr. The undigested material was removed by centrifugation (40,000 × g), and the supernate was dialyzed until salt free, against 1% acetic acid. During dialysis a precipitate may form which was again removed by centrifugation (40,000 × g), and the supernate was lyophilized. This supernate contained Col 1(I) when dermatosparactic skins were used or Col 1 to 3 when fetal calf skins were used. Purification of Col 1(I) and conversion of Col 1 to 3 into Col 1(III) was carried out as described in Section II.D. However, one additional purification step was required when skins were digested with bacterial collagenase. It is possible that the samples are cross-contaminated due to a small amount of P_N Type III collagen in dermatosparactic skins and a small amount of P_N Type I in fetal calf skins. The contamination can be removed by chromatography of the final product on a column (2.5 × 20 cm) of phosphocellulose (Whatman® P11 fibrous form) equilibrated at room temperature with 0.001 *M* sodium acetate buffer, pH 3.6, containing 6 *M* urea (deionized). The sample was eluted with a linear gradient of 0 to 0.2 *M* NaCl in a total volume of 2 ℓ.[25]

G. Isolation of P_N Type III Collagen Extension Peptide Fragments from Fetal Calf Skin Extracts

Another source of Col 1(III) was the 2.5 *M* NaCl supernate which remains after precipitating intact collagen out of a fetal calf skin extract (see Section II.A). When the supernate was dialyzed until salt free against 1% acetic acid, a copious white precipitate formed which was removed by centrifugation and the supernate was lyophilized. This material was purified using DEAE-cellulose and agarose chromatography as described in Section II.E. Two fractions were obtained, one containing material which appears to be slightly longer than Col 1(III) and the other containing a heterogeneous fraction containing various forms of the disulfide cross-linked extension peptide. Both these fractions can be converted to pure Col 1(III) by treating with bacterial collagenase under the conditions already described (see Section II.F).

III. RESULTS

A. Chromatographic Purification of Native Collagens

Figure 2 shows the curves obtained when native Type III collagens extracted from fetal calf skins are chromatographed on DEAE-cellulose. Under the conditions used

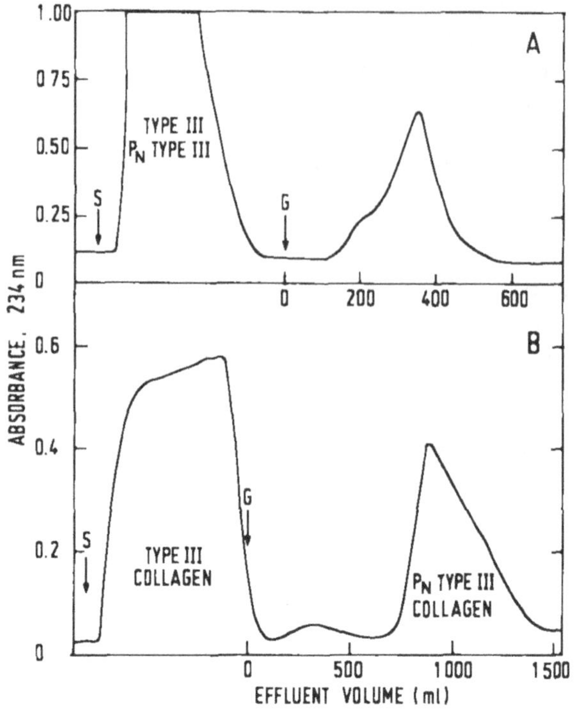

FIGURE 2. Elution profiles showing the purification and separation of Type III and P_N Type III collagens isolated from fetal calf skin on DEAE cellulose. In chromatogram A, the column was equilibrated at 5 to 10°C with 30 m*M* Tris-HCl, pH 8.6, containing 0.2 *M* NaCl and 2 *M* urea. Collagenous materials eluted before application of the gradient. Chromatogram B shows the separation of Type III from P_N Type III collagen when the column was equilibrated with 30 m*M* Tris-HCl, pH 8.6, containing 0.02 *M* NaCl and 2 *M* urea. S — application of the sample; G — application of the gradient.

for the first column, the native collagens were not retarded and were collected before applying the gradient. On the second column, Type III collagen eluted before the gradient was applied, and P_N Type III collagen was eluted with the gradient as an asymmetrical peak. Type I and pro-Type I, originating from dermatosparactic skin extracts, gave similar curves except that P_N Type I collagen eluted slightly earlier in the gradient of the second column than P_N Type III.

Figure 3 shows disk gel electrophoretic patterns of the DEAE fractions. As can be seen in the unreduced sample of Type III collagen, this fraction contained a trace of Type I collagen. This was nearly always the case, but the impurity was never more than about 5%. Further reprecipitations of the Type III fraction at 1.7 or 1.5 *M* NaCl before applying to the DEAE-cellulose columns did not remove the Type I collagen. It is also apparent from Figure 3 that the P_N collagens, particularly P_N Type III, are not homogeneous molecules. Not all three chains in the triple helix carry the aminoterminal extension peptides. The proportions of the $P_N\alpha$ to α-chains varied from one preparation to another in P_N Type III collagen, however, in P_N Type I collagen, $P_N\alpha 1$ and $P_N\alpha 2$ always predominated.

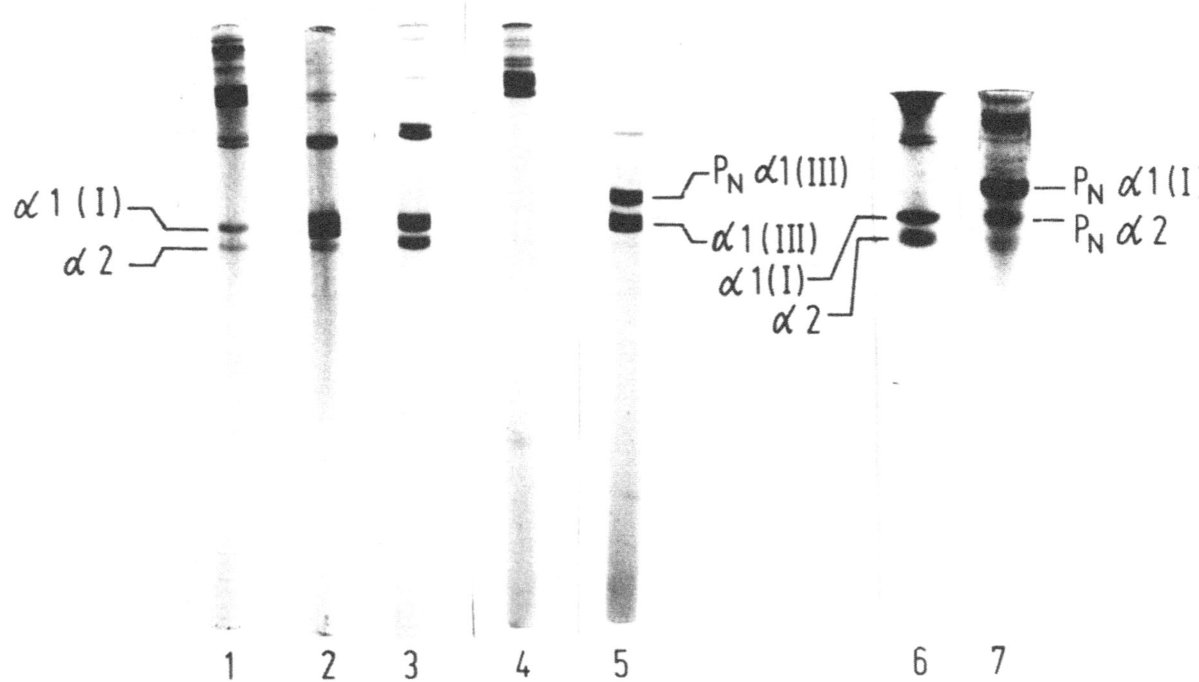

FIGURE 3. Electrophoresis gels (5%) stained with coomassie blue showing (1) fetal calf skin Type III collagen; (2) same as 1, but reduced; (3) fetal calf skin Type I collagen; (4) fetal calf skin P_N Type III collagen; (5) same as 4, but reduced; (6) dermatosparactic calf skin Type I collagen; (7) dermatosparactic calf skin P_N Type I collagen.

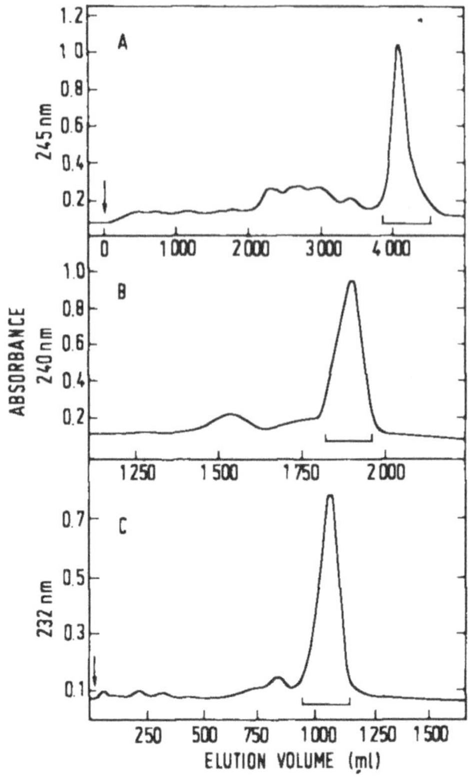

FIGURE 4. Elution profiles showing the purification of Col 1(I) from a collagenase digest of dermatosparactic calf skin. (A) Separation of the digest on DEAE cellulose; (B) fraction marked by a bar in A chromatographed on agarose 1.5m; (C) fraction from agarose 1.5m chromatographed on phosphocellulose. (From Wiestner, M., Krieg, T., Hörlein, D., Glanville, R. W., Fietzek, P. P., and Müller, P. K., *J. Biol. Chem.*, 254, 7016, 1979. With permission.)

B. Purification of Col 1(I), Col 1 to 3, and Col 1(III)

The first step in the purification of collagenase peptides, irrespective of whether skin or purified collagens were used as starting material, was always chromatography on DEAE-cellulose. The required peptide, due to the acidic character of the extension peptides, was always eluted last from the column. Figure 4 shows the purification of Col 1(I) from a collagenase digest of preextracted dermatosparactic calf skin. The major product of the digestion is clearly Col 1(I) which was obtained in a highly purified state by rechromatographing the DEAE-cellulose fraction first on agarose 1.5m and then phosphocellulose as shown.

The isolation of Col 1 to 3 from collagenase digests of either purified P_N Type III collagen or fetal calf skin, using DEAE-cellulose, gave a similar chromatogram to that shown in Figure 4A. Without further purification, this fraction was converted to Col 1(III) by a second treatment with collagenase as described in Section II.E. and purified using agarose 1.5m and phosphocellulose. Chromatograms were similar to those shown in Figures 4B and 4C for Col 1(I).

The isolation of Col 1 to 3 and other fragments of the P_N Type III extension

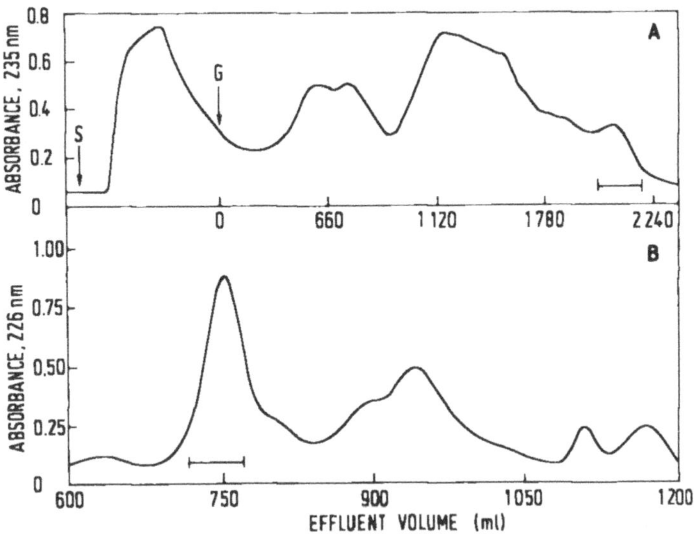

FIGURE 5. Elution profile showing the separation of P_N Type III collagen extension peptides from a fetal calf skin extract supernate. (A) DEAE cellulose (5 × 25 cm), gradient 0 to 0.4 *M* NaCl in 3 ℓ equilibration buffer as in Section II.D. S — sample; G — gradient. (B) agarose 1.5m (5 × 145 cm) equilibrated as in Section II.D. The material in the marked fractions of chromatogram A were applied. The bar indicates the fraction containing the P_N Type III collagen extension peptides.

Table 1
AMINO ACID COMPOSITIONS OF THE MAJOR BACTERIAL COLLAGENASE DERIVED PEPTIDES FROM P_N TYPE I AND P_N TYPE III COLLAGENS[20,21]

	Col 1 (I)	Col 1 (III)	Col 1 to 3
Hydroxyproline	—	—	9
Aspartic acid	15	11	16
Threonine	6	2	3
Serine	1	4	10
Glutamic acid	19	13	17
Proline	10	13	18
Glycine	10	8	25
Alanine	1	4	7
Cysteine	10	10	7
Valine	10	5	12
Isoleucine	3	4	7
Leucine	3	3	3
Tyrosine	1	1	4
Phenylalanine	—	1	1
Tyrosine	1	1	1
Histidine	1	1	1
Lysine	4	2	3[a]
Arginine	3	2	3
Total	98	85	147

[a] Value includes 0.6 residues hydroxylysine.

peptides, directly from the fetal calf skin extract supernate as described in Section II.G, resulted in more complex chromatograms. Figure 5A shows the separation of such an extract supernate on DEAE-cellulose (5 × 25 cm). Despite the large quantities of other proteins present, the collagen extension peptides could easily be identified. This fraction, chromatographed on agarose 1.5m gave the curve shown in Figure 5B. The fraction marked with a bar was Col 1 to 3. Other fractions in the chromatogram contained material related to Col 1 to 3, including a peptide which was a little larger than Col 1(III).

Table 1 gives the amino acid compositions of the peptides Col 1(I), Col 1(III), and Col 1 to 3 derived from the amino acid sequences of these peptides.[20,21]

IV. CONCLUSION

Until recently there were no chromatographic methods available to separate native Type I and Type III collagens. Type I can be obtained pure, direct from fractional salt precipitations, but Type III collagen, even after repeated precipitations, contains a small trace of Type I. A method called zone precipitation chromatography has now been reported, but this has not yet been used in my laboratory.[22] However, for raising antibodies, this impurity is not a serious problem as the antisera must be purified by passing over affinity columns of denatured Type I and Type III collagens to ensure monospecificity of the antisera (see Chapter 11, Volume I).

Collagen in dermatosparactic skins is less soluble than that in fetal calf skins, and therefore it is advisable to use the residue left after extraction of native collagens for preparation of Col 1(I). On the other hand, residues from fetal calf skin extracts should not be used for the production of Col 1(III) as the quantity of insoluble P_N Type III collagen remaining is relatively small. Dermatosparactic skins also contain very little extractable Type III or P_N Type III collagens in comparison to fetal calf skins.

The native P_N collagens are not homogeneous preparations. P_N Type III and frequently P_N Type I collagen molecules contain both intact P_N α-chains and α-chains. The conversion of P_N α-chains to α-chains may be due to nonspecific proteolytic cleavage which takes place during the preparation of the skins despite the presence of inhibitor cocktails. However, it is interesting to note that fragments of the extension peptides of P_N Type III collagen, similar to those isolated directly from the skin extract as described in Section II.G, have also been identified in human urine and ascites and serum.[23] This suggests that perhaps the fragments are natural breakdown products produced during the conversion of P_N Type III collagen to Type III collagen which have a regulatory function on collagen biosynthesis.[24]

ACKNOWLEDGMENTS

The author would like to thank Andrea Brandt, Dietrich Hörlein, Peter Bruckner, Rupert Timpl, and Peter Fietzek, whose original work made the preparation of this manuscript possible. Research carried out by the author was supported by the Deutsche Forschungsgemeinschaft, SFB 51, B/22.

REFERENCES

1. **Click, E. M. and Bornstein, P.,** Isolation and characterization of the cyanogen bromide peptides from the α1 and α2 chains of human skin collagen, *Biochemistry*, 9, 4699, 1970.
2. **Epstein, E. H.,** $[\alpha 1(III)]_3$ human skin collagen. Release by pepsin digestion and preponderance in fetal life, *J. Biol. Chem.*, 249, 3225, 1974.
3. **Chung, E. and Miller, E. J.,** Collagen polymorphism: characterization of molecules with a chain composition $[\alpha 1(III)]_3$ in human tissues, *Science*, 183, 1200, 1974.
4. **Fujii, T. and Kühn, K.,** Isolation and characterization of pepsin treated type III collagen from calf skin, *Hoppe-Seylers Z. Physiol. Chem.*, 356, 1793, 1975.
5. **Rauterberg, J. and Kühn, K.,** Acid soluble calf skin collagen. Characterization of the peptides obtained by cyanogen bromide cleavage of its α1 chain, *Eur. J. Biochem.*, 19, 398, 1971.
6. **Kang, A. H., Piez, K. A., and Gross, J.,** Characterization of the α chains of chick skin collagen and the nature of the NH_2 terminal crosslink region, *Biochemistry*, 8, 3648, 1969.
7. **Herrmann, H. and von der Mark, K.,** Isolation and characterization of type III collagen from chick skin, *Hoppe-Seylers Z. Physiol. Chem.*, 356, 1605, 1975.
8. **Bornstein, P. and Piez, K.,** The nature of the intramolecular cross links in collagen. The separation and characterization of peptides from the cross link region of rat skin collagen, *Biochemistry*, 5, 3460, 1966.
9. **Clark, C. C. and Bornstein, P.,** Cyanogen bromide cleavage of guinea pig skin collagen. Isolation and characterization of peptides from the α1 and α2 chains, *Biochemistry*, 11, 1468, 1972.
10. **Epstein, E. H., Scott, R. D., Miller, E. J., and Piez, K. A.,** Isolation and characterization of the peptides derived from soluble human skin and baboon skin collagen after cyanogen bromide cleavage, *J. Biol. Chem.*, 246, 1718, 1971.
11. **Bornstein, P. and Nesse, R.,** The comparative biochemistry of collagen: the structure of rabbit skin collagen and its relevance to immunochemical studies of collagen, *Arch. Biochem. Biophys.*, 138, 443, 1970.
12. **Timpl, R., Glanville, R. W., Nowack, H., Wiedemann, H., Fietzek, P. P., and Kühn, K.,** Isolation, chemical and electron microscopical characterization of neutral salt soluble type III collagen and procollagen from fetal bovine skin, *Hoppe-Seylers Z. Physiol. Chem.*, 356, 1783, 1975.
13. **Byers, P. H., McKenney, K. H., Lichtenstein, I. R., and Martin, G. R.,** Preparation of type III procollagen and collagen from rat skin, *Biochemistry*, 13, 5243, 1974.
14. **Lenaers, A., Ansay, M., Nusgens, B. V., and Lapiere, C.,** Collagen made of extended α chains, procollagen, in genetically defective dermatosparactic calves, *Eur. J. Biochem.*, 23, 533, 1971.
15. **Fjolstad, M. and Helle, O.,** A hereditary dysplasia of collagen tissue in sheep, *J. Pathol.*, 112, 183, 1974.
16. **Lapière, C. M., Lenaers, A., and Kohn, L. D.,** Procollagen peptidase: an enzyme excising the coordination peptides of procollagen, *Proc. Natl. Acad. Sci. U.S.A.*, 68, 3054, 1971.
17. **Timpl, R., Wick, G., Furthmayr, H., Lapière, C., and Kühn, K.,** Immunochemical properties of procollagen from dermatosparactic calves, *Eur. J. Biochem.*, 32, 584, 1973.
18. **Rohde, H., Becker, U., Nowack, H., and Timpl, R.,** Antigenic structure of the aminoterminal region of type I procollagen, *Immunochemistry*, 13, 967, 1976.
19. **Bruckner, P., Bächinger, H. P., Timpl, R., and Engel, J.,** Three conformationally distinct domains in the amino terminal segment of type III procollagen and its rapid triple helix coil transition, *Eur. J. Biochem.*, 90, 595, 1978.
20. **Hörlein, D., Fietzek, P. P., Wachter, E., Lapiere, C. M., and Kühn, K.,** Amino acid sequence of the aminoterminal segment of dermatosparactic calf skin procollagen type I, *Eur. J. Biochem.*, 99, 31, 1979.
21. **Brandt, A., Hörlein, D., Brückner, P., Timpl, R., Fietzek, P. P., and Glanville, R. W.,** Unpublished data, 1980.
22. **Ehrlic, H. P.,** Zone precipitation chromatography: its use in the isolation of different collagen types, *Prep. Biochem.*, 9, 407, 1979.
23. **Rohde, H., Vargas, L., Hahn, E., Kalbfleisch, H., Bruguera, M., and Timpl, R.,** Radioimmunoassay for type III procollagen peptide and its application to human liver disease, *Eur. J. Clin. Invest.*, 9, 451, 1979.
24. **Wiestner, M., Krieg, T., Hörlein, D., Glanville, R. W., Fietzek, P. P., and Müller, P. K.,** Inhibiting effect of procollagen peptides on collagen biosynthesis in fibroblast cultures, *J. Biol. Chem.*, 254, 7016, 1979.
25. **Hörlein, D.,** Personal communication, 1976.

Chapter 4

PREPARATION OF TYPE II COLLAGEN

William T. Butler and Charles A. Reese

TABLE OF CONTENTS

I. INTRODUCTION

The discovery of the distinct nature of chick sternal cartilage collagen by Miller and Matukas[1] was the key event in the numerous investigations that led to our recognition of the diversity of collagenous structures. The nomenclature for the chains of various members of the collagen families is derived from the observation that the α1(II)-chains of cartilage collagen elute from CM-cellulose columns in a position similar to α1(I)-chains of skin and bone (Type I) collagen and different from α2-chains. Subsequent to this discovery, Type II collagen was found in a variety of cartilaginous structures,[2–5] as well as in intervertebral disk,[6] the notochord,[7] and in the adult vitreous body.[8]

The preparation of native Type II molecules is complicated by the fact that cartilage collagen is highly insoluble. This problem has been circumvented in two ways. Small animals such as chicks can be rendered lathyritic by feeding them β-aminopropionitrile to inhibit the cross-linking process. Then the triple-helical Type II molecules can be extracted[9,10] and purified by differential salt precipitation. A second useful approach is to utilize limited pepsin cleavage of nontriple-helical ends where the cross-linkages were located to form truncated, soluble molecules with the triple-helical portions of the molecules still intact.[11] The Type II collagen can then be precipitated from solution and separated from the enzyme.[11–13]

The isolation of Type II collagen is further complicated by the observation that hyaline cartilage contains minor amounts of several collagenous chains different from α1(II).[14–16] Some of these additional chains probably arise from Type V-like macromolecules, while others are portions of disulfide-linked structures with unknown functions.[15,16] The procedure described here takes into account the existence of these minor collagenous constituents in the isolation of the major Type II collagen of chick sterna.

II. PROCEDURE

Chick sterna are obtained from young fryers at a local slaughter house shortly after they are sacrificed, quickly frozen at −80°C, and are stored at −18°C until utilized. Approximately 60 sterna (200 g wet wt.) are used for a preparation. The sterna are stripped of perichondrium by scraping with scalpels and blunt instruments (such as a weighing spatula). They are then diced into small pieces and homogenized in cold water, first in a Waring®-type blender and then for 5 min in a Polytron homogenizer. All the subsequent operations are at 4°C. After centrifugation to retrieve the small pieces of homogenized sterna, proteoglycans are removed by two extractions with 4 *M* guanidine HCl (GdmCl) in 50 m*M* Tris-HCl (pH 7.4) over a 24 hr period. The GdmCl extracts are discarded, and the residue is cleansed by rinsing with 1 to 2 ℓ of distilled water.

Next the collagenous molecules are solubilized by limited pepsin cleavage of bonds in nontriple-helical extremities. The sternal pieces are suspended in 400 mℓ of 0.2 *M* NaCl in 0.5 *M* acetic acid, containing 1 mg/mℓ of pepsin (Worthington). The digest is gently shaken for 20 hr. To insure a more complete digestion, the pieces are homogenized in the pepsin solution with the Polytron homogenizer three times during the last 10 hr of incubation. During the last 2 hr of digestion, the pepsin solution is diluted with 800 mℓ of 0.5 *M* acetic acid.

The digest is centrifuged at 30,000 × g to remove undigested material. The pH of the supernatant is then adjusted to 8.0 by the slow addition of 5 *N* NaOH while

Table 1
AMINO ACID COMPOSITION[a] OF CHICK STERNAL TYPE II COLLAGEN

Amino acid	Type II collagen[b] (residues per thousand)
4-Hydroxyproline	97
Aspartic Acid	46
Threonine	26
Serine	23
Glutamic Acid	90
Proline	125
Glycine	326
Alanine	115
Valine	15
Methionine	11
Isoleucine	7
Leucine	25
Tyrosine	2
Phenylalanine	12
Hydroxylysine	23
Lysine	12
Histidine	2
Arginine	43

[a] Samples of approximately 1 mg were hydrolyzed in constant boiling (6 N) HCl in an N_2 atmosphere at 108°C for 24 hr. The hydrolysates were analyzed with a Beckman 121m amino acid analyzer by a published procedure.[17]

[b] The Type II collagen used for this analysis was the pepsin-released material recovered by precipitation in 0.9 M NaCl, 0.5 M acetic acid as described in Section II.

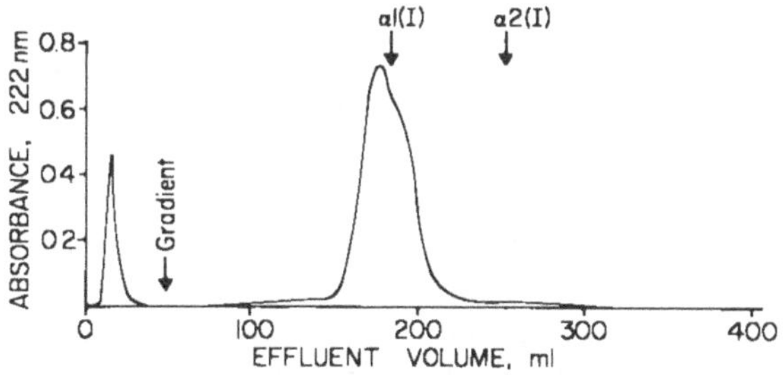

FIGURE 1. Separation of α1(II)-chains in the 0.9 *M* NaCl, 0.5 *M* acetic acid precipitate (see Section II) by CM-cellulose chromatography. The CM-cellulose column (1.5 × 10 cm) was equilibrated with 4 *M* urea, 0.02 *M* sodium acetate, pH 4.8 (starting buffer), at 42°C. Type II collagen (6 mg) was dissolved in starting buffer (6 mℓ) and applied to the column. The column was then eluted at 100 mℓ/hr with a linear gradient from 0.0 to 0.12 *M* NaCl (in starting buffer) over a total volume of 400 mℓ. Prior to further analysis, fractions containing the α1(II)-chains were pooled, lyophilized, dialyzed exhaustively against 0.1 *M* acetic acid, and lyophilized. Positions where α1(I)- and α2-chains elute are denoted.

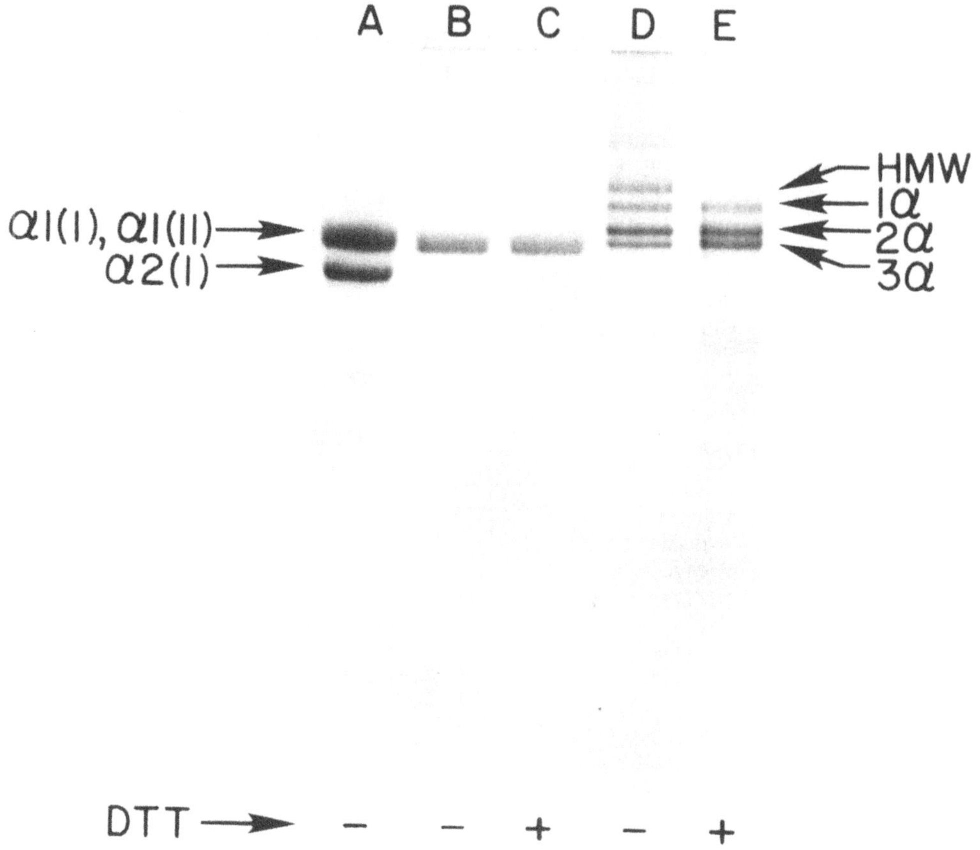

FIGURE 2. Slab gel electrophoresis of collagen chains. SDS-polyacrylamide gel electrophoresis was performed using a discontinuous system of buffers as described by Laemmli.[18] The separation gel was a 5 to 10% gradient gel that contained 8 *M* urea. The 3% stacking gel was in 4 *M* urea. Prior to electrophoresis, samples were boiled for 30 sec. Some samples were treated with dithiothreitol as indicated by a "+" below the appropriate lanes. Lane A, chicken Type I collagen; Lanes B and C, Type II collagen in the 0.9 *M* NaCl, 0.5 *M* acetic acid precipitate; Lanes D and E, minor collagenous components in a 1.2 *M* NaCl, 0.5 *M* acetic acid precipitate as reported elsewhere.[16] Arrows on the left- and righthand sides indicate the identities of the various components.

gently stirring it. After standing overnight, the Type II collagen is specifically precipitated from solution by dialysis against four changes of 0.9 *M* NaCl in 0.5 *M* acetic acid. The precipitated Type II collagen is recovered by centrifugation at 30,000 × g for 45 min. The supernatant can then be used for preparation of collagen structures with 1α-, 2α-, and 2α-chains[14,16] and of disulfide-linked collagenous components.[15,16] Type II collagen in the precipitate is redissolved in 0.5 *M* acetic acid, dialyzed exhaustively against 0.5 *M* acetic acid, and lyophilized.

III. CHARACTERIZATION OF TYPE II COLLAGEN

In order to determine the quality of the Type II collagen obtained in this way, we have employed several standard techniques. The amino acid analysis (see Table 1) shows a composition typical of collagen, with one third of the residues being glycine. However, the analysis shows distinctive features that differ from the analysis of other collagens. For example, the hydroxylysine content is unusually high (as is that of the glycosylated hydroxylysines). For a complete discussion of this point, see

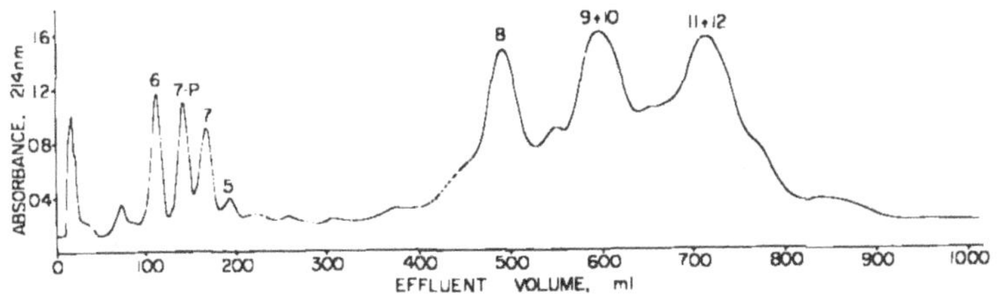

FIGURE 3. CM-cellulose chromatography of the CNBr peptides from pepsin-treated α1(II)-chains. About 30 mg of CNBr peptides were applied to a 1.5 × 10 cm column of CM-cellulose at 42°C that had been equilibrated with 0.02 *M* sodium formate buffer, pH 3.6 (starting buffer). The column was eluted at 60 mℓ/hr with 20 mℓ of starting buffer and then with a linear gradient formed from 500 mℓ each of starting buffer and of 0.02 *M* sodium formate buffer (pH 3.6) containing 0.17 *M* NaCl. (Numbering is from Miller, E. J., *Biochemistry*, 11, 4903, 1972.)

Miller.[5] Separation of the α1(II)-chains of denatured Type II collagen on CM-cellulose (see Figure 1) indicates the presence of a heterogeneous chromatographic peak that elutes just prior to α1(I)-chains. Furthermore, this analysis shows that α2-chains of Type I collagen and 1α- and 2α-chains of cartilage are absent. This chromatographic profile is typical of those reported earlier (e.g., see Miller[11] and Trelstad et al.[12]). Slab gel electrophoresis (see Figure 2) also indicates the lack of contamination of the preparation by other collagenous chains, and finally the CNBr peptide profile analyzed by slab gel electrophoresis (data not shown) and by CM-cellulose chromatography (see Figure 3) is typical of that reported for pepsin-treated Type II collagen.[11]

IV. SUMMARY

In summary, this simple procedure can be used to obtain Type II collagen of high purity in rather large quantities. Some of the earlier procedures utilizing supernatants containing Type II collagen rather than precipitated material may have been contaminated by minor amounts of collagenous components with 1α-, 2α-, and 3α-chains, now known to occur in cartilaginous structures.[14-16]

One other note of caution: amino acid sequence studies[17] have shown that preparations of α1(II)-chains contain more than one closely related structure. Thus, the Type II collagens obtained cannot be considered to be homogeneous in the strictest sense. The heterogeneity of α1(II)-chains seen on CM-cellulose chromatography (see Figure 1) is another indication of this fact; the nature of this heterogeneity is unclear.

REFERENCES

1. **Miller, E. J. and Matukas, V. J.,** Chick cartilage collagen: a new type of α1 chain not present in bone or skin of the species, *Proc. Natl. Acad. Sci. U.S.A.*, 64, 1264, 1969.
2. **Strawich, E. and Nimni, M. E.,** Properties of a collagen molecule containing three identical components extracted from bovine articular cartilage, *Biochemistry*, 10, 3905, 1971.
3. **Eyre, D. R. and Muir, H.,** Characterization of the major CNBr-derived peptides of porcine type II collagen, *Connect. Tissue Res.*, 3, 165, 1975.

4. **Miller, E. J. and Lunde, L. G.**, Isolation and characterization of the cyanogen bromide peptides from the α1(II) chain of bovine and human articular cartilage collagen, *Biochemistry*, 12, 3153, 1973.
5. **Miller, E. J.**, Biochemical characteristics and biological significance of the genetically-distinct collagens, *Mol. Cell Biochem.*, 13, 165, 1976.
6. **Eyre, D. R.**, Biochemistry of the intervertebral disc, *Int. Rev. Connect. Tissue Res.*, 8, 227, 1979.
7. **Miller, E. J. and Matthews, M. B.**, Characterization of notochord collagen as a cartilage-type collagen, *Biochem. Biophys. Res. Commun.*, 60, 424, 1974.
8. **Swann, D. A., Constable, I. J., and Harper, E.**, Vitreous structure. III. Composition of bovine vitreous collagen, *Invest. Ophthalmol.*, 11, 735, 1972.
9. **Miller, E. J.**, Isolation and characterization of a collagen from chick cartilage containing three identical α chains, *Biochemistry*, 10, 1952, 1971.
10. **Trelstad, R. L., Kang, A. J., Igarashi, S., and Gross, J.**, Isolation of two distinct collagens from chick cartilage, *Biochemistry*, 9, 4993, 1970.
11. **Miller, E. J.**, Structural studies on cartilage collagen following limited cleavage with pepsin, *Biochemistry*, 11, 4903, 1972.
12. **Trelstad, R. L., Kang, A. H., Toole, B. P., and Gross, J.**, Collagen heterogeneity. High resolution separation of native $[\alpha 1(I)]_2\alpha 2$ and $[\alpha 1(II)]_3$ and their component α chains, *J. Biol. Chem.*, 247, 6469, 1972.
13. **Miller, E. J., Harris, E. D., Jr., Chung, E., Finch, J. E., Jr., McCroskery, P. A., and Butler, W. T.**, Cleavage of type II and III collagens with mammalian collagenase: site of cleavage and primary structure at the NH_2-terminal portion of the smaller fragment released from both collagens, *Biochemistry*, 15, 787, 1976.
14. **Burgeson, R. E. and Hollister, D. W.**, Collagen heterogeneity in human cartilage. Identification of several new collagen chains, *Biochem. Biophys. Res. Commun.*, 87, 1124, 1979.
15. **Shimokomaki, M., Duance, V. C., and Bailey, A. J.**, Identification of a new disulphide bonded collagen from cartilage, *FEBS Lett.*, 121, 51, 1980.
16. **Reese, C. A. and Mayne, R.**, Minor collagens of chicken hyaline cartilage, *Biochemistry*, 20, 5443, 1981.
17. **Butler, W. T., Finch, J. E., Jr., and Miller, E. J.**, The covalent structure of cartilage collagen. Evidence for sequence heterogeneity of bovine α1(III) chains, *J. Biol. Chem.*, 252, 639, 1977.
18. **Laemmli, V. K.**, Cleavage of structural proteins during the assembly of the head of bacteriophage T4, *Nature*, 227, 680, 1970.

Chapter 5

ISOLATION AND CHARACTERIZATION OF PEPSIN SOLUBILIZED BASEMENT MEMBRANE COLLAGENS (TYPE IV) FROM HUMAN PLACENTA, BOVINE KIDNEY CORTICES, AND BOVINE LENS CAPSULE

S. N. Dixit and R. W. Glanville

TABLE OF CONTENTS

I. INTRODUCTION

Early chemical studies concerning the nature of the collagenous components in basement membrane structures, such as glomerular basement membrane, anterior lens capsule, Reicherts membrane, and Descements membrane, suggested that they contained a distinct collagen (Type IV) with three identical polypeptide chains represented by the molecular formular $[\alpha 1(IV)]_3$.[1,2] More recently, reports on glomerulus,[3-9] human placenta,[10-13] anterior lens capsule,[14-16] and murine tumor[17,18] basement membranes have established that they contain at least two distinct collagenous polypeptide chains. Biosynthetic studies substantiate these findings and indicate chain sizes of about 165,000 and 180,000 daltons.[19-21]

There are two major problems to be considered when isolating basement membrane collagens to use as antigens. First, basement membranes appear to contain more than one collagenous structure, and the relationship between these structures are still unknown. Second, as basement membranes are extremely insoluble, they have to be solubilized using proteolytic enzymes. Unlike interstitial collagens, basement membrane collagens are not resistant to proteolytic attack and are solubilized in the form of several fragments.[22] It is therefore difficult to determine the origin of such fragments which would be necessary for a meaningful interpretation of immunological data. These circumstances also make the preparation of a single intact native Type IV collagen molecule very difficult, with the result that all of the immunological work carried out to date has been done using pepsin fragments of basement membrane collagens. Thus, it is extremely important to be able to characterize these fragments and to know from which collagenous structure they derive.

In this article we describe the pepsin solubilization of human placenta, bovine kidney cortices, and bovine anterior lens capsule which are three of the tissues most commonly used as a source of basement membrane collagen. The purification and characterization of the solubilized native basement membrane pepsin fragments and their component chains are described.

Due to the lack of a uniform nomenclature for basement membrane collagens and their fragments at this time, we have used a nomenclature based upon one suggested by Crouch et al.[19]

The two intact chains with molecular weight of about 180,000 and 165,000 daltons that have been identified in biosynthetic systems are called α1(IV) and α2(IV), respectively. The pepsin fragments derived from these chains are then indicated by the name of the chain of origin, followed by the molecular weight of the fragment in kilo-daltons, e.g. α1(IV)95K. Table 1 lists the fragments that have been described in the literature to date with the nomenclature used by the authors. There are probably only five major components that have been isolated from placenta, kidney, and lens capsule tissues, the apparent molecular weight differences being due to experimental error. For the purpose of this report, they will be called α1(IV)140K, α2(IV)120K, α1(IV)95K, α2(IV)95K, and α2(IV)75K.

II. METHODS

A. Preparation of Tissues

Human placental tissue from normal births was deep frozen immediately after delivery and used for the preparation of basement membrane collagens. The umbilical cord was removed, and the chorion/amnion membrane was cut away around the circumference of the chorionic plate. The tissue was minced and washed exten-

Table 1
NOMENCLATURE OF PEPSIN-DERIVED FRAGMENTS OF BASEMENT MEMBRANE COLLAGENS FROM VARIOUS TISSUES

Tissue	α1 (IV)	α2 (IV)	Ref.
Placenta	140K (140)	70K-II (68)	13
	100K (100)		
	70K-I (70)		
	C' (140)	D (95)	12
	C (95)	80K (80)	
	$50K_1$ (50)	$50K_2$ (40)	
	α (IV) (95)	α' (IV) (70)	11
	α'' (IV) (95)		
	170K (170)	70K (70)	10
	100K (100)		
Lens capsule	C-1 (140)	D (95)	15
	C (95)	75K (75)	
	50K (50)		
		D (95)	
	C (95)	80K (80)	16
	$50K_1$ (50)	$50K_2$ (40)	
	α (IV) (95)		2
Kidney	A (93)	C (140)	6
	B (93)	D (140)	
	140 K (140)	80K (80)	7
	100K (100)		
	C-1 (140)	D-1 (120)	9
	C (95)	D (75)	
	50_1 (50)	50_2 (40)	
	α (IV) (108)		1
Mouse tumor	P1 (60)	P2 (48)	18
	P3A (72)		
Chick	F1 (53)	F2 (50)	24

Note: Numbers in parentheses are the reported molecular weights.

sively with 0.4 *M* sodium acetate solution to remove the blood components. A final wash was made with 0.5 *M* formic acid and tissue was ready for pepsin digestion.

For the cortices and anterior lens capsule, the bovine kidneys and eyes were obtained from a local slaughter house and transported on ice to the laboratory. Kidneys were defatted and freed of vascular tissues and the medula by dissection. Cortex slices were washed thoroughly with cold 0.5 *M* acetic acid and were homogenized (1 kg/2 ℓ wet weight in a Waring® blender for 2 min at 4°C. In the case of lens capsule, the eyes were cut open and the anterior lens capsules were dissected away, cleaned, and collected. The adhering cells were further removed by sonification (Branson Model 185, Plainview, Long Island, N.Y.) for 5 to 10 min, washed several times with cold water, and lyophilized.

B. Pepsin Solubilization and Purification of Native Collagen

1. Solubilization at 20°C

The washed placenta (2 kg) was suspended in 6 ℓ of 0.5 *M* formic acid. After addition of 1 g of pepsin (Boehringer Mannheim, 2500 θ/mg), the suspension was digested 16 to 20 hr. After centrifugation the residue was discarded. All following experiments were carried out at 5°C. The collagens were precipitated by addition of solid NaCl to a concentration of 6%. The precipitate was collected by centrifugation and resuspended in 2 ℓ of 0.3 *M* Tris-HCl buffer, pH 7.6, containing 0.2 *M* NaCl. The pH was readjusted to 7.6 with NaOH solution and dialyzed against the same buffer. During the dialysis, a precipitate was formed which was removed by centrifugation. Types I, III, IV, and V collagens which remained in solution were partially separated by fractional salt precipitation. To the supernate, a 5 *M* NaCl solution was added dropwise with stirring to give a final concentration of 1.8 *M* NaCl which precipitated Types III and IV collagens. At this NaCl concentration, Type I and V remained in solution. The 1.8 *M* NaCl precipitate was redissolved and precipitated once more at 1.8 *M* NaCl, centrifuged at 50,000 × g for 1 hr, and then dissolved in 0.5 *M* Tris-HCl, pH 8.0, at room temperature.

Mercaptoethanol was added to give a final concentration of 0.02 *M*, and the solution was left at 20°C for 16 to 20 hr. The free sulfhydryl groups were carboxymethylated by the addition of iodoacetate to a final concentration of 0.018 *M* and incubated at 20°C for 1 hr. The sample was then dialyzed against 0.5 *M* formic acid with several changes to remove the reagents. A second pepsin digest at 20°C for 16 to 20 hr was initiated by the addition of 1 mg pepsin per gram wet weight original 1.8 *M* NaCl precipitate. At the end of incubation, the collagen was precipitated by addition of NaCl to a concentration of 6%.

The precipitated collagen was dissolved in 0.03 *M* Tris-HCl, pH 8.6, containing 2 *M* urea and 0.2 *M* NaCl, the pH was adjusted to 8.6 to inactivate pepsin, and the solution was dialyzed against the same buffer at 4°C with two changes of the dialyzing buffer. The sample was passed through a DEAE-column (2.5 × 10 cm) equilibrated with the above Tris-HCl buffer at 15°C. The fraction which eluted unretarded from the column was collected and dialyzed against a 0.04 *M* sodium acetate buffer, pH 4.8, containing 2 *M* urea and then applied to a column (2.5 × 10 cm) of CM-cellulose equilibrated at 15°C with the same acetate buffer. The column was eluted using a linear gradient of NaCl from 0 to 0.2 *M* in 1000 mℓ of buffer.

For the preparation of collagen from anterior lens, the tissue was homogenized in 0.5 *M* formic acid (500 lenses in 500 mℓ) with an ultraturrax and 50 mg pepsin was added. Other experimental details were identical to those described for placenta above.

2. Solubilization at 4°C

The experimental details for placenta and anterior lens capsule were the same as those described above except reduction, alkylation, and repepsinization steps were not performed. The pepsin to substrate ratio for anterior lens capsule was 1 to 10.

The basement membrane collagens from bovine cortices were prepared as follows. The homogenized cortices were digested with pepsin (1 to 400 wet weight) at pH 2.5 in 0.5 *M* acetic acid for 62 hr with gentle stirring. The residue was collected by centrifugation (18,000 × g for 30 min) and discarded. Collagens were precipitated from the supernate by addition of NaCl to a concentration of 10%. The precipitate was redissolved in potassium phosphate buffer, pH 7.6 (I = 0.4 *M*), the pH was readjusted to 7.6 with 0.5 *M* NaOH, and the solution was dialyzed against the phosphate buffer for 24 hr with two changes. Type I and Type III collagens were precipitated by incubating the sample at 37°C for 16 hr.[23] After removing the gelled material by centrifugation at room temperature, the basement membrane collagen was precipitated by adding NaCl to 30% (w/v). The precipitate was dissolved in 0.01 *M* Tris-HCl buffer, pH 7.4, containing 0.1 *M* NaCl and dialyzed extensively against 0.01 *M* Tris-HCl buffer, pH 7.4 containing 0.02 *M* NaCl. Precipitated material was removed by centrifugation and discarded. The supernate which contained native basement membrane collagen was chromatographed on DEAE-cellulose and CM-cellulose as described above for placenta and lens capsule material.

C. Separation and Purification of Pepsin Fragments

Attempts to separate two distinct molecules, $[\alpha 1(IV)]_3$ and $[\alpha 2(IV)]_3$, in native state have been unsuccessful. A procedure to separate them is still under investigation.

The various chain components α1(IV)140K, α2(IV)120K, α1(IV)95K, and α2(IV)75K were separated by a combination of molecular sieve and CM-cellulose chromatography. The procedures described here are for bovine cortices solubilized at 4°C. Similar results have been obtained using placenta[12,13] and anterior lens capsule.[14–16]

The lyophilized native collagen obtained from bovine cortices was denatured by heating at 45°C for 10 to 15 min in 0.01 *M* Tris-HCl, pH 7.4, containing 1 *M* $CaCl_2$. The sample was applied to an agarose A5m (200 to 400 mesh, BioRad® Laboratories) column (4 × 50 cm) which was equilibrated at room temperature with 0.01 *M* Tris-HCl, pH 7.4, containing 1 *M* $CaCl_2$.

The separately pooled high molecular weight fractions and α size fraction were purified individually to isolate the α1(IV)140K-, α2(IV)120K-, α1(IV)95K-, α2(IV)95K-, and α2(IV)75K-chain components.

1. α1(IV)140K- and α2(IV)120K-Chains

The lyophilized high molecular weight fractions were reduced with dithiothreitol and applied to a CM-cellulose column equilibrated with 0.02 *M* acetate, 0.005 dithiothreitol, 1 *M* urea, pH 4.8, at 44°C. The sample (200 mg) was dissolved in 20 mℓ of 0.1 *M* dithiothreitol, 8 *M* urea, pH 8.5, and was warmed at 45°C for 2 hr with gentle stirring. After application of the sample, the column was eluted with equilibrated buffer until the original baseline was achieved, and then a linear gradient of 0 to 0.1 *M* NaCl in 1 ℓ of equilibration buffer applied. Further purification of the material in the CM-cellulose fractions was achieved by rechromatography on an agarose A5m column (4.0 × 120 cm) equilibrated at room temperature with 0.01 *M* Tris-HCl 1 *M* $CaCl_2$, pH 7.4. The agarose fractions were desalted into 0.1 *M* acetic acid using BioGel® P-2 (100 to 200 mesh) and lyophylized.

Attempts to isolate these chains from placenta or lens capsule material proved to be more difficult.[12–14,16] The α2(IV)120K component from these tissues has not yet been isolated and characterized.

2. α1(IV)95K-, α2(IV)95K-, and α2(IV)75K-Chains

The α size components from kidney cortices were chromatographed on a CM-cellulose column (2.5 × 10 cm) equilibrated with 0.02 *M* acetate, 1 *M* urea, pH 4.8, maintained at 44°C. After application of the sample, the column was eluted with a linear gradient of 0 to 0.1 *M* NaCl in 1 ℓ of starting buffer. The separately pooled fractions were desalted on BioGel® P-2 (100 to 200 mesh, BioRad® Laboratories) into 0.1 *M* acetic acid and lyophilized.

Similar results were obtained using identical experimental conditions from anterior lens capsule[14,16] and placenta[15,16] where pepsin digestion was performed at 4°C. In

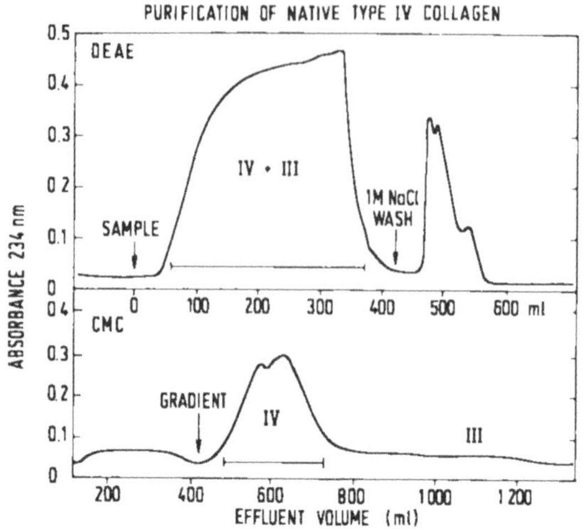

FIGURE 1. Chromatographic purification of native Type IV collagen from human placenta. Separation of native collagen components from acidic impurities using DEAE-cellulose; separation of native Type IV from native Type III collagen on CM cellulose.

experiments where pepsin digestion was carried out at 20°C followed by reduction and alkylation and repepsination, α2(IV)95K was converted to α2(IV)75K during the preparation of native material, and higher yields of both α1(IV)95K and α2(IV)75K were obtained.[11] The α1(IV)140K- and α1(IV)120K-chains were absent in such preparations.

3. *α1(IV)50K$_1$ and α2(IV)50K$_2$ Fragments*

The 50K fragments from kidney cortices,[9] placenta,[12] and anterior lens capsule[15-16] were resolved into α1(IV)50K$_1$ and α2(IV)50K$_2$ components by CM-cellulose chromatography. The experimental conditions are identical to that described for in the above section.

III. RESULTS

A. Characterization of Native Collagen Preparations

Native collagen preparations, prepared either by pepsin solubilization at 20 or 4°C, were chromatographically purified by passing over DEAE-cellulose and eluting from a CM-cellulose. Elution profiles obtained with placental material from a 20°C pepsin solubilization are shown in Figure 1. Similar profiles were obtained for lens capsule and kidney cortex material. The temperature at which pepsin solubilization was carried out had little or no effect on the elution position of basement membrane collagens. Collagenous material was not retarded by DEAE-cellulose, but a relatively small amount of acidic impurity was bound and thus removed (see Figure 1A). Small amounts of Type III collagen which may still be present in the collagenous DEAE cellulose fraction were removed using CM-cellulose. Basement membrane collagens were eluted as an asymmetrical peak at the beginning of the gradient followed by Type III collagen (see Figure 1B).

Disk gel electrophoresis patterns of the CM-cellulose fraction from placenta and lens capsule collagen, from both 4 and 20°C pepsin solubilization, are shown in Figure 2. The corresponding fractions from kidney cortexes were similar to these

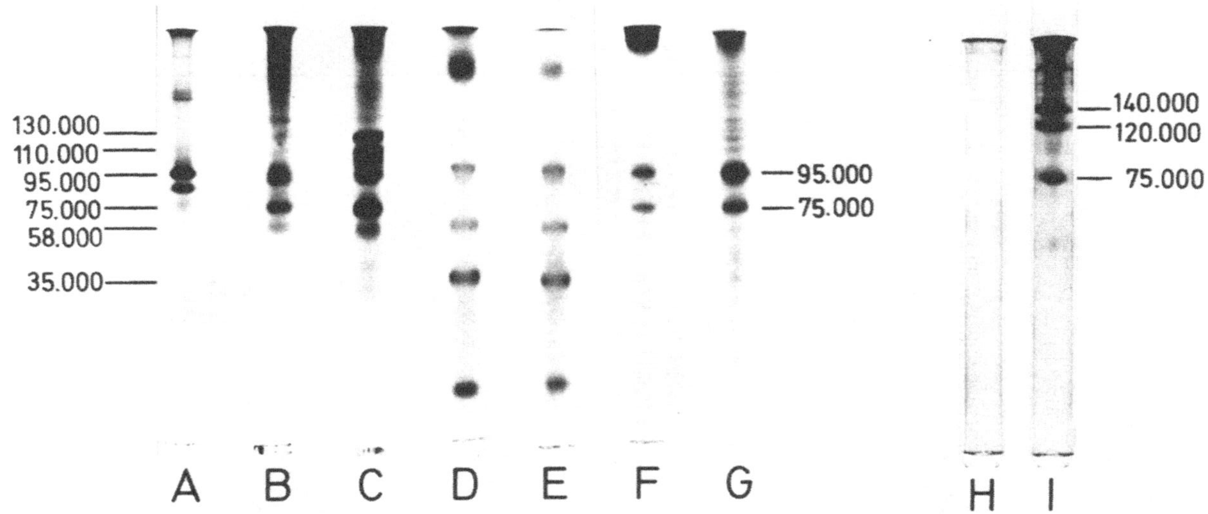

FIGURE 2. SDS disk gels (5%) stained with Coomassie blue showing purified native basement membrane collagen preparations isolated from bovine lens capsule and human placenta by pepsin solubilization at 4 and 20°C. (A) Type I collagen standard; (B) and (C) lens capsule, 4°C; (D) and (E) lens capsule, 20°C; (F) and (G) placenta, 20°C; (H) and (I) placenta, 4°C; In gels C, E, F, and I, the samples were reduced. (Gels H and I are not directly comparable to gels A to G.)

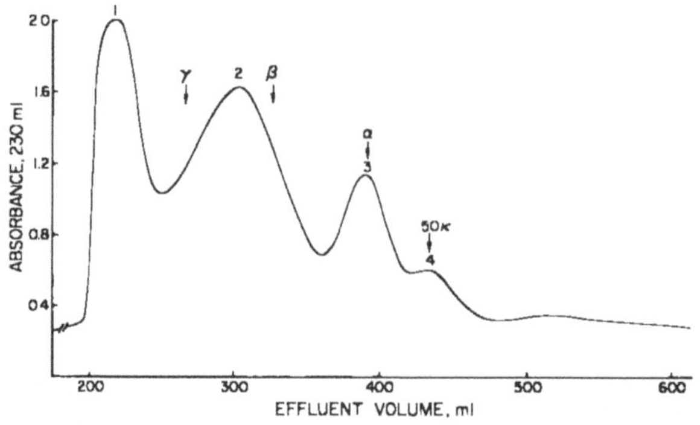

FIGURE 3. Molecular sieve chromatography of kidney cortex basement membrane collagen on agarose A5m. The elution positions of the γ (285,000 daltons), β (190,000 daltons), and α (95,000 daltons) components of Type I collagen are indicated.

Table 2
AMINO ACID COMPOSITIONS OF α1 (IV) 140K- AND α2 (IV) 120K-CHAINS FROM HUMAN PLACENTA (HP) BOVINE LENS CAPSULE (BL) AND BOVINE CORTICES (BC) IN RESIDUES PER 1000 AMINO ACIDS

	α1 (IV) 140K			α2 (IV) 120K
Amino acid	HP[a]	BL	BC	(BC)
3-Hyp	1	5	9	9
4-Hyp	118	122	128	114
Asp	45	45	49	48
Thr	20	19	22	24
Ser	36	36	36	38
Glu	88	90	87	72
Pro	88	68	56	58
Gly	324	342	349	337
Ala	30	28	30	36
Val	32	22	24	23
½ Cys	2	2	2	3
Met	14	12	13	11
Ile	32	24	30	32
Leu	50	54	49	58
Tyr	8	3	4	6
Phe	28	28	28	35
Hyl	49	60	52	45
Lys	7	9	6	5
His	6	8	6	6
Arg	22	23	21	44
Glc-Gal-Hyl	43	52	48	38
Gal-Hyl	2	6	2	3

[a] Taken from Reference 12.

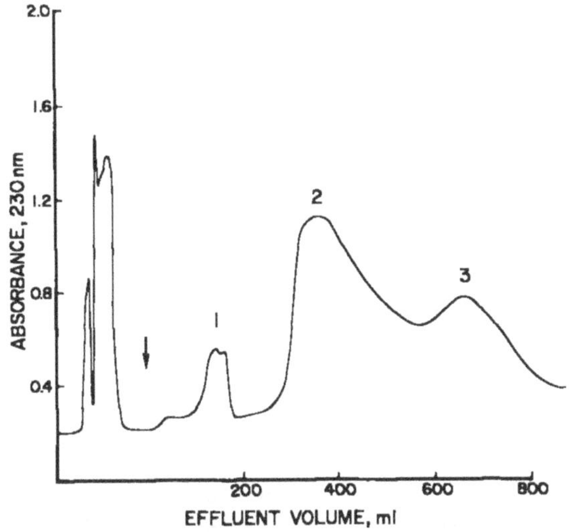

FIGURE 4. Separation of kidney cortex α1(IV)140K (Fraction 2) from α2(IV)120K and α2(IV)75K (Fraction 3) on CM-cellulose. The arrow indicates the start of the gradient.

Table 3
AMINO ACID COMPOSITIONS OF THE α1 (IV) 95K-CHAIN FROM HUMAN PLACENTA (HP), HUMAN GLOMERULUS (HG), BOVINE LENS CAPSULE (BL), BOVINE CORTICES (BC) AND PORCINE CORTICES (PC) IN RESIDUES PER 1000 AMINO ACIDS

Amino acid	HP	HG[a]	BL	BC	PC[a]
3-Hyp	nd	7	4	8	7
4-Hyp	146	126	126	123	126
Asp	48	41	52	50	52
Thr	18	16	20	20	17
Ser	34	36	38	40	34
Glu	77	85	75	82	73
Pro	79	61	64	58	64
Gly	318	342	352	354	359
Ala	31	30	35	35	35
Val	30	25	26	26	21
Met	15	13	11	13	13
Ile	34	28	27	25	23
Leu	54	53	48	48	55
Tyr	5	4	2	2	4
Phe	24	26	28	27	25
Hyl	61	57	58	52	60
Lys	5	6	7	7	8
His	6	6	6	5	7
Arg	20	24	21	21	21
Glc-Gal-Hyl	nd	nd	48	48	48
Gal-Hyl	nd	nd	5	1	2

Note: n.d. = not determined.

[a] Taken from Reference 8.

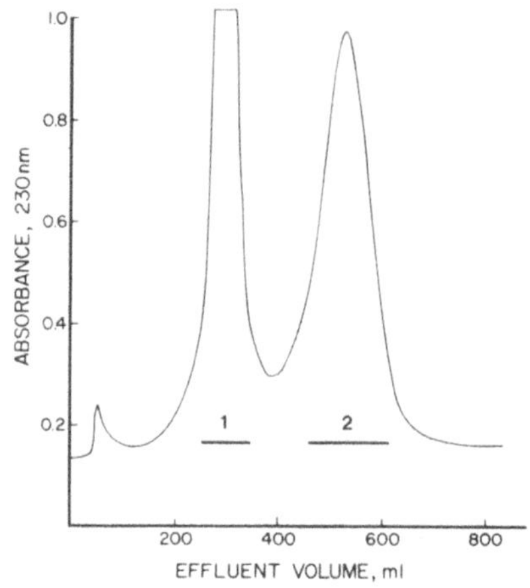

FIGURE 5. Separation of α1(IV)95K (Peak 1) and α2(IV)95K (Peak 2) on CM-cellulose.

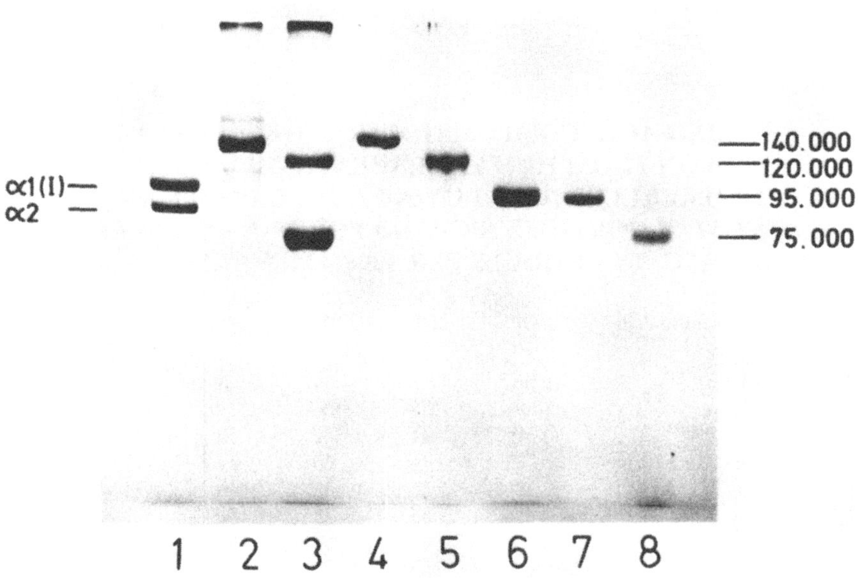

FIGURE 6. SDS slab gel (7.5%) stained with Coomassie blue showing in Lane 1 standard α1(I) and α2; Lane 2, Fraction 2 from Figure 4; Lane 3, Fraction 3 from Figure 4; Lane 4, purified α1(IV)140K; Lane 5, purified α2(IV)120K; Lane 6, α1(IV)95K from CM-cellulose (see Figure 5, Peak 1); Lane 7, α2(IV)95K from CM-cellulose (see Figure 5, Peak 2); Lane 8, purified α2(IV)75K.

from lens capsule. Clearly there are differences not only between the products derived from lens capsule and placenta when the same preparation procedure was used, but also between the products from one tissue when pepsin solubilization was carried either at 4 or 20°C. In Figure 2, gels D and E, the products of a lens capsule 20°C pepsin treatment and in F and G the same product derived from placenta are shown. Gels E and G show reduced products. The only common component is an α-chain size fragment of approximately 95,000 daltons. The placenta fragments are α1(IV)95K and α2(IV)75K and have both been reasonably well characterized.[11-13]

Table 4
AMINO ACID COMPOSITION OF THE α2 (IV) 75K-CHAIN FROM HUMAN PLACENTA (HP), BOVINE LENS CAPSULE (BL), BOVINE CORTICES (BC), AND PORCINE CORTICES (PC) IN RESIDUES PER 1000 AMINO ACIDS

Amino acid	HP	BL	BC	PC[a]
3-Hyp	nd	6	8	6
4-Hyp	127	117	115	126
Asp	52	50	46	48
Thr	28	24	23	19
Ser	26	38	37	36
Glu	62	70	72	58
Pro	69	58	60	64
Gly	313	338	340	337
Ala	49	38	37	48
Val	24	26	24	37
½ Cys	—	1	2	2
Met	12	12	11	9
Ile	42	31	30	21
Leu	59	58	56	60
Tyr	6	2	3	2
Phe	37	35	34	35
Hyl	42	42	43	40
Lys	6	4	4	6
His	8	6	6	8
Arg	45	44	45	38
Glc-Gal-Hyl	nd	28	34	28
Gal-Hyl	nd	4	2	2

Note: n.d. = not determined.

[a] Taken from Reference 28.

The reducible material at the top of gel F is mainly 7-S collagen which on reduction gives rise to several fragments that can be seen as a background in gel G. Lens capsule material appears to contain very little or no reducible material. The lens capsule fragments produced by a 20°C pepsin solubilization have not yet been characterized. Gels B, C, H, and I show basement membrane collagen solubilized using pepsin at 4°C. In this case, lens capsule material contains 95,000- and 75,000-dalton components before reduction (gel B) and after reduction (gel C) contains an additional distinct component of about 130,000 daltons plus two other diffuse bands of about 100,000–120,000 daltons. In contrast placental material did not enter the gel in the unreduced form (gel H) and on reduction showed three distinct components with molecular weight of 140,000, 120,000, and 75,000 daltons. Comparing the products of a 20 to a 4°C pepsin solubilization form in one tissue, it can be seen that at 20°C the 100,000 to 140,000 mol wt products are absent (compare gels C and E, I, and G).

B. Characterization of Pepsin Fragments

In order to divide the sample into reducible components and nonreducible components, the first step in the isolation of the fragments of basement membrane collagen chains was usually a molecular weight fractions as shown in Figure 3. Here the sample was divided into four fractions. The material in fractionization 1 and 2 after reduction and carboxymethylation was separated on CM-cellulose giving the elution profile shown in Figure 4. Disk gel electrophoresis patterns of the material in fractions 2 and 3 of Figure 4 are shown in Figure 6, lanes 2 and 3, respectively. Rechromatography of fraction 2 on agarose A5m gave α1(IV)140K, and rechromatography of fraction 3 gave α2(IV)120K and α2(IV)75K fragments. Gel electrophoresis patterns of the purified components are shown in Figure 6, and their amino acid composition is shown in Table 1.

The α-chain size components from kidney cortex, when chromatographed on CM-cellulose, gave rise to α1(IV)95K and α2(IV)95K as shown in Figure 5. Disk gel electrophoresis patterns of these two fragments are shown in Figure 6. On reduction the α2(IV)95K fragment is converted to α2(IV)75K (see Figure 6, lanes 7 and 8), a fragment similar to that obtained by reducing fractions 1 and 2 in Figure 3. Table 2 shows the amino acid compositions of α1(IV)95K fragments and Table 3 shows the α2(IV)75K fragments isolated from kidney cortex, lens capsule, and human placenta.

The agarose fraction containing material with a molecular weight of approximately 50,000 daltons (see Figure 3, peak 4) was also resolved into two components by chromatographing on CM-cellulose and are refered to as $50K_1$ and $50K_2$ (not shown).

IV. DISCUSSION

In general, the methods used to prepare native basement membrane collagen from the various tissues considered here are similar. The methods used to remove the interstitial collagens and Type V (AB collagen) vary, but appear to be equally effective. The most important differences are in the conditions used for pepsin treatment. The source and activity of pepsin used, the enzyme/substrate ratio, enzyme concentration, temperature, and digestion medium are all critical factors which are not always adequately reported in the literature.

Selection of starting material will depend upon the problem to be investigated, but when species is unimportant, bovine lens capsule is probably the cleanest and most homogenous tissue. A disadvantage of this tissue is the number of lenses that have to be processed to obtain sufficient quantities of end product (usually 500 to 1000 lenses). Conversely human placenta is readily available in large quantities, but is a very heterogeneous tissue. Another tissue which has been investigated extensively, but not discussed here, is a mouse tumor.[17,18] The major fragments, P1 and P2, produced by pepsin solubilization of this tissue are smaller than α2(IV)75K. However, this appears to be due only to small sequence differences that are normally found in collagens isolated from different species and not to a fundamentally different structure. More recently, similar-sized fragments have also been isolated from chick tissues.[24]

It is clear from the data presented here and reported elsewhere that it has not been possible to prepare a pure homogeneous native basement membrane collagen molecule from placenta, kidney, or lens capsule. Preparations usually contain non-reducible high molecular weight material, particularly when the pepsin solubilization is carried out at 20°C, and a mixture of chains, the molecular weights and relative proportions of which depend upon the starting material and conditions of pepsin solubilization. As the chains derive from two different molecules and the high molecular weight material derives possibly from 7-S collagen[25] (this is certainly true for material derived from placenta), antisera raised against native preparations will contain antibodies directed against a number of structures in basement membranes.[26] More specific antisera can therefore only be produced by purifying antisera to native basement membrane collagens by passing through affinity columns containing purified fragments.[27] The fragments most suited for this purpose are α1(IV)95K and α2(IV)75K, as they can be isolated from all three tissues described here and have been fairly well characterized.

To isolate α1(IV)95K and α2(IV)75K, pepsin solubilization at 20°C is probably preferable. Using this method, α1(IV)140K and α2(IV)120K are apparently converted to α1(IV)95K and α2(IV)75K, respectively, and thus the separation of the required fragments from the pepsin digest and from one another is made easier, and the yields are greater. However, α1(IV)95K and α2(IV)75K are not end products of this reaction, and so the digestion conditions have to be controlled carefully so that these two fragments are produced as major products.[22] The α2(IV)75K fragment is particularly labile and is very quickly degraded when the sample is denatured at neutral pH if traces of nonspecific neutral proteases, that originate from the starting material, are still active.

ACKNOWLEDGMENTS

During the preparation of this article, S. N. Dixit was supported by National Institutes of Health grants AM 20385 and 16506. Original work on human placental basement membrane collagen was supported by a grant from the Deutsche Forschungsgemeinschaft, Sonderforschungsbereich 51, B/22.

REFERENCES

1. **Kefalides, N. A.,** Structure and biosynthesis of basement membranes, *Int. Rev. Connect. Tissue Res.*, 6, 663, 1973.
2. **Dehm, P. and Kefalides, N. A.,** The collagenous component of lens basement membrane; the isolation and characterization of an α chain size collagenous peptide and its relation to newly synthesized lens components, *J. Biol. Chem.*, 253, 6680, 1978.
3. **Hudson, B. G. and Spiro, R. G.,** Studies on the native and reduced alkylated renal glomerular basement membrane, *J. Biol. Chem.*, 247, 4229, 1972.
4. **Sato, T. and Spiro, R. G.,** Studies on the subunit composition of the renal glomerular basement membrane, *J. Biol. Chem.*, 251, 4062, 1976.
5. **Freytag, J. W., Ohno, M., and Hudson, B. C.,** Bovine renal glomerular basement membrane: assessment of proteolysis during isolation, *Biochem. Biophys. Res. Commun.*, 72, 796, 1976.
6. **Daniels, J. R. and Chu, G. H.,** Basement membrane collagen of renal glomerulus, *J. Biol. Chem.*, 250, 3531, 1975.
7. **Tryggvason, K. and Kivirikko, K. I.,** Heterogeneity of pepsin-solubilized human glomerular basement membrane collagen, *Nephron*, 21, 230, 1978.
8. **Dixit, S. N.,** Isolation and characterization of two α chain size collagenous polypeptide chains C and D from glomerular basement membrane, *FEBS Lett.*, 106, 379, 1979.
9. **Dixit, S. N.,** Type IV collagens: isolation and characterization of two structurally distinct collagen chains from bovine kidney cortices, *Eur. J. Biochem.*, 106, 563, 1980.
10. **Bailey, A. J., Sims, T. J., Duance, V. C., and Light, N. D.,** Partial characterization of a second basement membrane collagen in human placenta, *FEBS Lett.*, 99, 361, 1979.
11. **Glanville, R. W., Rauter, A., and Fietzek, P. P.,** Isolation and characterization of a native placental basement collagen and its component α chains, *Eur. J. Biochem.*, 95, 383, 1979.
12. **Kresina, T. F. and Miller, E. J.,** Isolation and characterization of basement membrane collagen from human placental tissue. Evidence for the presence of two genetically distinct chains, *Biochemistry*, 18, 3089, 1979.

13. **Sage, H., Woodbury, R. G., and Bornstein, P.,** Structural studies on human type IV collagen, *J. Biol. Chem.*, 254, 9893, 1979.
14. **Dixit, S. N.,** Isolation and characterization of two collagenous components from anterior lens capsule, *FEBS Lett.*, 85, 153, 1978.
15. **Dixit, S. N. and Kang, A. H.,** Anterior lens capsule collagens: cyanogen bromide peptides of the C chain, *Biochemistry*, 18, 5686, 1979.
16. **Gay, S. and Miller, E. J.,** Characterization of lens capsule collagen: evidence for the presence of two unique chains in molecules derived from major basement membrane structures, *Arch. Biochem. Biophys.*, 198, 370, 1979.
17. **Timpl, R., Martin, G. R., Bruckner, P., Wick, G., and Wiedemann, H.,** Nature of the collagenous protein in a tumor basement membrane, *Eur. J. Biochem.*, 84, 43, 1978.
18. **Timpl, R., Bruckner, P., and Fietzek, P. P.,** Characterization of pepsin fragments of basement membrane collagen obtained from a mouse tumor, *Eur. J. Biochem.*, 95, 255, 1979.
19. **Crouch, E., Sage, H., and Bornstein, P.,** Structural basis for apparent heterogeneity of collagens in human basement membranes: type IV procollagen contains two distinct chains, *Proc. Natl. Acad. Sci. U.S.A.*, 77, 745, 1980.
20. **Tryggvason, K., Robey, P. G., and Martin, G. R.,** Biosynthesis of type IV procollagens, *Biochemistry*, 19, 1284, 1980.
21. **Alitalo, K., Vaheri, A., Krieg, T., and Timpl, R.,** Biosynthesis of two subunits of type IV procollagen and of other basement membrane proteins by a human tumor cell line, *Eur. J. Biochem.*, 109, 247, 1980.
22. **Rauter, A. and Glanville, R. W.,** Pepsin fragments of human placental basement membrane collagens showing interrupted triple helical amino acid sequences, *Eur. J. Biochem.*, submitted.
23. **Trelstad, R. L. and Lawley, K. R.,** Isolation and initial characterization of human basement membrane collagens, *Biochem. Biophys. Res. Commun.*, 76, 376, 1977.
24. **Mayne, R. and Zettergren, J. G.,** Type IV collagen from chick muscular tissues; isolation and characterization of pepsin resistant fragments, *Biochemistry*, 19, 4065, 1980.
25. **Risteli, J., Bächinger, H. P., Engel, J., Furthmayr, H., and Timpl, R.,** 7S collagen: characterization of an unusual basement membrane structure, *Eur. J. Biochem.*, 108, 239, 1980.
26. **Timpl, R., Glanville, R. W., Wick, G., and Martin, G. R.,** Immunochemical study on basement membrane (type IV) collagens, *Immunology*, 38, 109, 1979.
27. **Risteli, J., Schuppan, D., Glanville, R. W., and Timpl, R.,** Immunochemical distinction between two different chains of type IV collagen, *Biochem. J.*, 191, 517, 1980.
28. **Dixit, S. N. and Kang, A. H.,** Basement membrane collagens: cyanogen bromide peptides of the D chain from porcine kidney, *Biochemistry*, 19, 2692, 1980.

Chapter 6

THE PREPARATION OF TYPE V (AB) COLLAGEN*

Joseph A. Madri

TABLE OF CONTENTS

* Supported in part by U.S. Public Health Service Fellowship 5 F32 HL 05559 and National Cancer Institute Contract 1-CB-84255-37.

I. INTRODUCTION

Since its recent discovery by Burgeson et al. and Chung et al., the possible inclusion of Type AB collagen as a "basement membrane" and/or "cell surface associated" collagen and its suggested involvement in epithelial cell migration and hemostasis have prompted many studies on this collagen type.[1-7] This type of collagen has been termed Type V recently, and chain designations B, A, and C are used interchangeably with $\alpha 1(V)$, $\alpha 2(V)$, and $\alpha 3(V)$.

The stoichiometry of αA- or $\alpha 2(V)$- to αB- or $\alpha 1(V)$-chains (and possibly αC- or $\alpha 3(V)$-chains) is still an open question, as evidence has been adduced for the existence of separate molecules with the compositions of $(\alpha A)_3$ or $[\alpha 2(V)]_3$, $(\alpha B)_3$ or $[\alpha 1(V)]_3$, $\alpha A(\alpha B)_2$ or $\alpha 2(V)[\alpha 1(V)]_2$, $(\alpha A)_2\alpha B$ or $[\alpha 2(V)]_2\alpha 1(V)$, and possibly $(\alpha A)_x(\alpha B)_y(\alpha C)_z$ or $[\alpha 2(V)]_x\,[\alpha 1(V)]_y\,[\alpha 3(V)]_z$.[1,5,8-11] Possible explanations for these varied findings are tissue source differences, age of tissue(s) extracted, conditions of pepsin extraction, salt precipitation, and chromatography.[1,8-11]

Similarly, evidence has been presented for and against the inclusion of this collagen type in the category of basement membrane collagens. Its amino acid composition is similar to that of Type IV collagen,[1,9] and several studies have localized it to basement membranes using standard immunofluorescence techniques with affinity-purified antibodies.[5,6,12,13] In addition Type V collagen has been identified in glomerular and alveolar basement membranes utilizing ultrathin sectioning technology coupled with affinity purified antibody probes.[3,4] Arguments against its inclusion as a basement membrane collagen include isolation from tissues not rich in basement membranes and its rather ubiquitous distribution in the interstitium of various organs.[5,7,8] However, its localization on cell surfaces of various cultured cells, including epidermal cells,[4] endothelial cells,[15] and thymic epithelial cells,[16] suggests that it may be a component of the basal lamina surrounding certain cells and therefore can be considered a "basement membrane" collagen (assuming a broad definition of the term basement membrane). Thus the localization of Type V collagen in basement membranes in some organs (kidney, lung) and as a cell surface associated collagen in other tissues (thymus, epidermis, media of vessels) is compatible with this notion. In any case, the presence of a particular collagen in traditional basement membranes should not necessarily exclude it from being present in the "interstitium" of certain organs and the reverse could apply also.

II. PREPARATIVE METHODS

Several tissues have been used successfully as sources of type V collagen,[1,3,5,9,10,12,13,15,19,27] including lung, placental membranes, placental villi, skin, and bone.

A. Washing

Regardless of the tissue source, an important first step is a thorough washing of the tissue sample. For noncartilagenous and nonosseous tissues, this step should include an initial mincing using a meat grinder or blender, followed by exhaustive washing with water at 4°C. The water wash is usually followed by salt extraction with 1 *M* NaCl, 50 m*M* Tris-HCl, pH 7.5, and acid extraction with 0.5 *M* acetic acid.[1,5,6,8] These washing steps are necessary to remove contaminating blood components, as well as other noncollagenous connective tissue proteins. If cartilagenous tissues are used, they should also be extracted with 4.0 *M* guanidinium hydrochloride in 50 m*M* Tris-HCl, pH 7.5, to remove contaminating glycosaminoglycans. Osseous tissues can be decalcified initially using 0.5 *M* acetic acid prior to the washing protocol.

In addition to adequate washing, another important consideration is to minimize the time from acquisition of the tissues to washing. This is especially true in extractions involving parenchymal, blood-rich tissues, such as placenta, kidney, lung, and liver. If these tissues are allowed to stand (even at 4°C) for any length of time, breakdown products of the blood components will become a significant contamination problem.

B. Storage of Washed Tissues

Only after the tissues are adequately washed should they be stored. They may be stored frozen at −70°C in a small amount of water or diluted acetic acid. Alternatively, tissues may be stored conveniently in the lyophilized state at −30°C.

C. Enzymatic Digestion

Pepsin digestion is common to all the procedures of extracting Type V collagen from tissues. Variable enzyme to substrate ratios have been used. A commonly employed ratio is 1 g of crystalline pepsin A (Worthington) added to 10 g (dry weight) of tissue previously homogenized and suspended in 1 to 2 ℓ of 0.5 *M* acetic acid.[5,6,8,11] The mixture is incubated for 24 hr at 4°C with gentle, constant stirring. The pepsin digest is then clarified by centrifugation for 1 hr at 10,000 × g at 4°C, and undigested material in the precipitate is discarded. The supernatant fluid (containing soluble collagen types) can be stored frozen at −30°C, or one may proceed with differential salt precipitation to isolate and purify Type V collagen.

D. Isolation and Purification of Type V Collagen

These first three general steps in the isolation and purification of Type V collagen are fairly conventional and common to most procedures. However, the differential salt precipitation procedures allowing for the separation of Type V collagen from other collagen types can be done following two major protocols (1) differential salt precipitation under conditions of acid pH[2,8,11]; and (2) differential salt precipitation under conditions of neutral pH.[1,3-6,10] Further separation of Type V collagen involves ion exchange chromatography of the native collagens. Each of these protocols will be discussed in detail.

1. Differential Salt Precipitation Under Acid pH

This approach was first proposed by Chung et al.[2] and Rhodes and Miller[8] and subsequently used by Sage and Bornstein[11] to isolate Type V collagen. In this protocol, the pepsin digest supernatant fluid is adjusted to 0.7 *M* NaCl by the addition of solid crystalline NaCl with constant stirring at 4°C. After all the added salt is dissolved, the solution is left standing for 24 hr at 4°C, during which time a white fluffy precipitate forms. This precipitate can be removed by centrifugation at 10,000 × g for 1 hr at 4°C and is largely composed of the interstitial collagens Types I and III. The supernatant fluid also contains collagen Types IV, V, as well as other less well-characterized species.[17,18] By addition of crystalline NaCl to a final concentration of 1.2 *M* NaCl collagen Types IV and V are precipitated. Several alternative procedures are available which allow for the separation of Type IV collagen from Type V collagen. One method used in this laboratory is solubilization of the 1.2 *M* NaCl acid pH precipitate in 1 *M* NaCl, 50 m*M* Tris-HCl, pH 7.5, followed by pH adjustment with 1 *M* Tris® base to pH 7.5. After dialysis against 1 *M* NaCl, 50 m*M*

Tris-HCl, pH 7.5, solid NaCl is added to the collagen solution to a final concentration of 1.7 *M*. After the added NaCl is dissolved, the solution is left standing overnight at 4°C. The precipitate that forms is enriched for Type IV collagen and is collected by centrifugation. The supernatant fluid containing Type V collagen is made 4.5 *M* with respect to NaCl by addition of solid NaCl, or it can be exhaustively dialyzed against water to precipitate Type V collagen. The precipitation cycle may be repeated to further enrich for Type V collagen. The precipitated Type V collagen is then dissolved in 0.01 *M* acetic acid, dialyzed against 0.01 *M* acetic, and stored frozen at −30°C.

Another method that can be used to separate Types IV and V collagen after fractional salt precipitation in acid pH conditions involves the use of DEAE cellulose (Whatman® DE52) chromatography (See Section 3.3).

2. Differential Salt Precipitation Under Neutral pH

This approach was first proposed by Burgeson et al.[1] and used subsequently by Madri and Furthmayr,[5,6] Roll et al.,[3] and Hong et al.[10] Other groups have employed this approach to separate individual collagens after initial fractional salt precipitation at acid pH.[8,11,18,19]

In this protocol, the pepsin digest supernatant fluid is adjusted to 1 *M* NaCl by addition of solid crystalline NaCl with constant stirring at 4°C. After the added salt is dissolved, the solution is left standing for 24 hr at 4°C. The resultant precipitate, containing (the interstitial) collagen Types I and III, as well as Types IV and V, is collected by centrifugation and dissolved in 1 *M* NaCl, 50 m*M* Tris-HCl, pH 7.5. The pH is adjusted with 1 *M* Tris® base. After dialysis against 1 *M* NaCl, 50 m*M* Tris-HCl, pH 7.5, the solution is brought up to 1.7 *M* NaCl by addition of solid NaCl and is left standing at 4°C overnight. The resultant precipitate is collected by centrifugation. This fraction contains largely Type III collagen and smaller amounts of Type IV.[5,6,11] The supernatant fluid is then brought up to 2.6 *M* NaCl as before. The precipitate formed is predominantly Type I, with small amounts of Type IV.[5] The supernatant fluid contains Type V which can be precipitated by adding solid NaCl to a final concentration of 4.5 *M* or by exhaustive dialysis against water. As with the acid pH precipitation, this precipitation cycle may be repeated to further enrich for Type V collagen. The precipitated collagen is solubilized in 0.01 *M* acetic acid, dialyzed against 0.01 *M* acetic acid, and stored at −30°C.

3. Ion-Exchange Chromatography

A third approach used in conjunction with differential salt precipitation to separate Type V collagen from other collagen types is ion-exchange chromatography. Bentz et al.[9] and Glanville and Kühn[20] first described the use of DEAE-cellulose (Whatman® DE52) in separating Type V collagen from noncollagenous contaminants and other collagens (Types I, III, and IV). DEAE-cellulose is equilibrated at 15°C with 2.0 *M* deionized urea, 0.02 *M* NaCl, 50 m*M* Tris HCl, pH 8.6. The collagen solution is equilibrated in the DEAE-cellulose buffer and applied to the column (2.5 × 12 cm). The collagen Types I, II, and IV are not bound to the column under these conditions and are eluted with starting buffer. After the $OD_{230\ nm}$ returns to baseline, a 2000-mℓ linear gradient of 0.02 to 0.32 *M* NaCl is run through the column. Type V collagen is eluted in the early part of the gradient (200 to 400 mℓ or 0.05 to 0.08 *M* NaCl) (see Figure 1).

The other ion exchange system which has been used to purify Type V collagen is carboxymethyl (CM)-cellulose (Whatman® CM52). Welsh et al.[19] equilibrated CM-cellulose and the collagen solution in 6.0 *M* deionized urea, 0.02 *M* NaCl, 20

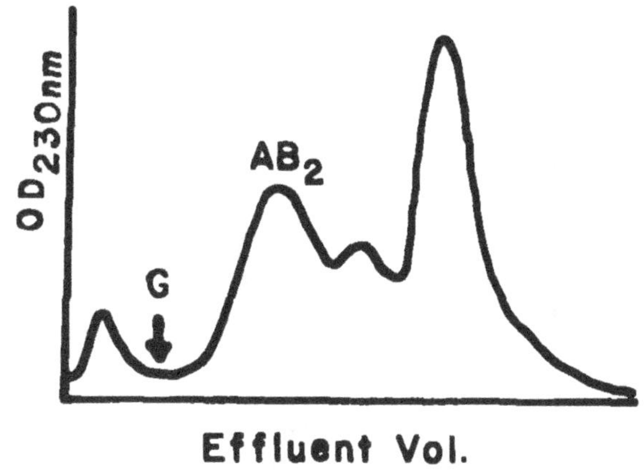

FIGURE 1. DEAE cellulose chromatography of native placental collagen. The peak containing A- and B-chains is labeled AB_2. The gradient elution begins at the arrow labeled "G." (Redrawn from Bentz, H., Bächinger, H. P., Glanville, R. and Kühn, K., *Eur. J. Biochem.*, 92, 563, 1978.)

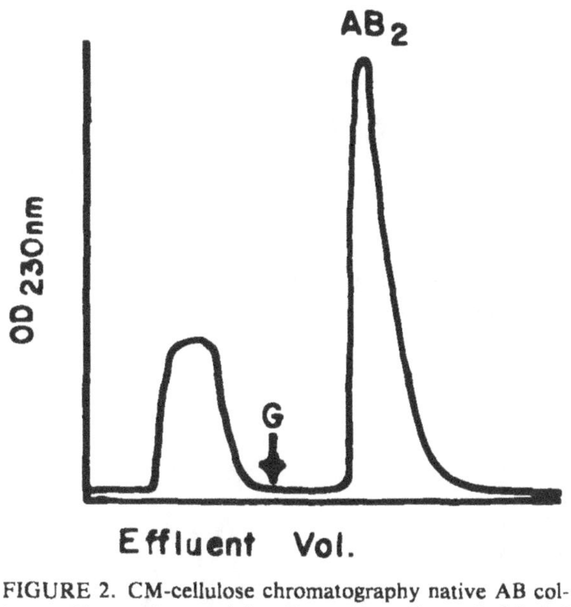

FIGURE 2. CM-cellulose chromatography native AB collagen. The peak containing Type AB collagen is labeled AB_2. The gradient elution begins at the arrow labeled "G". (Redrawn from Welsh, G., Gay, S., Rhodes, R. K., Pfister, R., and Miller, E. J., *Biochem. Biophys. Acta*, in press.)

m*M* Tris-HCl, pH 8.0, at 4°C. The collagen solution was passed over the column (2.5 × 8.0 cm), and the column was washed with starting buffer until the $OD_{230\,nm}$ returned to baseline. At this time, an 800-mℓ linear gradient of 0.02 to 0.23 *M* NaCl was run through the column. While Type I collagen is not bound to the column, Type V is and can be eluted early in the NaCl gradient (50 mℓ or 0.04 *M* NaCl) (see Figure 2).

Ayad et al.[18] have used CM-cellulose also, but with a different buffer system

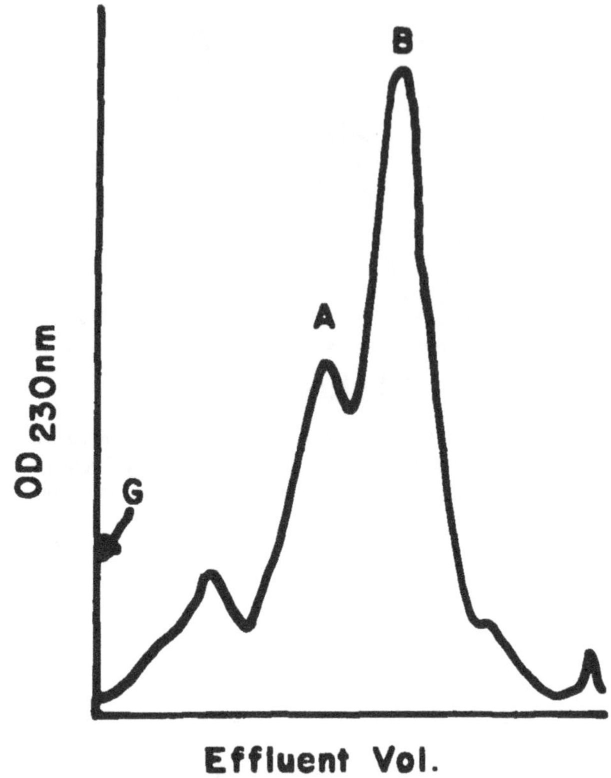

FIGURE 3. (A) CM-cellulose chromatography of Type AB collagen chains. Elution positions of α2(V)- and α1(V)-chains are indicated by A and B. The gradient begins at the arrow labeled "G". (Redrawn from Burgeson, R. E., El Adli, F. A., Kaitila, I. I., and Hollister, D. W., *Proc. Natl. Acad. Sci. U.S.A.*, 73, 2579, 1976.)

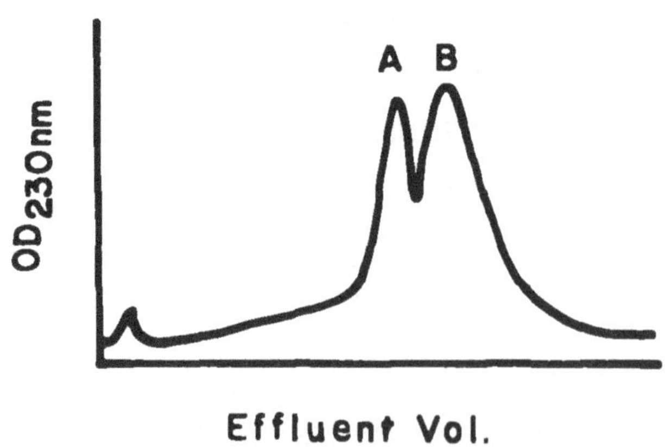

FIGURE 3. (B) CM cellulose chromatography of Type AB collagen chains. Elution positions of α2(V)- and α1(V)-chains are indicated by A and B. (Redrawn from Chung, E., Rhodes, R. K., and Miller, E. J., *Biochem. Biophys. Res. Commun.*, 71, 1167, 1976.)

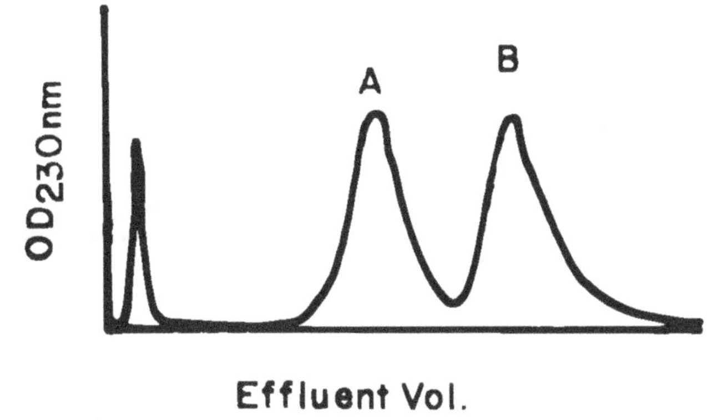

FIGURE 4. Phosphocellulose chromatography of Type AB collagen chains. Elution positions of α2(V)- and α1(V)-chains are indicated by A and B. (Redrawn from Rhodes, R. K. and Miller, E. J., *Biochemistry*, 17, 3442, 1978.)

consisting of 2.0 *M* deionized urea, 40 m*M* sodium acetate, pH 4.8, with a linear gradient of 0 to 0.2 NaCl. In this system, Type IV collagen elutes at 0.04 NaCl, while Type V as well as the interstitial collagens Types I, II, and III elute at 0.09 to 0.1 *M* NaCl.

E. Isolation of the α-Chain Components of Type V Collagen

Several chromatographic systems have been developed to separate the individual chains that make up Type V collagen. The first chromatographic system developed involved the use of CM-cellulose.[1,2]

1. CM-cellulose

Burgeson et al.[1] used a starting buffer consisting of 0.04 *M* sodium acetate, pH 4.8, with 6.0 *M* deionized urea. The column (dimensions not given) was developed with a superimposed linear gradient of 300 mℓ from 0 to 80 m*M* NaCl at 42°C (see Figure 3A).

Chung et al.[2] also used a CM-cellulose column (0.9 × 9.0 cm), but their starting buffer consisted of 0.02 *M* sodium acetate, pH 4.8, and 1.0 *M* deionized urea. Elution was accomplished with a 200-mℓ linear gradient from 0 to 0.12 *M* NaCl (see Figure 3B).

Using these methods, the αA- or α2(V)- and αB- or α1(V)-chains are incompletely separated, and rechromatography is necessary in order to obtain pure α-chains. Another potential complication that may occur in these systems is the inability to separate α1(V)- and αC- or α3(V)-chains in samples of Type V collagen which contains the so-called α3(V)-chain.[11]

2. Phosphocellulose

Rhodes and Miller[8] subsequently described a chromatographic system utilizing phosphocellulose (Whatman® P11). The starting buffer was 0.03 *M* dibasic sodium phosphate adjusted to a pH of 6.3 with phosphoric acid. The collagen sample is dissolved in starting buffer and applied to the column after washing. The column is then washed with starting buffer $OD_{230\,nm}$ returns to baseline a 400-mℓ linear gradient of 0 to 0.2 *M* NaCl is run through the column at 43°C. The separation of αA- and αB-chains apparently is more complete then that observed using CM-cellulose columns (see Figure 4).

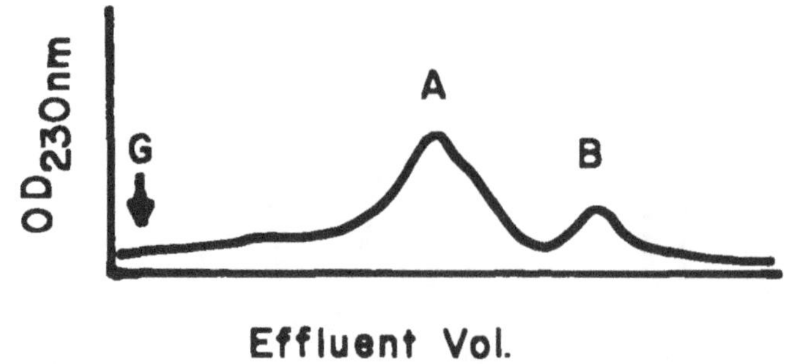

FIGURE 5. (A) Phosphocellulose chromatography of AB and C collagen chains. Elution positions of α2(V)-, α1(V)-, and α3(V)-chains are indicated by A and B. The gradient begins at the arrow labeled "G".

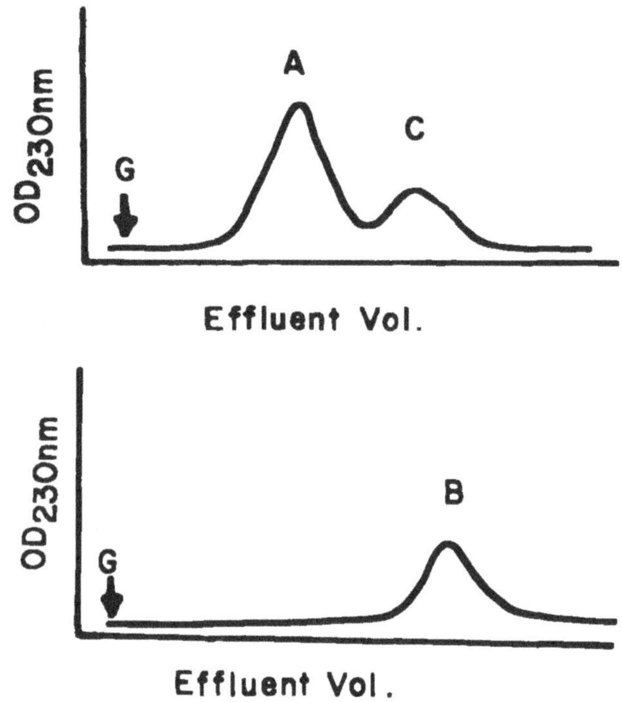

FIGURE 5B&C. CM-cellulose chromatography of αA, αB, and αC collagen chains previously fractionated by phosphocellulose chromatography. (B) αA plus αC collagen chains (fraction "A" of the phosphocellulose column); (C) αB collagen chain (fraction "B" of the phophocellular column). Elution positions of αA-, αB-, and αC-chains are indicated by A, B, and C. The gradient begins at the arrow labeled "G". (Redrawn from Sage, H. and Bornstein, P., *Biochemistry*, 18, 3815, 1979.)

3. Phosphocellulose/CM-cellulose

Sage and Bornstein[11] rechromatographed the αA- and αB-chains obtained by initial separation on phosphocellulose on CM-cellulose columns using the conditions established by Burgeson et al.[1] and showed that the αA- and αC-chains which coeluted in the phosphocellulose system could be separated. This combination of ion exchange

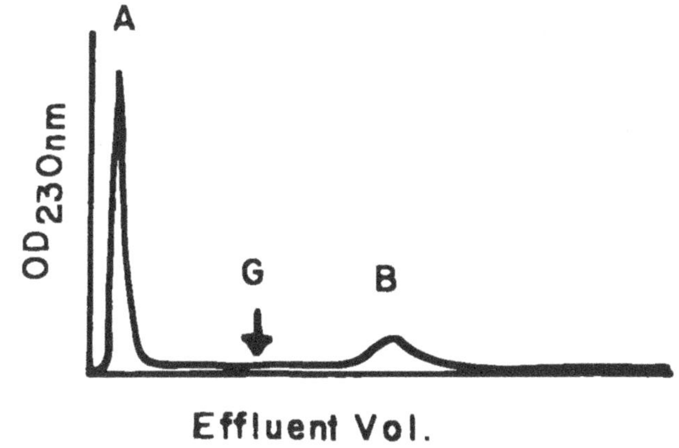

FIGURE 6. Hydroxylapatite chromatography of Type AB collagen chains. The elution positions of the α2(V)- and α1(V)-chains are indicated by A and B. The gradient begins at the arrow labeled "G". (Redrawn from Hong, B. S., Davison, P. F., and Cannon, D. J., *Biochemistry*, 10, 4278, 1979.)

columns also allows for the separation of αA-, αB-, and αC-chains (see Figure 5A,B,C).

4. Hydroxylpatite

Hong et al.[10] have established a hydroxylapatite chromatography system for separating α1(V)- and α2(V)-chains. The column (0.9 × 12.0 cm) (DNA grade BioGel® HTP, BioRad® Laboratories) was equilibrated with 0.005 *M* sodium phosphate buffer, pH 6.3, containing 4.0 *M* deionized urea and 0.1 m*M* phenylmethylsulfonyl fluoride. The sample is dissolved in the starting buffer, heated to 100°C for 2 min, then cooled to 20°C, and applied to the column. The column is washed with starting buffer. The α2(V)-chains are not bound and are eluted in the washing step. When the $OD_{230\,nm}$ returns to baseline, a 200-mℓ linear gradient from 0 to 0.5 *M* NaCl in starting buffer is run through the column. The α1(V)-chains elute at approximately 0.15 *M* NaCl. This system allows for good separation of the α1(V)- and α2(V)-chains. Since the authors did not test this separation system with Type V (AB) collagen preparations that contain α3(V)-chains (placental villi), the chromatographic behavior of the αC-chain is unknown (see Figure 6).

5. DEAE-cellulose

Bentz et al.[9] have developed a DEAE-cellulose (Whatman® DE52) chromatography system for the separation of α1(V)- and α2(V)-chains. The sample is equilibrated in 4.0 *M* deionized urea, 0.025 *M* Tris-HCl, pH 8.6, and applied to the DEAE column (2.5 × 11.0 cm) equilibrated in the same buffer at 42°C. The column is washed with starting buffer, and α2(V)-chains were found not to bind to the column and were eluted in the wash. When the $OD_{230\,nm}$ returned to baseline, a 1000-mℓ linear gradient of 0 to 0.1 *M* NaCl in starting buffer was passed through the column. The α1(V)-chains eluted at approximately 0.03 *M* NaCl. This system also allows for good separation of α2(V)- and α1(V)-chains, but preparations containing α3(V)-chains were not reported, and the chromatographic behavior of the α3(V)-chains in this system is unknown (see Figure 7).

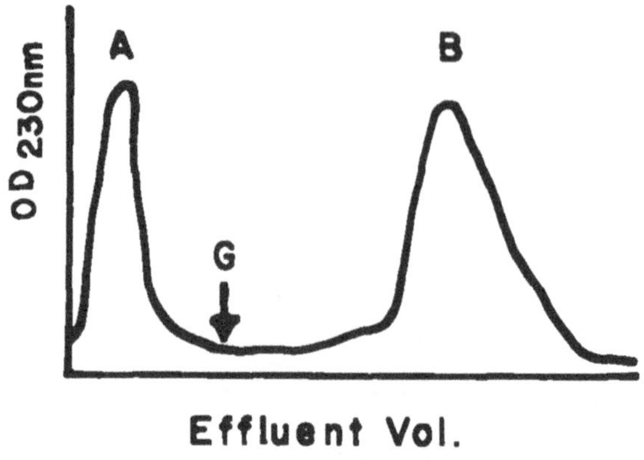

FIGURE 7. Separation of the A and B chains by chromatography on DEAE-cellulose. The start of the gradient is marked by arrow labeled "G". The elution positions of α2(V)- and α1(V)-chains are indicated by A and B. (Redrawn from Bentz, H., Bächinger, H. P., Glanville, R., and Kühn, K., *Eur. J. Biochem.*, 92, 563, 1978.)

F. Characterization of the α-Chain Components of Type V Collagen

Several groups have analyzed the A- and B-chains utilizing cyanogen bromide hydrolysis coupled with SDS polyacrylamide gel analysis, CM-cellulose column profiles, and amino acid composition the classical methods of analyzing chains.[1,2,8,9,10,11] In addition Sage and Bornstein[11] have used limited mast cell protease digestion of denatured αA-, αB-, and αC-chains and generated unique one-dimensional "maps" using SDS polyacrylamide electrophoresis. The data support the notion that the αA- and αB-chains are chemically distinct and that the αC-chain is distinct from both the αA- and αB-chains. Previous studies suggested that the αC-chain represented a form of the αA-chain which migrates slower than the majority of αA-chain material due to either variable posttranslational modifications or to incomplete proteolysis with pepsin.[8]

To further characterize the chain components of Type V collagen and to better ascertain whether the α3(V)-chain found in Type V collagen preparations from selected tissues (placental villi) is related to the α2(V)- or α1(V)-chain, we have developed a two-dimensional mapping technique based on the techniques of Elder et al. and Cleveland et al.[21,22] However, since the collagens contain only a few tyrosine residues, we elected to radiolabel the collagen samples by using the Bolton-Hunter reagent. One mg samples of Type V collagen isolated from placental villi and placental membranes were iodinated as previously described.[23] The Bolton-Hunter reagent (New England Nuclear) solubilized in benzene having a specific activity of 4000 Ci/mmol was evaporated by a gentle stream of N_2 for 30 min. One mg of collagen (previously dissolved in 1 mℓ of 0.05% acetic acid and dialyzed in 0.1 *M* calcium acetate, pH-adjusted to 8.5, with 0.2 *M* $NaBO_3$, pH 9.0) was added to the dried Bolton-Hunter reagent. The tube was capped and shaken on ice for 2 hr. The reaction was then quenched by the addition of 0.2 mℓ of 0.2 *M* glycine in 0.1 *M* calcium acetate, pH 8.5, with shaking at 4°C for 30 min. The solution was then exhaustively dialyzed against 0.05% acetic acid at 4°C and lyophilized. The radiolabeled samples were run on 5% SDS polyacrylamide slab gels.[24] The gel slice containing the α2(V)-, α1(V)-chains and (in the samples isolated from placental villi) the α3(V)-chains were cut out of the gel, dialyzed against 25% isopropyl alcohol and 10% methanol to remove SDS, and the acrylamide slices were lyophilized. These samples were then digested with chymotrypsin (code CDS, Millipore® Freehold, N.J.) or *Staphylococcus aureus* V.8 protease (Miles® Laboratories, England) in 0.05 *M* ammonium bicarbonate, pH 8.0, at a concentration of 50 μg enzyme per milliliter buffer (at 1 mℓ per gel slice). After 24 hr at 37°C, the supernatant fluids were lyophilized and analyzed on cellulose coated TLC plates (10 × 10 cm) (EM Laboratories, Elmsford, N.Y.). Samples are dissolved in 20 μℓ of buffer consisting of acetic acid to formic acid to water in a 15 to 5 to 80 ratio.

Aliquots (100 cpm/dalton) are spotted on the TLC plates and electrophoresed at 22°C at 1000 W for 1 hr. The plates are dried and then chromatographed in the second dimension in buffer consisting of butanol to pyridine to acetic acid to water in a 32.5:25:5:20 ratio plus 7% 2,5-diphenylazazole to increase sensitivity. The plates are dried and placed in X-ray cassettes with Kodak Royal X-O-Mat Film® and exposed at −70°C and finally developed in a commercial x-o-mat processor. The peptide map of the α2(V)-chain is distinct from that of α1(V)- or α3(V)-chains. The maps of α1(V)- and α3(V)-chains are also distinct. We conclude from these experiments (using the two above-mentioned enzymes and chloramine-T labeling, as well as the Bolton-Hunter labeling technique) that the α2(V)-, α1(V)-, and α3(V)-chains are chemically distinct (data not shown).[25] Amino acid compositional analyses of the α1(V)-, α2(V)-, and α3(V)-chains is also suggestive of their distinct nature (see Table 1).[11]

G. Evaluation of the α-Chain Composition of Type V Collagen

Since its discovery, there has been controversy over the molecular composition of Type V (AB) collagen. Rhodes and Miller[8] presented evidence for the existence of two separate molecular species with a chain composition $(\alpha A)_3$ or $[\alpha 2(V)]_3$ and $(\alpha B)_3$ or $[\alpha 1(V)]_3$ based on thermal denaturation studies. Bentz et al.,[9] on the other hand, concluded from their studies on DEAE column-purified Type V (AB) collagen isolated from tissue known to contain no detectable αC- or α3(V)-chain that Type V (AB) collagen has a chain composition $\alpha A(\alpha B)_2$ or $\alpha 2(V)[\alpha 1(V)]_2$. It may be that the differences in tissue source and the resulting preparations could account for the differences in the results.

We have attempted to approach this problem using a two-dimensional gel electrophoresis system developed in this laboratory.[25,26] In this system, the collagen sample is electrophoresed in the native state (first dimension), followed by electrophoresis in a reduced, denatured state (second dimension). In this system, all the collagen types migrate as native monomers and/or multiples of the native monomers in the first dimension and as reduced, denatured chain components in the second dimension. Thus, the particular chain composition of a collagen in the native state can be evaluated. Placental membrane Type V (AB) collagen (as well as Type I and III collagen) migrate as a single band in the first dimension, suggesting a single molecular species. In the second dimension, the α2(V)- and α1(V)-chains migrate in a line perpendicular to each other, indicating that they originated from a single point in the first dimension gel. If two or more molecular species are present, they would have different locations in the native gel (as observed for Type IV samples)[26] and the α-chain components would be in a nonperpendicular array in the second dimension.[25] This is the case for placental villi Type V collagen preparations. The collagen in this preparation migrates similarly to the Type V collagen isolated from placental membranes. However, in addition to this band, there is another fainter

Table 1
AMINO ACID RESIDUES/1000

Amino acid residue	α1 (V)	α2 (V)	α3 (V)
Hyl	46	23	43
Lys	15	13	15
His	6.3	9.9	14
Arg	42	52	42
3-Hyp	0.9	1.1	0.9
4-Hyp	99	107	91
Asx	50	50	42
Thr	22	26	19
Ser	23	34	34
Glx	100	88	98
Pro	125	105	99
Gly	325	325	332
Ala	41	57	49
Val	21	31	29
Ile	20	18	20
Leu	44	39	56
Cys	1.2	—	1.3
Met	6.8	10	8.1
Tyr	2.5	2.1	2.4
Phe	11	12	9.2
GlcGal Hyl	9.5	18	17
Gal Hyl	4.7	12	6.5

band which migrates slower than the Type V collagen in the first dimension. This has been putatively identified as being composed of α3(V)-chains using migration patterns in the second dimension (data not shown).[25]

H. Preparation of Biosynthetically Labeled Type V Collagen in Tissue Culture

Another approach that may allow for a better understanding of the chain composition of Type V collagen is the analysis of Type V collagen produced in a tissue culture system. To date several investigators have developed tissue culture systems in which this collagen is synthesized as a major collagenous product. In one case, aortic smooth muscle cells were cultured and found to produce Type V collagen in a 1 to 2 α-chain ratio of α2(V) and α1(V) within certain cell passage numbers.[27] We have observed that calf aortic endothelial cells produce Type V collagen with the same ratio in cell culture for several passages.[15] Recently, Kumamoto and Fessler have developed a tissue culture system using chick embryo muscle tissue for obtaining procollagen forms of Type V collagen.[29]

III. SUMMARY

The preparative procedures that our laboratory has found most useful for the isolation and purification of Type V collagen are outlined in Tables 2 and 3. We have found placental membranes (amnion and chorion) and pulmonary parenchyma to be the best tissue sources for large amounts of Type V collagen. Placental villi, while an excellent source of Type V collagen, also contains significant amounts of α3(V)-chains.

Table 2
PREPARATION OF TYPE V COLLAGEN: NEUTRAL SALT FRACTIONATION

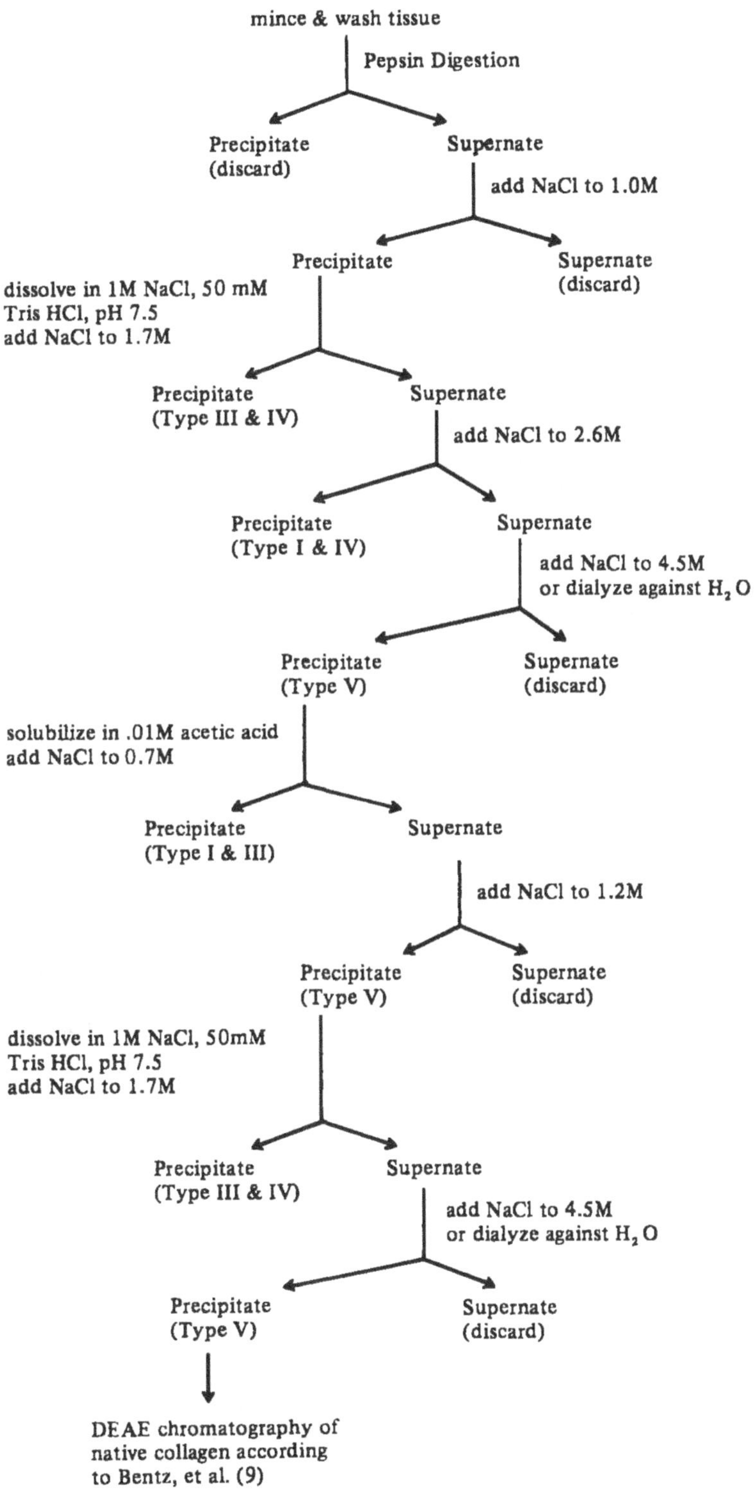

Table 3
PREPARATION OF TYPE V COLLAGEN: ACID SALT FRACTIONATION

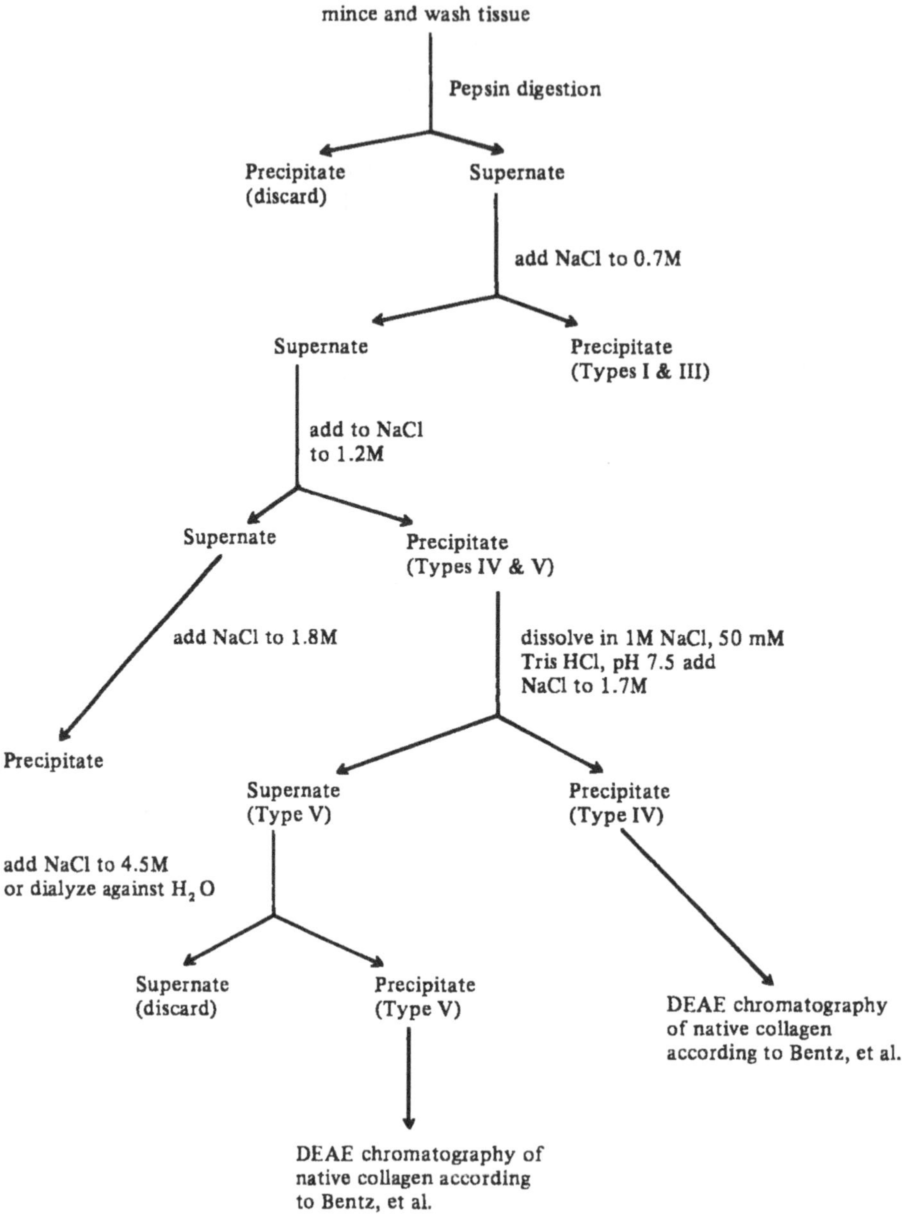

REFERENCES

1. **Burgeson, R. E., El Adli, F. A., Kaitila, I. I., and Hollister, D. W.,** Fetal membrane collagens: identification of two new collagen alpha chains, *Proc. Natl. Acad. Sci. U.S.A.*, 73, 2579, 1976.
2. **Chung, E., Rhodes, R. K., and Miller, E. J.,** Isolation of three collagenous components of probable basement membrane origin form several tissues, *Biochem. Biophys. Res. Commun.*, 71, 1167, 1976.
3. **Roll, F. J., Madri, J. A., Albert, J., and Furthmayr, H.,** Co-distribution of collagen types IV and AB_2 in basement membranes and mesangium of the kidney: an immunoferritin study of ultrathin frozen sections, *J. Cell. Biol.*, 85, 597, 1980.

4. **Stenn, K. S., Madri, J. A., and Roll, F. J.,** Migrating epidermis produces AB_2 collagen and requires continual collagen synthesis for movement, *Nature*, 277, 229, 1979.
5. **Madri, J. A. and Furthmayr, H.,** Isolation and tissue localization of type AB_2 collagen from normal lung parenchyma, *Am. J. Pathol.*, 94, 171, 1979.
6. **Madri, J. A. and Furthmayr, H.,** Collagen polymorphism in the lung: an immunochemical study of pulmonary fibrosis, *Human Pathol.*, 11, 353, 1980.
7. **Furthmayr, H., Roll, F. J., Madri, J. A., and Foellmer, H. G.,** Composition of basement membranes viewed with the electron microscope, in *New Trends in Basement Membrane Research*, Raven Press, New York, in press, 1981.
8. **Rhodes, R. K. and Miller, E. J.,** Physiochemical characterization and molecular organization of the collagen A and B chains, *Biochemistry*, 17, 3442, 1978.
9. **Bentz, H., Bächinger, H. P., Glanville, R., and Kühn, K.,** Physical evidence for the assembly of A and B chains of human placental collagen in a single triple helix, *Eur. J. Biochem.*, 92, 563, 1978.
10. **Hong, B. S., Davison, P. F., and Cannon, D. J.,** Isolation and characterization of a distinct type of collagen from bovine fetal membranes and other tissues, *Biochemistry*, 18, 4278, 1979.
11. **Sage, H. and Bornstein, P.,** Characterization of a novel collagen chain in human placenta and its relation to AB collagen, *Biochemistry*, 18, 3815, 1979.
12. **Bailey, A. J., Shellswell, G. B., and Duance, V. C.,** Identification and change of collagen types in differentiating myoblasts and developing chick muscle, *Nature*, 278, 67, 1979.
13. **Duance, V. C., Dubowitz, V., Black, C. M., and Huges, G. R. V.,** Polymyositis—an immunofluoresence study on the distribution of collagen types, *Muscle Nerve*, in press.
14. **Madri, J. A., Roll, F. J., Wyatt, R., and Furthmayr, H.,** The distribution of the connective tissue components of the murine lung: an immunoelectron microscopic study, in preparation.
15. **Madri, J. A., Dreyer, B., Pitlick, F., and Furthmayr, H.,** The collagenous components of the subendothelium: correlation of structure and function, *Lab. Invest.*, 43, 303, 1980.
16. **Jenson, J., Janeway, C., and Madri, J. A.,** Unpublished observations, 1981.
17. **Furuto, D. K. and Miller, E. J.,** Isolation of a unique collagenous fraction from limited pepsin digests of human placental tissue, *J. Biol. Chem.*, 255, 290, 1980.
18. **Ayad, S., Abedin, M. Z., Shuttleworth, C. A., and Weiss, J. B.,** Bovine placental collagens. I. Origin of types IV and V (AB) collagens and the presence of an unusual disulphide bond-rich collagen, submitted.
19. **Welsh, C., Gay, S., Rhodes, R. K., Pfister, R., and Miller, E. J.,** Collagen heterogeneity in normal rabbit cornea. I. Isolation and biochemical characterization of the genetically distinct collagens, *Biochem. Biophys. Acta*, in press.
20. **Glanville, R. W. and Kühn, K.,** Preparation of two basement membrane collagens from human placenta, *Front. Matrix Biol.*, 7, 19, 1979.
21. **Elder, J. H., Pickett, R. A., Hampton, J., and Lerner, R. A.,** Radioiodination of proteins in single polyacrylamide gel slices, *J. Biol. Chem.*, 252, 6510, 1977.
22. **Cleveland, D. W., Fischer, S. G., Kirschner, M. W., and Laemmli, U. K.,** Peptide mapping by limited proteolysis in sodium dodecylsulfate and analysis by gel electrophoresis, *J. Biol. Chem.*, 252, 1102, 1977.
23. **Roll, F. J., Madri, J. A., and Furthmayr, H.,** A new method of Iodinating collagens for use in radioimmunoassay, *Anal. Biochem.*, 96, 489, 1979.
24. **Furthmayr, H. and Timpl, R.,** Characterization of collagen peptides by sodium dodecylsulfate-polyacrylamide electrophoresis, *Anal. Biochem.*, 41, 510, 1971.
25. **Madri, J. A., Foellmer, H. G., and Furthmayr, H.,** Type V collagens of the human placenta: trimerαchain composition, untrastructural morphology and peptide analysis, *Collagen Rel. Res.*, 2, 1, in press, 1981.
26. **Foellmer, H. G., Madri, J. A., Lwebuga-Mukasa, J., and Furthmayr, H.,** Type IV collagen of the human placenta: native trimerαchain composition, molecular organization, and domain structure, in preparation.
27. **Mayne, R., Vail, M. S., and Miller, E. J.,** Characterization of the collagen chains synthesized by cultured smooth muscle cells derived from Rhesus monkey thoracic aorta, *Biochemistry*, 17, 446, 1978.
28. **Bornstein, P., Sage, H.,** Structurally distinct collagen types, *Annu. Rev. Biochem.*, 49, 957, 1980.
29. **Kumamoto, C. A., and Fessler, J. H.,** Biosynthesis of A, B procollagen, *Proc. Natl. Acad. Sci. U.S.A.*, 77, 6434, 1980.

Chapter 7

PREPARATION OF ANTISERA SPECIFIC FOR THE AMINO AND CARBOXYL PROPEPTIDES OF TYPE I AND TYPE II PROCOLLAGENS

D. M. Pesciotta, S. Curran, and B. R. Olsen

TABLE OF CONTENTS

I. INTRODUCTION

During the past decade, several laboratories have employed antisera against different types of collagens as tools for histological, chemical, cell biological, and pathological studies on connective tissues.

The preparation, characterization, and application of such antisera have been described extensively in several reviews.[1-3] As discussed in detail in Chapter 2, collagens are relatively weak immunogens, and large amounts of antigens are needed to produce antisera of high titers against collagen antigenic determinants. In contrast, when the precursor forms of collagens, procollagens, are used as antigens, antisera of high titers containing antibodies against the amino (N) and carboxyl (C) propeptides of procollagen are easily prepared with relatively small amounts of antigen. In fact, the strong immunogenicity of procollagen propeptides was noted in the early studies on procollagen.[4-6] For this reason, antisera specific for procollagen propeptides were useful in initial studies elucidating the synthesis and structure of procollagen.

Procollagens differ from collagens in that they have propeptide extensions at the amino and the carboxyl termini of the collagen α-chains[7-9] (see Figure 1). These propeptides are proteolytically removed from procollagen molecules by procollagen N- and C-proteases during the physiological conversion of procollagen to collagen in tissues.[10-12] Several types of data suggest that the conversion of procollagen to collagen occurs by endoproteolytic scission of the propeptides leaving the N- and C-propeptides as intact molecules in the extracellular connective tissue matrix from which they can be recovered relatively easily. Procollagen precursors of most types of collagens have been isolated from several systems. However, only the Type I and Type II procollagens and their propeptides have been isolated in quantities to permit a detailed chemical characterization of both the N- and C-propeptides.[13-20] In addition, a truncated form of Type III procollagen, pN (III)-collagen and its N-propeptide have been isolated from extracts of fetal bovine skin.[21,22] The preparation of pN (III)-collagen will be discussed in Chapter 3. Here we will describe the isolation of Type I and Type II procollagens and their propeptides. In addition, we will discuss the development and use of specific antisera against these propeptides.

II. ISOLATION AND PURIFICATION OF CHICK TYPE I AND TYPE II PROCOLLAGENS

A. Type I Procollagen

Chick Type I procollagen can be conveniently isolated from the media of chick tendon fibroblasts in suspension culture. The cells are isolated from embryonic chick leg tendons by digestion with collagenase-trypsin according to the method of Dehm and Prockop.[23] In a typical experiment, leg tendons are dissected from 12 dozen 17-day-old chick embryos. The eggs are opened, and the embryos are separated from the yolk sac. The chick embryos are held upside down, and the legs are cut off with a pair of scissors. The legs are temporarily stored in a modified Krebs medium II containing 111.2 m*M* (0.9%) NaCl, 15.7 m*M* Na phosphate, pH 7.4, 5.4 m*M* KCl, 4.0 m*M* $NaHCO_3$, 1.3 m*M* $MgCl_2$, and 1.6 m*M* KH_2PO_4. The medium is supplemented with 13 m*M* glucose immediately prior to use. Toes with attached tendons are dissected by holding the thighs and gently pulling the toes with forceps. The toes with tendons attached are stored in the modified Krebs medium. Tendons are removed by cutting at the junction between the toe and the tendon with a scalpel blade. During the dissection period the tendons are stored in the Krebs medium.

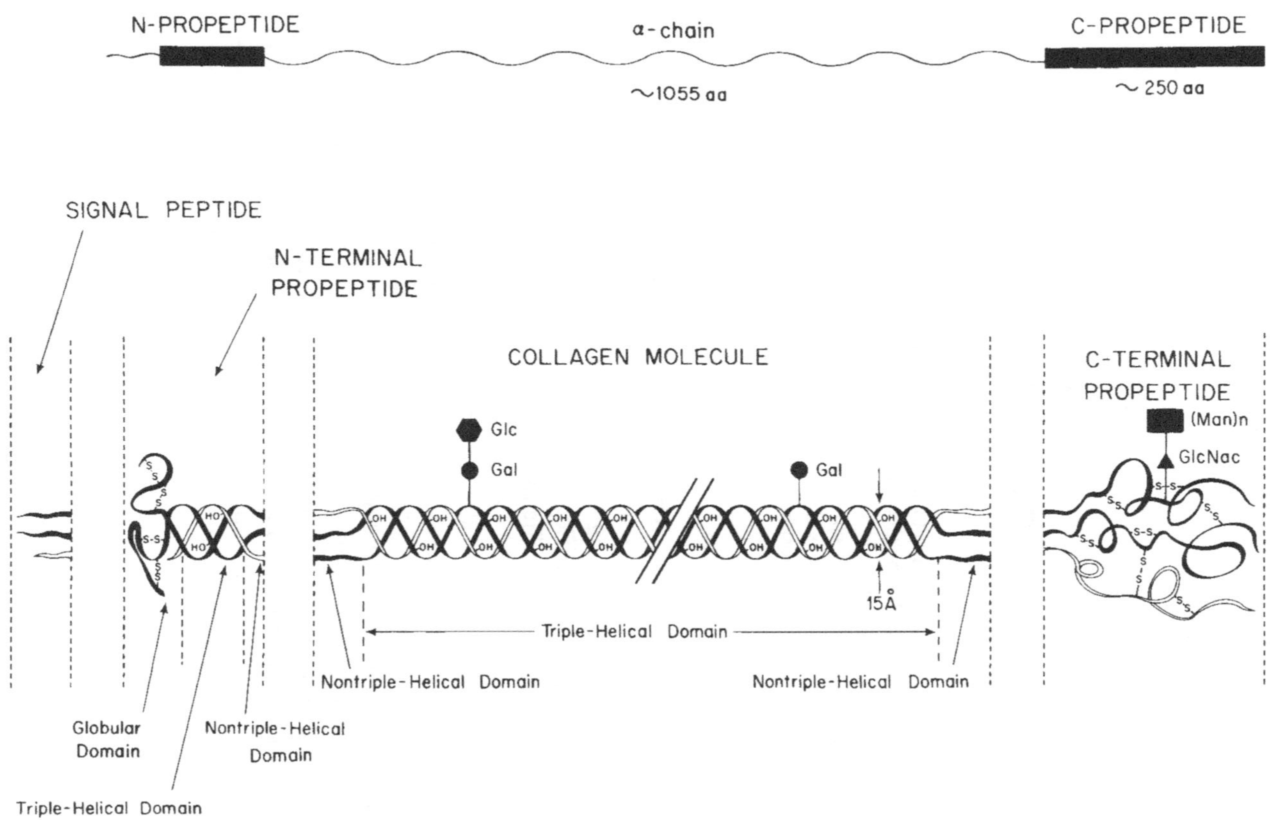

FIGURE 1. Diagram showing the structure of preproα-chains and pro-collagen Type I. Symbols: Glc (glucose); Gal (galactose); Man (mannose), and GlcNac (N-acetylglucosamine). During synthesis of procollagen polypeptide chains on membrane-bound ribosomes, the chains are synthesized as preproα-chains (top part of figure) with hydrophobic signal peptides at their amino termini (see Vol. 2, Chapter 1). The signal peptides are rapidly cleaved off the preproα-chains in the rough endoplasmic reticulum. (From Olsen, B. R., *Cell Biology of the Extracellular Matrix*, Hay, E. D., Ed., Plenum Press, 1981, 139. With permission.)

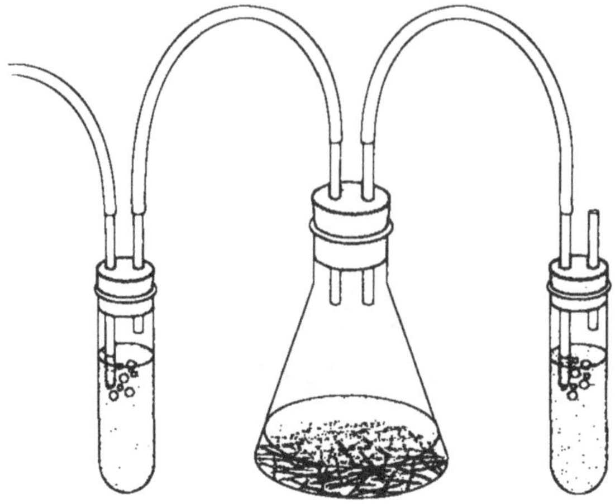

FIGURE 2. Arrangement used during tendon incubation. Tube A contains Krebs/glucose medium. Flask B contains Dulbecco modified Eagle medium containing trypsin, bacterial collagenase, and chick embryo tendons, and Tube C is a water trap. Gas is slowly bubbled through Tube A so as to create an atmosphere of 5% CO_2, 95% air within Flask B.

The tendons are drained of excess Krebs medium, and the wet weight is determined. Twelve dozen chick embryos usually yield about 10 g of tendons. For every gram of tendon, 3 mℓ of modified Eagles medium (Gibco®) is placed in a 500-mℓ Erlenmeyer flask, and 3.3 mg of purified bacterial collagenase (Worthington®) per gram of tendon is added to the flask. The medium is equilibrated at 37°C in a shaking water bath under an atmosphere of 5% CO_2, 95% air. After equilibration for 10 min, the tendons are added to the flask, followed immediately by 0.33 mℓ/g tendon of trypsin (Gibco®, 2.5% trypsin in physiological saline). The flask is gently shaken by gyratory motion for 45 to 60 min under 5% CO_2, 95% air at 37°C (see Figure 2). The digestion is considered complete when no large pieces of tendon are visible.

The digest is filtered into two 50-mℓ plastic conical tubes through a Millipore® (Swinnex-25 mm) filter holder fitted with two sheets of lens paper (see Figure 3). The tubes are centrifuged at room temperature in a GLC-Sorvall centrifuge for 6 min at 1800 r/min (see Figure 4). The supernatant is removed by aspiration, and each cell pellet is resuspended in 10 mℓ Krebs medium containing 10% fetal calf serum. The resuspension is accomplished by gently vortexing the tubes. The cells are sedimented by centrifugation at room temperature at 1800 r/min for 3 min. The washing procedure is repeated three times. After the final wash, the cell pellets are combined and resuspended into approximately 20 mℓ of the Krebs solution. The cell suspension is allowed to settle by gravity for 3 min to remove large aggregates of cells, and the suspension is decanted into a plastic tube. The cells are counted in a hemocytometer by diluting 0.1 mℓ of the cell suspension into 10 mℓ of Krebs solution. This dilution gives a convenient cell density for counting. In a typical experiment, 12 dozen embryos yield about 2×10^9 cells. The cells are suspended in the Krebs-glucose medium to give a final concentration of 7.5×10^6 cells per milliliter of medium. The appropriate isotopes are added to the suspension. For most labeling experiments,

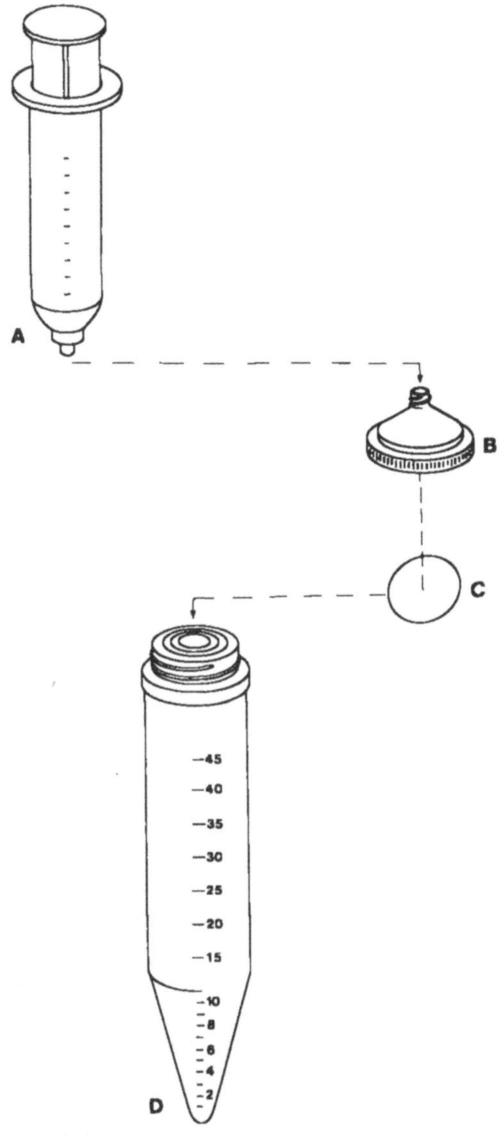

FIGURE 3. Assembly of Swinnex filter holder. A 10-mℓ plastic syringe (A) is inserted into the top of a Swinnex® filter holder (B). Two sheets of lens paper (C) are fitted into the filter holder, and the holder is attached to a 50-mℓ plastic conical tube. The digested tissue is poured into the syringe and filtered through the lens paper.

[^{14}C]-proline is used. Usually 50 μCi of [^{14}C]-proline per 10^9 cells yields a good labeling of the procollagen. The cell suspension is incubated in a 1-ℓ Erlenmeyer flask for 4 to 6 hr at 37°C in a shaking water bath (see Figure 4). The top of the flask is covered with parafilm which is punctured several times with a fine needle to allow equilibration with air inside the flask during the incubation.

After incubation, the cells are separated from the medium by centrifugation at room temperature for 10 min at 3000 r/min. The supernatant is placed on ice, and

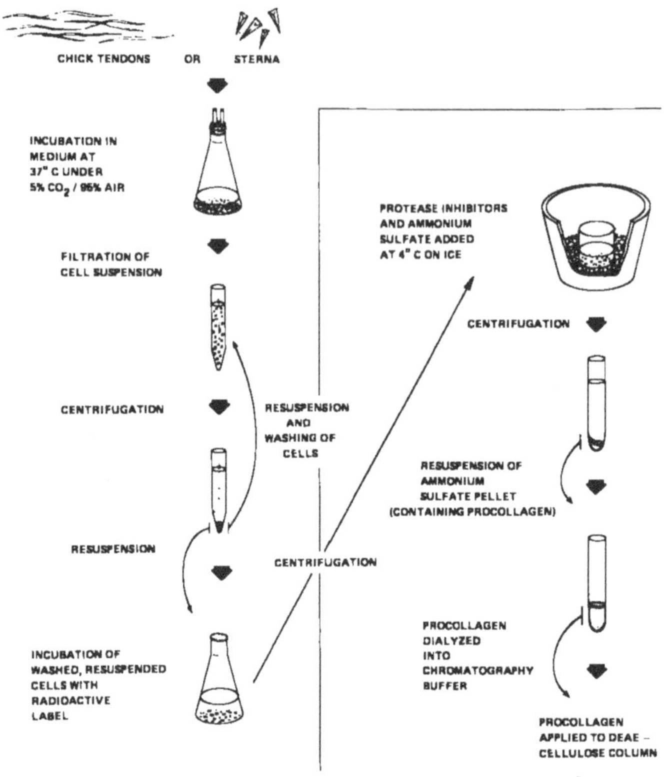

FIGURE 4. Flow chart for the preparation of procollagen from chick tendons or chick sterna.

protease inhibitors are added to give a final concentration of 25 m*M* Na_4-EDTA, 10 m*M* *N*-ethylmaleimide, 1 m*M* *p*-aminobenzamidine, and 1 m*M* phenylmethylsulfonyl fluoride.[24] Then 176 mg of $(NH_4)_2SO_4$ per milliliter of medium is slowly added while stirring. Precipitation with $(NH_4)_2SO_4$ is allowed to proceed at 4°C overnight.

The solution is centrifuged at 12,500 r/min for 30 min at 4°C in a Sorvall RC-2B centrifuge. Plastic tubes or siliconized glass tubes should be used to avoid any loss of the procollagen onto the glass surfaces. The procollagen-containing pellets are resuspended in 0.1 *M* Tris-HCl, pH 7.4 (4°C), and 0.4 *M* NaCl. The total volume should not exceed 15 mℓ. The procollagen is extracted from the pelleted material by stirring at 4°C for 4 to 5 hr. The suspension is then centrifuged at 4°C for 30 min at 12,500 r/min. The supernatant is dialyzed overnight against 0.1 *M* Tris-HCl, pH 8.6 (at 4°C), containing 2 *M* deionized (or ultrapure) urea (see Figure 4).

The procollagen is purified on a column of DEAE-cellulose (see Figure 5). Usually, a 1.5 × 10 cm column is used.[25] The column is equilibrated with the 0.1 *M* Tris-HCl, pH 8.6 (at 4°C), 2 *M* urea buffer. The sample is pumped onto the column at 4°C with a peristaltic pump using a flow rate of 60 to 80 mℓ/hr. After washing with an additional 25 mℓ of 0.1 *M* Tris-HCl, pH 8.6 (at 4°C), 2 *M* urea, the procollagen is eluted with a 400-mℓ linear gradient of 0 to 0.12 *M* NaCl in the 0.1 *M* Tris-HCl, 2 *M* urea buffer, and 5 mℓ fractions are collected. The column fractions are assayed by taking 50 μℓ aliquots for scintillation counting. The fractions containing procollagen are pooled. The procollagen is dialyzed against 0.1 *M* Tris-HCl, pH 7.4 (at 4°C), 0.4 *M* NaCl and stored frozen at −20°C in aliquots.

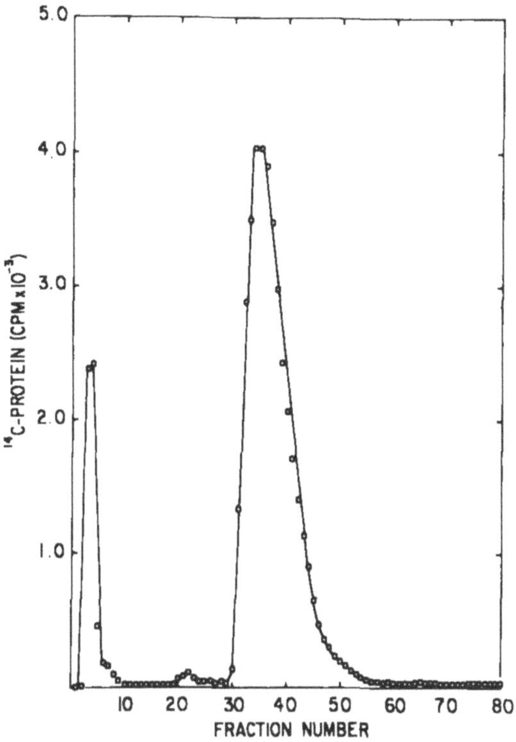

FIGURE 5. Isolation of chick Type I procollagen by DEAE-cellulose chromatography. The medium of tendon fibroblasts in suspension culture was prepared as described in the text and chromatographed on a DEAE-cellulose column at 4°C. Procollagen was eluted from the column using a linear salt gradient.

B. Type II Procollagen

Chick Type II procollagen is isolated from chondroblasts in suspension culture. The chondroblasts are isolated from sterna by a modification of the method outlined above for the isolation of fibroblasts from chick leg tendons.[26–28]

Sterna are dissected from 17-day-old chick embryos by cutting through the skin along the anterior median line with the sharp tip of a small forceps. The skin is folded back, and the sternal cartilage is exposed. The sternoclavicular junction is severed with the forceps, and the sternum is pulled out. During the dissection, the sterna are stored in modified Krebs II medium. In a typical experiment, sterna from 12 dozen embryos are used.

The sterna are incubated in 15-mℓ modified Eagles medium (Gibco®) in a 500-mℓ Erlenmeyer flask. To the medium, 4.0 mg of collagenase per 30 sterna and 0.4 mℓ 2.5% trypsin per 30 sterna are added. Usually the collagenase and trypsin are added to the medium and the medium is equilibrated at 37°C under 10% CO_2, 90% air before the sterna are added. The flask is gently shaken by gyratory motion for 2 hr at 37°C under 10% CO_2, 90% air. After the incubation, the cell suspension is filtered through lens paper, and the cells are washed with 10% fetal calf serum in Krebs medium as described for the tendon fibroblasts. It has been observed that the chondroblasts are more difficult to resuspend than the fibroblasts. Therefore the sternal cells are pelleted by centrifugation at 1400 r/min, as compared with 1800 r/min for

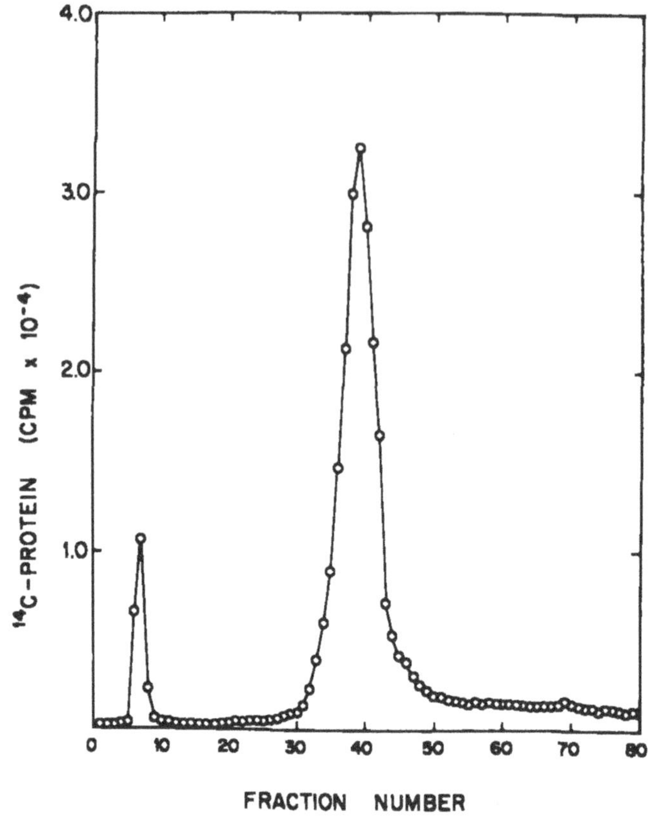

FIGURE 6. Isolation of chick Type II procollagen by DEAE-cellulose chromatography. The medium of sternal cells in suspension culture was prepared as described in the text and chromatographed on a DEAE-cellulose column at 4°C. Procollagen was eluted from the column using a linear salt gradient of 0 to 0.12 *M* NaCl in a 0.1 *M* Tris-HCL, pH 8.0, buffer containing 2 *M* urea and 5 m*M* Na_4-EDTA.

the tendon fibroblasts. For incubation of the chondroblasts, the cells are incubated at a concentration of 7.5×10^6 cells per milliliter in a Krebs-glucose medium containing 50 μg/mℓ of β-aminopropionitrile fumarate and 50 μg/mℓ of sodium ascorbate. If Type II procollagen is to be used as a radioactive marker, 20% fetal calf serum is usually added to the Krebs medium during the incubation to minimize proteolytic degradation of the procollagen during its isolation. However, if procollagen is prepared for antibody production, it is recommended that the fetal calf serum be ommitted, even if this may result in some degradation of the procollagen.

After incubation for 2 hr at 37°C, the chondroblasts are separated from the medium, and protease inhibitors are added to the medium as described for Type I procollagen. The procollagen is precipitated from the medium by adding 176 mg/mℓ of $(NH_4)_2SO_4$, and the precipitate is collected by centrifugation as described above. The pelleted Type II procollagen is resuspended by stirring in 15 mℓ of 0.1 *M* Tris-HCl, pH 7.4 (4°C), 0.4 *M* NaCl, 25 m*M* EDTA, 10 m*M* *N*-ethyl maleimide, 1 m*M* *p*-aminobenzamidine, and 1 m*M* phenylmethylsulfonyl fluoride. After centrifugation at 12,500 r/min for 30 min, the supernatant is dialyzed at 4°C against 0.1 *M* Tris-HCl, pH 8.6, containing 2 *M* deionized (or ultrapure) urea and 5 m*M* EDTA. The procollagen is purified on a DEAE-cellulose column as described for Type I procollagen except that 5 m*M* EDTA is added to the elution buffer to minimize proteolytic degradation of the procollagen during the isolation procedure (see Figure 6). Column fractions are assayed by scintillation counting, and fractions containing procollagen are pooled. It has been noted that dialysis of the pooled fractions against Tris-NaCl prior to storage leads to degradation and loss of the Type II procollagen. Therefore, to minimize losses, the procollagen in the pooled fractions is precipitated by adding 176 mg $(NH_4)_2SO_4$ per milligram of solution. After stirring at 4°C overnight, the procollagen is precipitated by centrifugation and the pellet is stored frozen at −20°C.

III. ISOLATION AND PURIFICATION OF CHICK TYPE I AND TYPE II PROPEPTIDES

A. Type I Propeptides

Chick Type I propeptides can be isolated in relatively large amounts from leg tendons in tissue culture.[19] During incubation of the tendons, the procollagen that is synthesized and secreted by the tendon fibroblasts is converted to collagen by the procollagen N- and C-proteases in the tissue. The collagen is mostly incorporated into fibrils in the tendons, while the propeptides diffuse into the medium. Also, small amounts of intact procollagen and *p*-collagen intermediates are found in the medium. The system described below allows the isolation of the N terminal propeptide of proα1(I)-chains and the disulfide-linked Type I C-propeptide by two column chromatographic steps. Attempts to isolate the N-propeptide of proα2-chains in this system have been unsuccessful.

Leg tendons from 17-day-old chick embryos are dissected as described above. During the dissection, the tendons are kept in modified Krebs medium II supplemented with glucose. In a typical experiment, tendons from 12 dozen embryos are used. If necessary, the dissection can be scaled up, and 36 dozen embryos can be used for one experiment. The tendons are placed in 100 mℓ of Dulbecco modified Eagle medium per 12 dozen embryos in a 1-ℓ Erlenmeyer flask (sterilized). Sodium ascorbate is added to a final concentration of 40 μg/mℓ and β-aminopropionitrile furmarate is added to 64 μg/mℓ. To label the propeptides, 100 μCi of a mixture of [^{14}C]-labeled amino acids (New England Nuclear®, 15 different amino acids) is added. To minimize bacterial contamination, penicillin and streptomycin are added to final concentrations of 100 units per milliliter and 100 μg/mℓ, respectively. The tendons are incubated at 37° under an atmosphere of 10% CO_2, 90% air while shaking. After 24 hr of incubation, the medium is separated from the tendons by centrifugation at 30,000 × g at 4°C for 30 min. To the supernatant are added protease inhibitors (25 m*M* Na_4-EDTA, 10 m*M* *N*-ethylmaleimide, 1 m*M* *p*-aminobenzamidine, and 1 m*M* phenylmethylsulfonyl fluoride), and the medium is dialyzed overnight at 4°C against 0.05 *M* Tris-HCl, pH 8.6 (23°C), 2 *M* deionized (or ultrapure) urea. For every 100 mℓ of medium, three changes of 1 ℓ of buffered urea should be used for dialysis. It is not necessary to remove completely the phenol red indicator from the medium by dialysis prior to the last column chromatography step. The dialyzed medium is fractionated on a 2.5 × 15 cm DEAE-cellulose column equilibrated with 0.05 *M* Tris-HCl, pH 8.6 (23°C), and 2 *M* urea at 4°C (see Figure 8). (Note: It is important to use a column of this size because the phenol red present in the medium binds tightly to the column, thus reducing its capacity for proteins and peptides.) The column is eluted with a 1000-mℓ linear gradient of 0 to 0.3 *M* NaCl in the 0.05 *M* Tris-HCl, pH 8.6, 2 *M* urea buffer. The flow rate is 100 to 120 mℓ/hr, and 10-mℓ

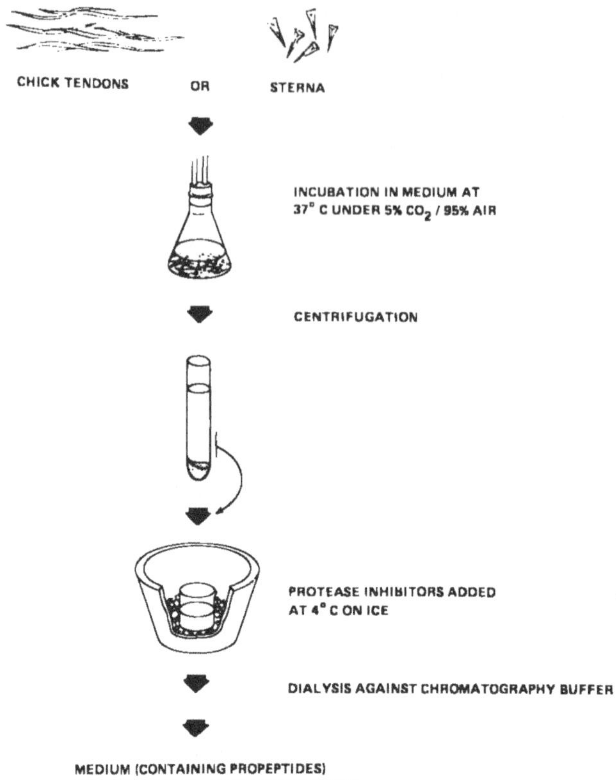

FIGURE 7. Diagram showing the preparation of Type I and Type II procollagen propeptides.

fractions are collected. The fractions are assayed by taking 100-μℓ aliquots for scintillation counting.

Using these column conditions, five distinct peaks of radioactive material are eluted with the gradient. The C-propeptide is contained within the third peak, and the N-propeptide is eluted together with the phenol red indicator dye within the fourth and fifth peaks. The separately pooled peaks are desalted on a 4.2 × 45 cm polyacrylamide gel filtration column (BioGel® P-2, 200 to 400 mesh, BioRad® Corp.) equilibrated with 0.2 *M* ammonium bicarbonate. This desalting column separates the N-propeptide from the phenol red indicator dye. The recovery of radioactive peptides is essentially quantitative. The fractions from the P-2 column are assayed by liquid scintillation counting; 50 μℓ aliquots are counted. The peak fractions are frozen and lyophilized.

Further purification of the propeptides is achieved by ion-exchange chromatography on carboxymethyl cellulose (Whatman® CM52) in the presence of 6 *M* urea.[19] The lyophilized C-propeptide is dissolved in a 0.05 *M* sodium acetate buffer containing 6 *M* deionized (or ultrapure) urea. This buffer is prepared by mixing appropriate aliquots of 1 *M* sodium acetate buffer, pH 3.6 (55.6 mℓ glacial acetic acid and 10.07 g sodium acetate trihydrate in 1 ℓ), and 8 *M* urea. The final unadjusted pH is 4.2. The C-propeptide sample is dialyzed against this buffer overnight at 4°C to remove excess salt. The sample is pumped onto a 1.5 × 10 cm CM-cellulose column equilibrated with the 6 *M* urea, 0.05 *M* sodium acetate buffer at a flow rate of 80 mℓ/hr at room temperature. The column is eluted with a 600-mℓ linear gradient of

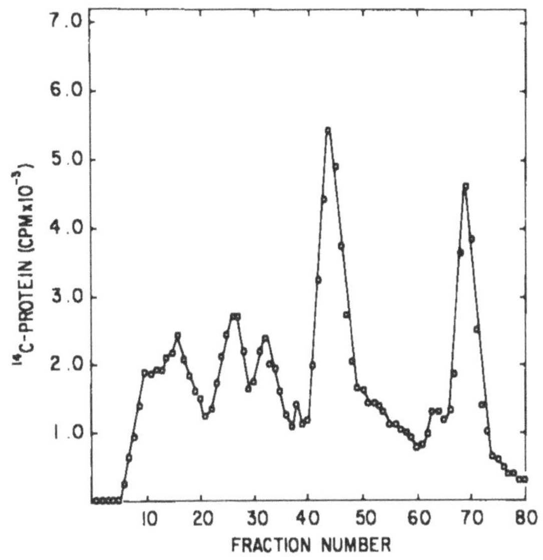

FIGURE 8. DEAE-cellulose chromatography of the tendon medium propeptides. Tendons were incubated as described in the text, and the medium was chromatographed on a 2.5 × 15 cm DEAE-cellulose column at 4°C. The C-propeptide is contained within the peak eluting between Fractions 40 and 50. The proα1(I) N-propeptide elutes between Fractions 65 and 75.

0 to 0.2 *M* NaCl and 5.0 mℓ fractions are collected and assayed for radioactivity. Optical density at 230 nm is monitored continuously throughout the column elution to follow the removal of contaminating nonradioactive proteins. The C-propeptide peak is desalted on P-2 and lyophilized as described above.[19]

The N-propeptide is dissolved in a 0.01 *M* sodium acetate buffer, pH 3.6, containing 6 *M* urea (prepared as above), dialyzed, and applied to a 1.5 × 10 cm CM-cellulose column[20] (see Figure 9). The N-propeptide is eluted at room temperature with a 400-mℓ linear salt gradient of 0 to 0.3 *M* NaCl, and 5-mℓ fractions are collected. The elution profile shows four distinct peaks of radioactive material. The N-propeptide is contained within the second peak. The N-propeptide fraction is desalted on P-2 and lyophilized as described above. The lyophilized propeptides are stored at −20°C.

B. Type II Propeptides

Type II propeptides are isolated from the media of chick sterna in tissue culture by a modification of the organ culture system described above for tendon. Sterna are dissected as described above and kept in modified Krebs medium II supplemented with glucose.

For the Type II C-propeptide, the isolation procedure is similar to that used for the Type I C-propeptide.[29] After the 24-hr incubation, the medium is separated from the sterna by centrifugation for 10 min at 2,500 r/min in a Sorvall GLC centrifuge at room temperature.

Protease inhibitors are added to the medium which is then dialyzed overnight at 4°C against 0.01 *M* Tris-HCl buffer, pH 8.6 (23°C), containing 2 *M* urea. The dialyzed medium is fractionated on a 2.5 × 15 cm DEAE-cellulose column equilibrated with

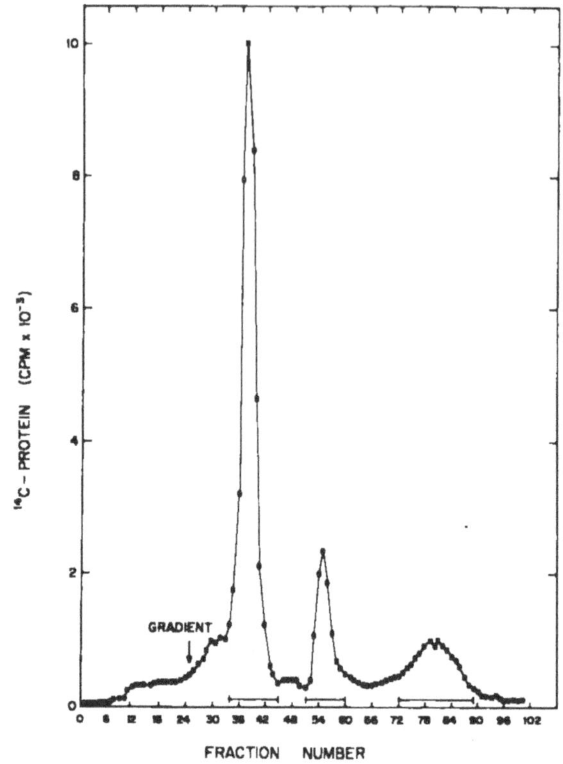

FIGURE 9. Purification of the proα1(I) N-propeptide by chromatography on CM-cellulose. Before the sample was applied to the column, particulate material was removed by centrifugation. The chromatographic conditions were as described in the text. (From Pesciotta, D. M., Silkowitz, M. H., Fietzek, P. P., Graves, P. N., Berg, R. A., and Olsen, B. R., *Biochemistry,* 19, 2447, 1980. With permission.)

0.01 *M* Tris-HCl buffer, pH 8.6 (23°C), containing 2 *M* urea (see Figure 10). The column is eluted with an 800-mℓ linear gradient of 0 to 0.35 *M* NaCl at a flow rate of 120 mℓ/hr, 8-mℓ fractions are collected. The fractions are assayed for radioactivity by taking 100-μℓ aliquots for scintillation counting. The C-propeptide is contained within the first major peak. It is desalted on P-2 and lyophilized as described above.

For purification on CM-cellulose (see Figure 11), the lyophilized C-propeptide is dissolved in 0.05 *M* sodium acetate, pH 3.6, containing 8 *M* urea. To ensure good recovery of the C-propeptide from the CM-cellulose column, 1 mg of bovine serum albumin is added as a carrier to the propeptide before dialysis against the sodium acetate buffer. The dialysis is carried out overnight at 4°C. The subsequent column chromatographic step separates the bovine serum albumin and the C-propeptide.

The C-propeptide sample is applied to a 1.5 × 5 cm CM-cellulose column at a flow rate of 120 mℓ/hr at room temperature. The column is eluted with a 1000-mℓ linear salt gradient of 0 to 0.2 *M* NaCl, 8-mℓ fractions are collected and assayed for radioactivity. Five distinct peaks of radioactivity are eluted. The fifth peak contains the C-propeptide. To minimize losses, the Type II C-propeptide is usually desalted by dialysis against 0.2 *M* ammonium bicarbonate instead of by gel filtration on P-2.

The Type II N-propeptide is isolated by a modification of the procedure described

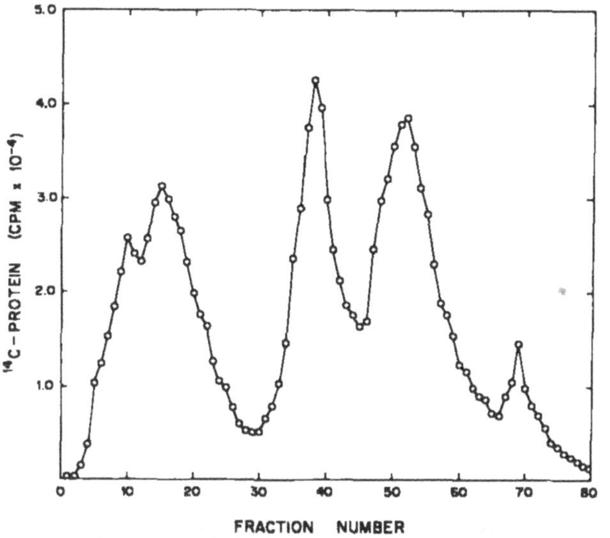

FIGURE 10. DEAE-cellulose chromatography of the sternal medium propeptides. Sterna were incubated as described in the text. The sternal medium was chromatographed on a 2.5 × 15 cm DEAE-cellulose column at 4°C. After application of the sample, the propeptides were eluted with a linear salt gradient of 0 to 0.35 *M* NaCl as described in the text. The gradient elution started at Fraction 25.

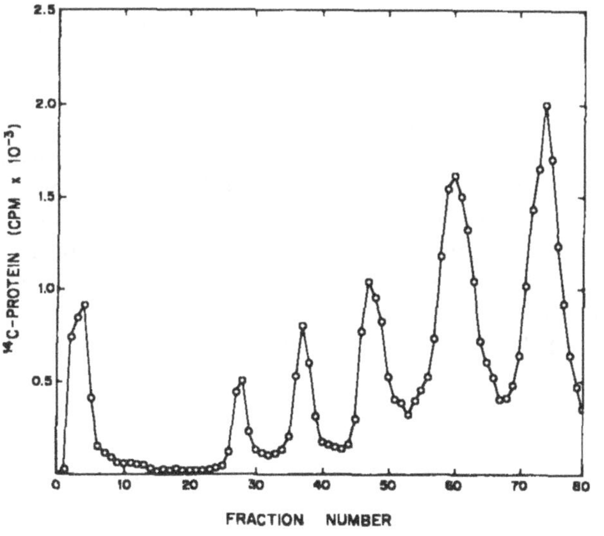

FIGURE 11. Further purification of the sternal propeptides by chromatography on a CM-cellulose column at 23°C. Propeptide fractions from the DEAE-cellulose column were chromatographed as described in the text. One mg of bovine serum albumin was added as a carrier for the propeptides before application to the CM-cellulose column. The gradient elution started at Fraction 10.

above for the Type II C-propeptide. The sterna are incubated for 4 to 5 hr in Krebs medium II supplemented with 13 m*M* glucose and 5 m*M* $CaCl_2$. The incubation time is relatively short, since it has been observed that the Type II N-propeptide is particularly susceptible to proteolytic degradation during longer incubation times. The sterna are separated from the medium after the incubation as described above for the Type II C-propeptide. Protease inhibitors are added, and the medium is dialyzed against 0.01 *M* Tris-HCl buffer, pH 8.6 (23°C), containing 2 *M* urea. The dialyzed medium is fractionated using the DEAE-cellulose column and the conditions described above for the Type II C-propeptide. The N-propeptide is eluted in the breakthrough volume. It is desalted, lyophilized, and further purified using the CM-cellulose column described for the Type II C-propeptide. The second peak of radioactivity eluted by the gradient contains the N-propeptide (see Figure 11).

IV. PREPARATION AND CHARACTERIZATION OF ANTISERA AGAINST PROCOLLAGEN PROPEPTIDES

Antisera against the C-propeptides of Type I and Type II procollagens have been prepared in rabbits by injecting a mixture of Freund's adjuvant and the purified propeptide. The initial injection usually contains 0.5 to 1.0 mg of propeptide suspended in 0.15 *M* NaCl mixed with an equal volume of Freunds complete adjuvant. For booster injections (about 2 weeks apart), 0.25 to 0.5 mg of peptide is used in a mixture with Freund's incomplete adjuvant. The first injection is intradermal; the booster injections are intramuscular. The rabbits are bled from an ear vein before the injections and about 10 days after each booster.

For testing the antisera, we have used direct binding radioimmune assays with sheep antirabbit IgG antiserum in the second precipitation step.[20,30] The assays are performed by adding various amounts of antiserum to 2000 to 5000 cpm of Type I or Type II ^{14}C-procollagen, prepared as described above, in 0.1 mℓ 0.1 *M* Tris-HCl, pH 7.5 (23°C), 0.15 *M* NaCl, 0.1% ovalbumin, and 0.01% sodium azide. The total amount of rabbit IgG in each tube is kept constant by adding appropriate amounts of nonimmune rabbit serum. The antisera are usually diluted 1 to 10 or 1 to 100 with the Tris, NaCl, ovalbumin-solution described above. The exact dilution depends, of course, on the titer of the antiserum. After incubation of the antigen with antiserum at room temperature for 1 hr with constant shaking, sheep antirabbit IgG serum is added, and the tubes are incubated at 4°C for 24 hr while shaking. The amount of sheep antiserum is predetermined to ensure complete precipitation of the rabbit IgG. After incubation, the tubes are centrifuged for 10 min at 2000 r/min in a PR-6 International centrifuge. The pellets are washed twice with 0.15 *M* NaCl and then dissolved in 1 mℓ of 0.2 *N* HCl for counting in 15 mℓ Aquasol II (New England Nuclear®).

In some experiments we have attempted to replace the sheep antirabbit IgG antiserum with Protein A-Sepharose® to precipitate the immuncomplexes. However, in our hands, the use of sheep antirabbit IgG antiserum has resulted in more complete precipitations of ^{14}C-procollagen than the use of Protein A-Sepharose®.

The reactivity of different antisera with different propeptides have been compared using inhibition assays.[20,31] For the inhibition assays, the inhibitors are incubated with the antiserum for 24 hr at 4°C before the radioactively labeled antigen is added. Various amounts of inhibitor are added to a constant amount of antiserum in 0.1 mℓ 0.1 *M* Tris-HCl, pH 7.5 (23°C), 0.15 *M* NaCl, 0.1% ovalbumin, and 0.01% sodium azide. The amount of antiserum is predetermined as the amount of serum that will precipitate 80% of the radioactively labeled antigen. The antisera are diluted with

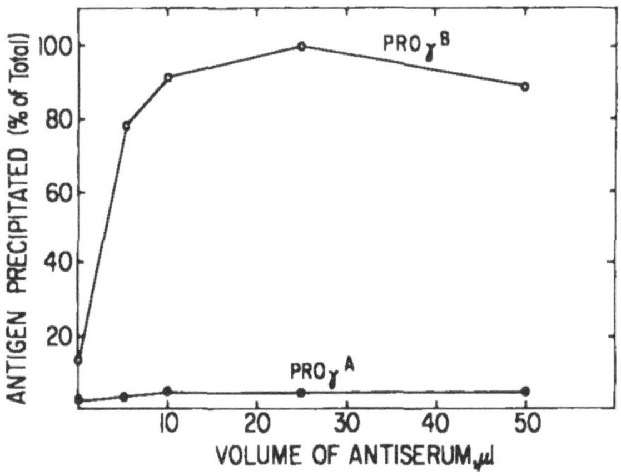

FIGURE 12. Titration curves using antiserum against the Type I procollagen C-propeptide. Antigens were (○——○) carboxyterminal fragment proγ^B obtained by cleavage of procollagen with tadpole collagenase; (●——●) amino terminal fragment proγ^A obtained by cleavage of procollagen with tadpole collagenase. The antiserum was diluted 1 to 10, and the assay was performed as described in the text. (From Olsen, B. R., Guzman, N. A., Engel, J., Condit, C., and Aase, S., *Biochemistry*, 14, 3030, 1977. With permission.)

the Tris, NaCl, ovalbumin-solution described above. After addition of a constant amount of labeled antigen (usually 3,000 to 10,000 cpm), the incubation is continued at 4°C for an additional 24 hr. The antigen-antibody complexes are precipitated by the addition of sheep antirabbit IgG antiserum and incubated at 4°C for 24 hr. The isolation, washing, and counting of the immunoprecipitates are as described above for the direct binding immunoassays.[20]

The antibodies that are raised in rabbits against the Type I C-propeptide as described here do not cross-react with the Type I N-propeptides or with Type I collagen from which the propeptides have been removed by treatment with pepsin (see Figures 12 and 13).[19] The antisera precipitate intact Type I procollagen[19] and are therefore useful in immunological studies of procollagen.[32,33] The antibodies show cross-reactivity with Type I procollagens from rat[31] and humans.[33] This cross-reactivity in fact makes the chick propeptide isolation procedure as described here an attractive system for the production of antihuman Type I procollagen antibodies.

The antiserum raised against the chick Type I C-propeptide clearly distinguishes between chick Type I and Type II procollagens. This is evident both in a direct binding assay (see Figure 14), and in inhibition assays (see Figure 15). However, inhibition assays demonstrate that the antisera against the chick Type I C-propeptide contain antibodies that show partial cross-reactivity with determinants in the chick Type II C-propeptide. It should be noted that the precipitation of Type I procollagen by the anti-Type I C-propeptide antibodies is inhibited by the purified Type II C-propeptide at concentrations that are more than tenfold higher than those required with the Type I C-propeptide (see Figure 15).

Antibodies raised in rabbits against the proα1(I) N-propeptide have been shown to react with intact procollagen.[20] They show no cross-reactivity with the chick Type I C-propeptide.[20] The antibodies raised against the chick Type II C-propeptide show a weak but distinct cross-reactivity with Type I procollagen. The results of the direct

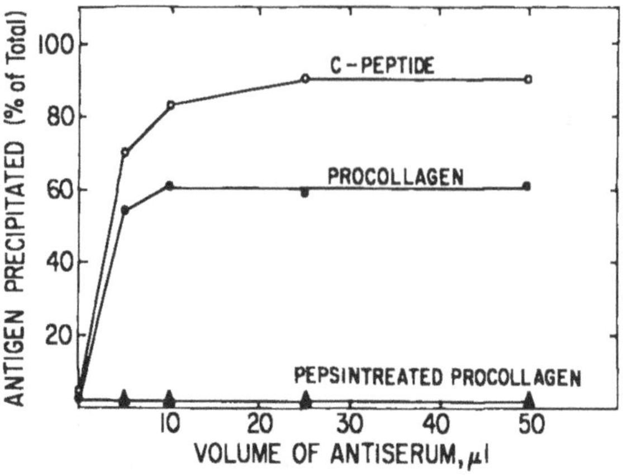

FIGURE 13. Titration curves using antiserum against the Type I procollagen C-propeptide. Antigens were (○——○) C-propeptide; (●——●) Type I procollagen; (▲——▲) pepsin-treated Type I procollagen. For precipitation of C-propeptide, undiluted antiserum was used; for precipitation of procollagen and pepsin-treated procollagen, the antiserum was diluted 1 to 10. The assay was performed as described in the text. (From Olsen, B. R., Guzman, N. A., Engel, J., Condit, C., and Aase, S., *Biochemistry*, 14, 3030, 1977. With permission.)

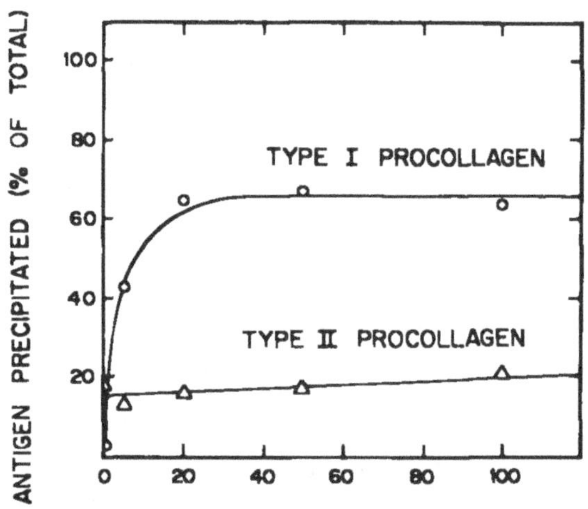

FIGURE 14. Titration curves using antiserum against the Type I procollagen C-propeptide. Antigens were (○——○) Type I procollagen; (△——△) Type II procollagen. The antiserum was diluted 1 to 100. The assay was performed as described in the text.

binding immunoassays suggest that the major fraction of the antibodies is specific for the Type II C-propeptides and that the cross-reactivity with Type I procollagen is due to a minor subpopulation of antibody molecules (see Figure 16).

The isolation of propeptides of Type I and Type II procollagens as described here allows the production of antibodies that are specific for the N- and C-propeptides. Also, the availability of relatively large amounts of propeptides allows the prepa-

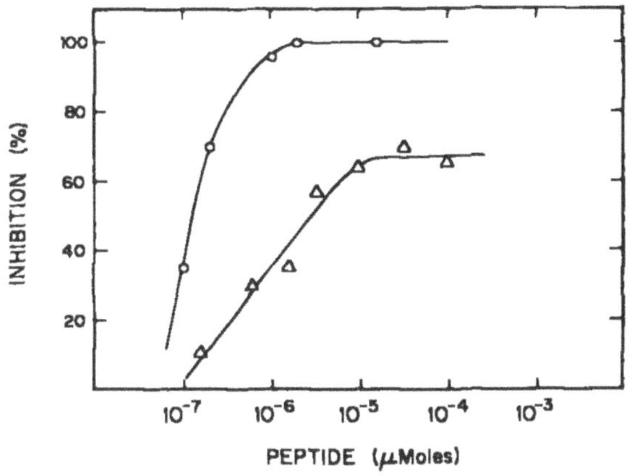

FIGURE 15. Inhibition assay with Type I [^{14}C] procollagen and different propeptide inhibitors. For the assay, sufficient antiserum was used to bind 80% of the [^{14}C] procollagen. Inhibitors used were the Type I procollagen C-propeptide (O——O) and the Type II procollagen C-propeptide (Δ——Δ). The number of micromoles of propeptide inhibitors were calculated from the weights of peptides and a molecular weight of 100,000 daltons for each of the two C-propeptides. The assay was performed as described in the text.

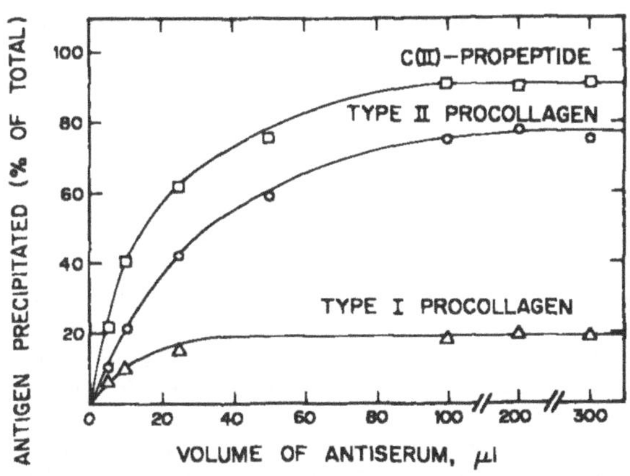

FIGURE 16. Titration curves using antiserum against the Type II procollagen C-propeptide. Antigens were (O——O) Type II procollagen; (Δ——Δ) Type I procollagen; (□——□) Type II procollagen C-propeptide. For the precipitation of the C-propeptide, undiluted antiserum was used; for the precipitation of the two types of procollagens, the antiserum was diluted 1 to 10. The assay was performed as described in the text.

ration of propeptide affinity columns for immunoadsorption of the antisera. Although not discussed here, we have prepared[36] such immunoadsorbents by linking the N- and C-propeptides of chick Type I procollagen to CNBr-activated agarose, and we have used them to purify the antibodies against each of the two propeptides. The procedure we have described for isolation and purification of propeptides from chick tendon and cartilage should also be applicable to tissues from other animals.

ACKNOWLEDGMENTS

Original contributions from this laboratory were supported in part by research grant AM 21471 from the National Institutes of Health of the U.S. Public Health Service. The authors express their sincere thanks to Nancy Kedersha and Frank David Murphy for excellent technical and artistic assistance.

REFERENCES

1. **Timpl, R., Wick, G., and Gay, S.,** Antibodies to distinct types of collagens and procollagens and their application in immunohistology, *J. Immunol. Methods,* 18, 165, 1977.
2. **Timpl, R.,** Immunological studies on collagen, in *Biochemistry of Collagen,* Ramachandran, G. N. and Reddi, A. H. Eds., Plenum Press, New York, 1976, 61.
3. **Furthmayr, H. and Timpl, R.,** Immunochemistry of collagens and procollagens, *Int. Rev. Connect. Tissue Res.,* 319, 7, 1976.
4. **von der Mark, K., Click, E. M., and Bornstein, P.,** The immunology of procollagen. I. Development of antibodies to determinants unique to the proα1 chain, *Arch. Biochem. Biophys.,* 156, 356, 1973.
5. **Timpl, R., Wick, G., Furthmayr, H., Lapière, C. M., and Kühn, K.,** Immunochemical properties of procollagen from dermatosparactic calves, *Eur. J. Biochem.,* 32, 584, 1973.
6. **Dehm, P., Olsen, B. R., and Prockop, D. J.,** Antibodies to chick-tendon procollagen. Affinity purification with the isolated disulfide-linked NH_2-terminal extensions and reactivity with a component in embryonic serum, *Eur. J. Biochem.,* 46, 107, 1974.
7. **Fessler, J. H. and Fessler, L. I.,** Biosynthesis of procollagen, *Annu. Rev. Biochem.,* 47, 129, 1978.
8. **Bornstein, P. and Traub, W.,** The chemistry and biology of collagen, in *The Proteins,* Neurath, H. and Hill, R. L., Eds., Academic Press, New York, 1979, 411.
9. **Prockop, D. J., Kivirikko, K. I., Tuderman, L., and Guzman, N. A.,** The biosynthesis of collagen and its disorders, *N. Engl. J. Med.,* 301, 13, 1979.
10. **Goldberg, B., Taubman, M. B., and Radin, B.,** Procollagen peptidase: its mode of action on the native substrate, *Cell,* 4, 45, 1975.
11. **Leung, M. K. K., Fessler, L. I., Greenberg, D. B., and Fessler, J. H.,** Separate amino and carboxyl procollagen peptidases in chick embryo tendon, *J. Biol. Chem.,* 254, 224, 1979.
12. **Tuderman, L., Kivirikko, K. I., and Prockop, D. J.,** Partial purification and characterization of a neutral protease which cleaves the N-terminal propeptides from procollagen, *Biochemistry,* 17, 2948, 1978.
13. **Furthmayr, H., Timpl, R., Stark, M., Lapière, C. M., and Kühn, K.,** Chemical properties of the peptide extension in the proα1 chain of dermatosparactic skin procollagen, *FEBS Lett.,* 28, 247, 1972.
14. **Becker, U., Timpl, R., Helle, O., and Prockop, D. J.,** NH_2-terminal extensions on skin collagen from sheep with a genetic defect in conversion of procollagen into collagen, *Biochemistry,* 15, 2853, 1976.
15. **Becker, U., Helle, O., and Timpl, R.,** Characterization of the amino-terminal segments in procollagen proα2 chain from dermatosparactic sheep, *FEBS Lett.,* 73, 197, 1977.
16. **Murphy, W. H., von der Mark, K., McEneany, L. S. G., and Bornstein, P.,** Characterization of procollagen-derived peptides unique to the precursor molecule, *Biochemistry,* 14, 3243, 1975.
17. **Rohde, H., Wachter, E., Richter, W. J., Bruckner, P., Helle, O., and Timpl, R.,** Amino acid sequence of the N-terminal non-collagenous segment of dermatosparactic sheep procollagen type I, *Biochem. J.,* 179, 631, 1979.
18. **Hörlein, D., Fietzek, P. P., Wachter, E., Lapière, C. M., and Kühn, K.,** Amino acid sequence of the aminoterminal segment of dermatosparactic calf skin procollagen type I, *Eur. J. Biochem.,* 99, 31, 1979.
19. **Olsen, B. R., Guzman, N. A., Engel, J., Condit, C., and Aase, S.,** Purification and characterization of a peptide from the carboxy-terminal region of chick tendon procollagen type I, *Biochemistry,* 14, 3030, 1977.

20. **Pesciotta, D. M., Silkowitz, M. H., Fietzek, P. P., Graves, P. N., Berg, R. A., and Olsen, B. R.,** Purification and characterization of the amino-terminal propeptide of proα1(I) chains from embryonic chick tendon procollagen, *Biochemistry*, 19, 2447, 1980.
21. **Nowack, H., Olsen, B. R., and Timpl, R.,** Characterization of the amino-terminal segment in type III procollagen, *Eur. J. Biochem.*, 70, 205, 1976.
22. **Bruckner, P., Bächinger, H. P., Timpl, R., and Engel, J.,** Three conformationally distinct domains in the amino-terminal segment of type III procollagen and its rapid triple helix $\rightleftharpoons$ coil transition, *Eur. J. Biochem.*, 90, 595, 1978.
23. **Dehm, P. and Prockop, D. J.,** Synthesis and extrusion of collagen by freshly isolated cells from chick embryo tendon, *Biochim. Biophys. Acta*, 240, 367, 1971.
24. **Olsen, B. R., Hoffman, H. P., and Prockop, D. J.,** Interchain disulfide bonds at the COOH-terminal end of procollagen synthesized by matrix-free cells from chick embryonic tendon and cartilage, *Arch. Biochem. Biophys.*, 175, 341, 1976.
25. **Hoffmann, H-P., Olsen, B. R., Chen, H-T., and Prockop, D. J.,** Segment-long-spacing aggregates and isolation of COOH-terminal peptides from type I procollagen, *Proc. Natl. Acad. Sci. U.S.A.*, 73, 4304, 1976.
26. **Dehm, P. and Prockop, D. J.,** Biosynthesis of cartilage procollagen, *Eur. J. Biochem.*, 35, 159, 1973.
27. **Uitto, J. and Prockop, D. J.,** Biosynthesis of cartilage procollagen. Influence of chain association and hydroxylation of prolyl residues on the folding of the polypeptides into the triple-helical conformation, *Biochemistry*, 13, 4586, 1974.
28. **Uitto, J.,** Biosynthesis of type II collagen. Removal of amino- and carboxy-terminal extensions from procollagen synthesized by chick embryo cartilage cells, *Biochemistry*, 16, 3421, 1977.
29. **Curran, S. and Prockop, D. J.,** Isolation and partial characterization of the carboxy-terminal propeptide of Type II procollagen from chick embryos, *Biochemistry*, in press, 1982.
30. **Nist, C., von der Mark, K., Hay, E. D., Olsen, B. R., Bornstein, P., Ross, R., and Dehm, P.,** Location of procollagen in chick corneal and tendon fibroblasts using ferritin-conjugated antibodies, *J. Cell Biol.*, 65, 75, 1975.
31. **Rohde, H., Nowack, H., Becker, U., and Timpl, R.,** Radioimmunoassay for the amino-terminal peptide of the procollagen proα1(I)-chain, *J. Immunol. Methods*, 11, 135, 1976.
32. **Monson, M. M. and Goodman, H. M.,** Translation of chick calvarial procollagen messenger RNAs by a messenger RNA dependent reticulocyte lysate, *Biochemistry*, 17, 5122, 1978.
33. **Peltonen, L., Palotie, A., and Prockop, D. J.,** A defect in the structure of type I procollagen in a patient who had osteogenesis imperfecta: excess mannose in the COOH-terminal propeptide, *Proc. Natl. Acad. Sci. U.S.A.*, 77, 6179, 1980.
34. **Cournil, I., Leblond, C. P., Pomponio, J., Hand, A. R., Sederlof, L., and Martin, G. R.,** Immunohistochemical localization of procollagens. I. Light microscopic distribution of procollagen I, III and IV antigenicity in the rat incisor tooth by the indirect peroxidase-antiperoxidase method, *J. Histochem. Cytochem.*, 27, 1059, 1979.
35. **Prockop, D. J. and Guzman, N. A.,** Collagen diseases and the biosynthesis of collagen, *Hosp. Pract.*, 12, 61, 1977.
36. **Olsen, B. R.,** Unpublished data, 1981.

Chapter 8

ISOLATION OF FIBRONECTIN FROM PLASMA AND CELLS

Kenneth M. Yamada

TABLE OF CONTENTS

I. INTRODUCTION

Fibronectins are adhesive glycoproteins present on cell surfaces and in extracellular matrices, as well as in blood, amniotic fluid, and cerebrospinal fluid.[1-7] There are at least two major classes of fibronectins which are termed cellular (cell surface) fibronectin and plasma fibronectin. Plasma fibronectin is also known as "cold insoluble globulin" or "α-2 opsonic glycoprotein." Amniotic cells produce a third type of fibronectin that is unusual in its high degree of glycosylation.[8]

Plasma and cellular fibronectins are very similar in amino acid and carbohydrate composition and in certain biological activities involving cell attachment to collagen and plastic, as well as in secondary and tertiary structure.[1-7] However, these forms of fibronectin clearly differ in certain other biological activities that involve cell-cell interactions, have slightly different subunit polypeptide sizes, and display major differences in solubility properties (see Tables 1 and 2; discussed in more detail in Reference 9).

In this chapter, procedures for the isolation of plasma fibronectin and cellular fibronectin will be presented separately. Each section will compare published methods in general terms, then a detailed, recommended protocol will be provided for the purification of each form of fibronectin.

II. PURIFICATION OF PLASMA FIBRONECTIN

A. Approaches

There are currently two different approaches to purifying plasma fibronectin. Plasma fibronectin ("cold insoluble globulin") was first purified to homogeneity by Mosesson and Umfleet.[10] This procedure involves glycine fractionation of plasma cryoprecipitate, ethanol precipitation, and DEAE column chromatography. A more rapid, modified purification procedure has been described.[11] The purity of fibronectin purified by these procedures has been well documented immunologically, e.g., as free of contamination by fibrinogen or factor XIII.[11] However, the preparations do contain significant amounts of polypeptides of lower molecular weight that are apparently proteolytic fragments of plasma fibronectin; similar fragments are reported to be present in blood prior to any purification.[12] For reasons that are not yet clear, there often appears to be less of this heterogeneity in plasma fibronectin prepared by affinity chromatography (see below). Although it is a well accepted procedure, there are two possible disadvantages to this classical purification scheme: The ethanol precipitation step may cause the loss of certain biological activities (e.g., opsonic activity),[13] and the procedure is more complex or lengthy than other recent procedures.

The second approach to purifying plasma fibronectin has been by means of its specific, high-affinity binding to denatured collagen.[14-19] This approach is currently the most popular, and one version will be described in detail below. After affinity chromatography of plasma on columns containing covalently coupled denatured collagen (usually gelatin), the noncovalently bound purified protein is eluted from the column by 1 *M* KBr, 4 to 8 *M* urea, or amines.[16-20]

These eluting molecules are than removed from the fibronectin preparation by various procedures such as dialysis. A comparison of the recovery and biological activity of fibronectin eluted in various ways from collagen in terms of ability to mediate the attachment of cells to collagen suggests that urea is the safest eluting

Table 1
COMPARISON OF MOLECULAR PROPERTIES OF FIBRONECTINS

Property	Cellular fibronectin	Plasma fibronectin
Monomer molecular weight	200,000—240,000	215,000—220,000
Subunit structure	Disulfide-linked dimers and multimers	Disulfide-linked dimers
Sedimentation constant	8.5 (pH 8) 7.6 (pH 11)	12.3–13.0 (pH 7) 8.0 (pH 11)
Partial specific volume	0.72	0.72
Frictional ratio	2.8 (pH 11)	2.8 (pH 11)
Stokes radius	110 Å at pH 11	—
Extinction coefficient at 280 nm ($E^{1\ cm}_{mg/ml}$)	1.2	1.3
Circular dicroism (far UV) minimum/maximum	212 nm/227 nm	213 nm/227 nm

Table 2
COMPARISON OF BIOLOGICAL ACTIVITIES OF FIBRONECTINS

Activity	Relative activity $\left(\frac{\text{cellular fibronectin}}{\text{plasma fibronectin}}\right)$
Cell attachment to collagen	1
Cell spreading on plastic	1
Opsonic activity in macrophages	1
Cell attachment of transformed Nil cells	1.5
Restoration of normal phenotype to transformed cells	50
Hemagglutination	150–250

molecule for recovering fully active fibronectin.[21] However, no complete comparative studies have been performed concerning the best pH for urea elution.

The fibronectin purified by the one-step gelatin-agarose procedure is relatively pure. However, substantial contaminants are sometimes present, e.g., fibrinogen. For immunological studies, some laboratories have absorbed the immune sera obtained using these preparations with plasma previously passed through gelatin affinity columns to remove fibronectin. Absorption with such flow-through plasma reportedly removes contaminating antigens.[17,22,23]

An alternative and probably preferable approach is to purify further the fibronectin preparations used as antigens. Approaches used include applying the preparations to the gelatin affinity columns a second time or gel filtration chromatography to obtain further purification according to molecular size.[9]

Heparin affinity columns can also be used to purify fibronectin.[24] However, many other plasma proteins can also bind to heparin, and heparin affinity columns appear to be the most useful when used as a second affinity step following a gelatin affinity column.[25] This second column reliably removes the common contaminants of gelatin-agarose purified fibronectin. A routine, reliable procedure that has been modified from the methods of Engvall et al.[16,17] combined with the more recent methods of Miekka et al.[25] is described below. The strategy is to purify the fibronectin by gelatin affinity chromatography, then to elute and further purify by heparin affinity chromatography in the presence of 4 *M* urea which prevents reassociation of fibronectin with contaminating proteins to which it may bind.

B. Protocols for Affinity Chromatographic Purification of Plasma Fibronectin

1. Preparation of Affinity Columns

a. Collagen (Gelatin) Affinity Columns

Porcine skin gelatin is washed in sequential extracting solutions to remove soluble contaminants.* For example, 2 g of dry gelatin is washed with four changes of deionized water at 0 to 4°C over an 18 hr period, with occasional shaking, centrifugation at approximately 1000 × g for 5 min, and decanting of the wash solution. The gelatin is then washed 2 hr with 3 *M* NaCl, 10 m*M* EDTA, then three more times with deionized water at 0 to 4°C. The washed gelatin is solubilized by heating to 100°C with stirring for 10 min. The concentration of gelatin can be estimated using a value for absorbance at 235 nm of $E_{1\ mg/m\ell}^{1\ cm} = 0.8$.

The gelatin can be coupled to agarose activated by cyanogen bromide following detailed published procedures[26] or by coupling to commercially available CNBr-activated Sepharose® 4B according to the procedures recommended by the manufacturer (e.g., Pharmacia® or Pierce®).

An example of an effective coupling protocol follows. A 40 mg/mℓ solution of gelatin stored at −20°C is warmed to room temperature, then melted by brief heating in a boiling water bath and cooled to approximately 40°C. CNBr-activated Sepharose® (Pharmacia) is washed in a Buchner funnel with a medium-porosity fritted glass bottom in 10^{-3} *M* HCl as recommended by the company. An appropriate volume of gelatin solution (e.g., 0.7 mℓ/10 g of moist CNBr-activated agarose beads) is added to the gel beads immediately after the addition of coupling buffer (e.g., 7 mℓ of 0.2 *M* $NaHCO_3$, 1 *M* NaCl, plus water to bring the total solution volume to 14 mℓ/10 g of moist gel). The final concentration of gelatin we currently use is 10 mg in 5 mℓ of liquid per 3.5 g of moist gel beads.** The mixture is vigorously shaken after these additions, then placed in a plastic bottle on an end-over-end mixer for 2 hr at room temperature.

The amount of gelatin that is finally covalently bound to the beads can be determined by the absorbance of the supernatant solution. More accurate estimates can be obtained by quantitative amino acid analysis of beads or by determining the amount of ^{125}I-labeled gelatin that remains coupled to well-washed gel beads.

The reaction is stopped by incubation in 1 *M* ethanolamine (2-amino-ethanol), pH 8, for 2 hr. The gelatin-agarose is then washed on a Buchner funnel with three sequential cycles of low and high pH washes, using 1 *M* NaCl, 0.1 *M* sodium acetate, pH 4, then 1 *M* NaCl, 0.1 *M* sodium borate, pH 8, in order to remove adsorbed but noncovalently bound gelatin. The washed gel beads can be stored at neutral pH, e.g., in phosphate or Tris® buffered saline at 4°C as a 50% (v/v) suspension. Sodium azide (0.02%) can be added to prevent the growth of microorganisms during long-term storage.

* Although it has not been proven that this gelatin washing step is important, it probably decreases levels of contaminating proteoglycans and proteases.

** The quantity of moist gel beads is determined as follows: beads suspended in the final wash solution are aspirated free of liquid on a Buchner funnel with a fritted glass bottom. Immediately after the liquid is removed, the funnel is twisted out of the suction flask, inverted, and tapped on a hard surface to spill the moist gel bead cake onto a piece of filter paper. The amount of moist gel can then be adjusted as desired on a balance.

The gelatin-agarose is poured into columns 4 to 10 cm high;* if glass columns are used, they should be presiliconized, e.g., with Aquasil® (Pierce). The gelatin-agarose columns are washed with 0.15 *M* NaCl, 10 m*M* Tris-Cl, pH 7.4 ("Tris® saline"), containing the anticoagulant used for the plasma to be utilized, e.g., citrate or 5 m*M* EDTA. All procedures are performed at room temperature.

A parallel precolumn is also prepared, consisting of a similar amount of agarose or Sepharose® CL 4B beads which is also washed with tris-saline solution containing anticoagulant. This precolumn serves to remove proteins or protein complexes that bind nonspecifically to agarose columns.

b. Heparin Affinity Columns

Heparin can be coupled to agarose as described by Fujikawa et al.[27] Commercially available CNBr-activated Sepharose® can also be used, and heparin is coupled as described[27] using a ratio of 1 g heparin (Sigma Type I)® to 50 g of moist gel beads in 0.1 *M* $NaHCO_3$, pH 8.3. Heparin-agarose can also be purchased from commercial sources such as Pharmacia®, Pierce®, or BRL®. These commercial affinity matrices can be used successfully, but they appear to contain less bound heparin than material prepared as described above.** Heparin columns are 4 to 10 cm high and are rinsed with 4 *M* urea, 10 m*M* Tris-Cl, pH 7.4, in preparation for use.

2. Sources of Plasma

Outdated human plasma can be obtained from blood banks,*** anticoagulated by either citrate or EDTA. Chicken, bovine, and other plasmas can be purchased from Pel-Freez Biologicals. Plasma that contains any visible clotted material appears to be less satisfactory, since subsequent contaminants often appear in the purified preparations, perhaps due to the binding of fibrin monomer to fibronectin and isolation of the complexes during affinity chromatography.

3. Procedure for Purification of Plasma Fibronectin

Plasma is brought to 0.5 to 1 m*M* in the protease inhibitor phenylmethylsulfonyl fluoride (Sigma, Pierce®, or BRL®).† The plasma is centrifuged at $\geq$10,000 × g for 15 min.

* The effects of column height on purification have not been tested, although longer gel beds substantially reduce flow rates when gravity feed is used. We have chosen conditions that yield large amounts of purified material in a short period of time, using less gelatin-agarose matrix than is required for quantitative recovery of fibronectin from the plasma. The final procedure involves compromises between optimal economy (if commercial CNBr-Sepharose® is used or if large-scale columns are not available), rapidity, and percent recovery.

** The effects of these differences in amounts of bound heparin on the effectiveness of heparin affinity columns have not been examined by our laboratory except with regard to the purification of a heparin-binding fragment of fibronectin. In this particular case, the commercial matrices that were tested provided substantially lower yields of a 50,000-dalton heparin-binding fragment of cellular fibronectin.

***Other sources include fresh-frozen plasma or human serum, although the latter source is less satisfactory because of the reduced amounts of plasma fibronectin usually found in serum after clotting.

† Companies sell PMSF preparations that differ in the appearance of the crystals, which is reflected in the ease of solubilization. In some cases, visible precipitates remain in solubilized stocks; we currently use material from BRL due to recent changes in the apparent purity of the PMSF from Sigma. Stock solutions of PMSF are prepared immediately before use, since hydrolytic breakdown of this compound occurs over a period of hours in aqueous solution. A stock of 0.2 *M* is prepared in 95% ethanol, and a 1 to 200 dilution is added to the plasma with very vigorous stirring. In our experience, even anticoagulated chicken plasma will sometimes unpredictably initiate clotting during column chromatography, resulting in poor flow rates and contaminated preparations. This problem is completely prevented by the addition of hirudin (Sigma) to inhibit thrombin activity; this inhibitor is added to both the plasma and to the prewashes of pre- and gelatin-agarose columns. The minimum amount needed has not been determined; we have arbitrarily used two to five units per milliliter.

The supernatant plasma is applied to the precolumn at room temperature, and the flow-through material is applied as it is collected to the gelatin affinity column. The ratio of plasma to gelatin-agarose is approximately 500 mℓ/50 mℓ of packed gel. In a typical small preparative run, 500 mℓ of plasma is applied to a 2.5 × 9 cm gelatin-agarose column over a period of 3 hr.

The column is rapidly washed with at least two column volumes of 0.5 *M* NaCl, 10 m*M* Tris-Cl, pH 7.4, without anticoagulant, then with three column volumes of Tris® saline. The bound fibronectin is eluted with 4 *M* urea, 10 m*M* Tris-Cl, pH 7.4, over a period of 30 to 45 min.* The elution consists of at least 1.5 column bed volumes, collecting fractions of roughly one quarter to one third of the column bed volume each (e.g., 15 mℓ per fraction).

The peak of eluted protein can be determined by the Coomassie blue protein assay (BioRad®) or by absorbance at 280 nm ($E^{1\ cm}_{1\ mg/m\ell} = 1.2$) with appropriate buffer blanks. In a simplified protocol, the entire eluate can be pooled. The pooled protein fractions are applied to the heparin-agarose column that has been preequilibrated with the 4 *M* urea solution. In a routine preparation, 30 to 50 mℓ is applied to a 2.5 × 4 cm column over a period of 2 hr. The ratio of original plasma to heparin affinity matrix is 25 mℓ/mℓ of packed gel.

The column is then washed with three column volumes of 4 *M* urea in 10 m*M* Tris®, then with 10 m*M* Tris-Cl, pH 7.4, to remove all urea. The fibronectin is then eluted with 0.5 *M* NaCl, 10 m*M* Tris-Cl, pH 7.4, and the protein peak is determined by chemical protein assays (BioRad®) Lowry et al.[28]) or by absorbance at 280 nm. The peak fractions are pooled and dialyzed against Tris® saline at 4°C for 18 hr.

The final, dialyzed fibronectin preparation is assayed for protein by the method of Lowry et al.,[28] with care to include the same concentration of Tris in the blanks; protein concentration can also be determined by absorbance at 280 nm. Rarely, a second cycle of affinity chromatography on heparin-agarose is needed to remove trace contaminants detected by SDS gel electrophoresis as described later.

4. Storage

Purified fibronectin can be stored in polyethylene containers at 0 to 4°C for at least several days with no apparent deterioration. Use of phosphate buffers or glass or wettable plastic containers may result in precipitation or nonspecific losses by binding to vessel walls. For routine long-term storage, we quick-freeze aliquots in polypropylene tubes, e.g., vials used for storage of materials in liquid nitrogen; excellent results are obtained by immersion in liquid nitrogen. Samples are stored at −70°C or colder; our laboratory stores most stocks in liquid N_2. Storage at −20°C is not satisfactory, since eventual loss of activity or even grossly visible aggregation is observed.

Samples are quickly defrosted by placing vials in room temperature water with occasional agitation. Occasionally, a small precipitate is visible which is either dissolved by heating to 37°C or by centrifugation.

An alternative method of storage is to dialyze the fibronectin against an alkaline buffer: 0.15 *M* NaCl, 1 m*M* $CaCl_2$, 10 m*M* CAPS (cyclohexylaminopropane sulfonic acid), pH 11.0. At this pH, there are few problems with fibronectin insolubility after

* The ionic strength of the column elution buffer is reduced to promote the later binding of fibronectin to heparin-Sepharose® which is sensitive to ionic strength. We omit the 1 *M* urea wash of Engvall and Ruoslahti, since some loss of fibronectin can occur that may be due to our higher ratios of fibronectin to gelatin.

freezing and defrosting.* However, the conformation of the fibronectin is slightly altered at alkaline pH.[29] Although this conformational change does not appear to affect the biological activities analyzed to date, it may slightly alter proteolytic cleavage products.[39]

5. Regeneration of Columns

The gelatin-agarose column can be reused after extensive washing with 8 *M* urea, 10 m*M* Tris-Cl, pH 7.4, then 0.15 *M* NaCl, 10 m*M* Tris-Cl, pH 7.4. The heparin-agarose column can be washed with 2 *M* NaCl, 10 m*M* Tris-Cl, pH 7.4, then 10 m*M* Tris-Cl, pH 7.4, and stored in the presence of 0.02% sodium azide.

III. PURIFICATION OF CELLULAR FIBRONECTIN

A. Approaches

Cellular fibronectin poses unusual problems in purification due to its insolubility at neutral pH after purification. It has been purified by at least three different methods. The first approach is to purify it by inducing its specific release from the cell surface with low concentrations of urea in serum-free medium.[30–32]

The mechanism of this process is still not fully understood, although up to half of the fibronectin on the cell surface can be recovered from chick fibroblasts. This procedure is relatively gentle, although it is possible that it selects for molecules that are not extensively cross-linked by disulfide bonds.

A second approach is to purify fibronectin that is released by cells into culture medium.[33–35] Many cell types tend to secrete large amounts of the fibronectin they synthesize into the culture medium, rather than retaining most of it on the cell surface. Spent culture medium is therefore a logical source of fibronectin which can be purified by gelatin affinity chromatography. It is important that the serum used for culturing the cells be depleted of plasma fibronectin by gelatin affinity chromatography ahead of time so as to avoid contamination of the secreted cellular fibronectin with plasma fibronectin. One possible disadvantage of this approach would be the binding of procollagen to the fibronectin, if the cell line secretes large amounts of both proteins. The fibronectin purified may also differ in its extent of disulfide cross-linking from the fibronectin remaining on cell surfaces.

A third approach is to extract cells with ATP and EDTA, then with a strong detergent. Fibronectin remains insoluble and is solubilized in boiling 8 *M* urea. The fibronectin is then purified by lectin affinity chromatography. This procedure has been used successfully with hamster fibroblasts, although there is variable contamination with cellular actin,[36] and the harsh conditions may destroy biological activity.

The procedure that currently appears to yield the purest product is urea extraction of chick fibroblasts. However, purification of cellular fibronectin from human and hamster fibroblasts by this procedure is more difficult, perhaps due to the slightly lower amounts of fibronectin in these cells.[37] It is apparent that purification of cellular fibronectin from any particular cell type may require a different one of the three approaches. For example, urea extraction is particularly ineffective with certain permanent cell lines that have low amounts of cell surface fibronectin, and

* It is important to avoid acidification of buffers by CO_2 during dialysis or storage; the CO_2 in room air apparently dissolves in pH 11 buffers and lowers the pH. We cover dialysis containers well and usually store samples kept for periods of longer than a month in liquid N_2 to avoid exposure of even frozen samples to air.

purification from culture medium is quite effective with the NIL cell line which appears to make much less procollagen than primary fibroblast cultures. For cultures in which the fibronectin contains bound procollagen contaminants, DEAE-column chromatography in the presence of 8 *M* urea may provide a means of purifying the fibronectin by preventing noncovalent associations with other macromolecules, as well as maintaining the protein in solution.

Since there is no evidence as yet that the biological activities or overall structure of cellular fibronectins differ substantially between species (e.g., see Reviews in References 1–7), chick embryo fibroblasts provide a convenient source for isolating the largest amounts of this protein; a routine protocol is provided below.

B. Protocols for Purification of Cellular Fibronectin

1. Cells

Chick embryo fibroblasts are cultured as described by Vogt.[38] Ten-day-old chicken embryos (e.g., Truslow Farms, Chestertown, Md.) are dissociated with trypsin as described by Vogt and placed into primary culture in 10% heat-inactivated calf serum in Ham's F10 medium (see full recipe below). Cells are allowed to remain confluent for 3 to 4 days without a medium change in order to select for fibroblasts. After passaging once, secondary cultures are trypsinized, and tertiary cultures are established in 850 cm^2 plastic disposable roller bottles (Falcon®) at concentrations of 2.5 to 4×10^7 cells in 100 mℓ of F10 medium containing 5% (not 10%) heat-inactivated calf serum, 10% tryptose phosphate broth, 0.5% beef embryo extract, 0.056% sodium bicarbonate, 50 units per milliliter penicillin, and 50 μg/mℓ streptomycin.[38]

The use of tertiary cultures substantially reduces contamination with nonfibroblastic cells and also appears to decrease the amount of procollagen production without decreasing quantities of fibronectin synthesized. The initial cell plating density in roller bottles must be adjusted periodically according to changes in the age or composition of the flocks of laying hens in order to achieve a heavily confluent monolayer before the cells undergo senescence. This type of culture appears to be quite sensitive to initial plating concentrations in determining whether cells form a dense monolayer before losing the capacity to replicate. Cells are fed on days 3, 4, and 5 of roller bottle culture and are usually urea-extracted on day 6. It is critical that the cell monolayers be heavily confluent before extraction.[30]

2. Procedure for Purification of Cellular Fibronectin

Media listed below should be warmed to 37°C before use:

1. Hanks balanced salt solution
2. Serum-free Dulbecco-Vogt modified Eagles medium containing 4 m*M* glutamine and freshly added 2 m*M* PMSF (add 1 mℓ stock per 100 mℓ of medium with vigorous stirring, using a stock of freshly made PMSF at 34.8 mg/mℓ of 95% ethanol)—the PMSF-containing medium should be incubated 15 min at 37°C to inactivate proteases
3. 1 *M* ultrapure urea in Dulbecco-Vogt medium with 2 m*M* PMSF and glutamine

Care should be taken to avoid exposure of preparations to proteases, e.g., by wearing disposable gloves and using only materials known to be protease-free.

Cell monolayers are rinsed four times with 25 to 50 mℓ of Hanks solution per roller bottle. A total of 25 mℓ of the serum-free Dulbecco-Vogt medium with PMSF is added to each roller bottle and incubated for 1 hr at 37°C with rotation at 1 r/min.

The solution is decanted, and the monolayers are rinsed once with PMSF-containing Dulbecco-Vogt medium left over from the original preparation. The bottles are drained well by decanting, and 20 mℓ of freshly prepared 1 *M* urea in Dulbecco-Vogt medium with PMSF is added. Bottles are incubated with rolling at 1 r/min for 2 hr at 37°C. Care should be taken at all steps to avoid drying of cell monolayers which will result in cell lysis.

After urea-extraction, the medium is decanted into protease-free polypropylene centrifuge tubes (50 mℓ with caps) and centrifuged for 15 min at 25,000 × g. The supernatant solutions are decanted and either frozen in a plastic bottle in dry ice-ethanol and stored at below −70°C or processed further. The cells can be used for a second extraction on the following day; the medium is changed to 0.5 *M* urea in Dulbecco-Vogt or F10 medium without PMSF for ≥30 min, then rinsed well with Dulbecco-Vogt or F10 and returned to regular culture medium.

The fibronectin extract is precipitated with 70% saturated ammonium sulfate (0.472 g of ultrapure ammonium sulfate crystals per milliliter of solution, dissolved at 4°C using a magnetic stirrer). After the ammonium sulfate is dissolved, the solution is allowed to stand without mixing for at least 1 hr at 4°C. The precipitate is collected by centrifugation at 25,000 × g for 15 min using capped tubes.

The supernatant solution is decanted and discarded, and the pink pellet is allowed to drain well by inverting the tubes for at least 15 sec. The walls of the tube are wiped with a kimwipe that is tightly wound around a pair of forceps in order to remove as much ammonium sulfate as possible. The pellet is then resuspended in the equivalent of 1 mℓ of 0.2 *M* CAPS buffer, pH 11.0, per 20 mℓ of original roller bottle extract. The pellet is rapidly resuspended by mixing with a plastic pipet tip and immediately adjusted to pH 11 using 5 N NaOH. (The pH meter electrode should be precleaned by soaking in 0.1 *N* HCl for at least several minutes and rapidly alternating between 0.1 *N* HCl and 0.1 *N* NaOH solutions to remove proteases and contaminating proteins.) Rapidity is critical in resuspending the fibronectin pellet and adjusting the pH after addition of the pH 11 buffer, since otherwise a clear gel that is surprisingly resistant to solubilization can form.

The resolubilized fibronectin is then dialyzed against buffer A (0.15 *M* NaCl, 1 m*M* $CaCl_2$, 10 m*M* CAPS, pH 11.0) with vigorous stirring, usually in a 4-ℓ beaker at 4°C covered with plastic wrap layered on the surface of the liquid and held in place with a sheet of aluminum foil that is molded to press the plastic wrap flat on the liquid surface. The purpose of this maneuver is to exclude air from contact with the alkaline liquid; otherwise, CO_2 from the air will enter the liquid and begin to neutralize the buffer with the formation of a calcium carbonate precipitate and insolubilization of the fibronectin.

We usually reprecipitate the fibronectin after it has been dialyzed for 16 to 18 hr by adding ammonium sulfate to 50% saturation (313 mg/mℓ of original solution). It is resuspended to approximately 5 mg/mℓ and dialyzed overnight again against buffer A. The concentrated fibronectin preparation is then centrifuged for 15 min at 25,000 × g and applied by upward flow to a 2.5 × 100 cm Sepharose® CL-4B column* preequilibrated with 0.1 *M* NaCl, 1 m*M* $CaCl_2$, 50 m*M* CAPS, pH 11.0. The cross-linked agarose column is stable to pH 11, and the higher buffer concentration allows for less pH fluctuation during fraction collection.

In a routine preparative run, 40 to 50 mg of fibronetin in 8 mℓ is chromatographed

* Certain lots of Sepharose® CL-4B are found to bind fibronectin nonspecifically, resulting in lower recovery of protein. This problem can be alleviated by pretreating the column with 1 mg/mℓ bovine serum albumin (crystalline), followed by extensive washing with column buffer.

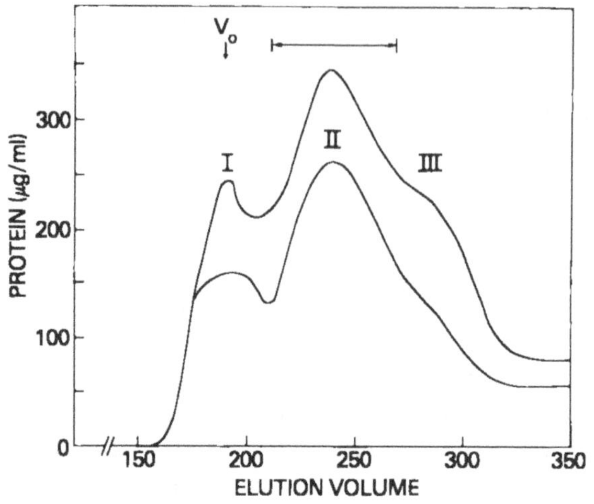

FIGURE 1. Elution profiles of cellular fibronectin. Two preparations of cellular fibronectin were applied to 2.5 × 100 cm columns of Sepharose® CL-4B and eluted as described in the text. Protein concentrations were determined on aliquots using column buffer blanks.[28] Peak I is the void volume (V_o) as determined by Blue Dextran 2000 (Pharmacia®). Peak II is the major peak, and it contains fibronectin dimer.[32] Peak III contains fibronectin monomer.[32] The horizontal arrows indicate the fractions to be pooled containing purified fibronectin dimer.

over a period of 18 to 24 hr, and 6-mℓ fractions are collected in polystyrene centrifuge tubes. The protein profile is determined on aliquots from each tube by the procedure of Lowry et al.;[28] example profiles are shown in Figure 1. The void volume peak (I) consists of fibronectin multimers, as well as residual collagenous and proteoglycan contaminants. The major peak is the fibronectin dimer peak. A trailing peak or shoulder (III) consists of fibronectin monomer which has an elution volume similar to that of intact fibrinogen used as a size marker. We usually pool and save only the fibronectin dimer peak.

The pooled peak is neutralized with 1 *N* HCl, and the fibronectin is precipitated as described earlier for unfractionated fibronectin. The pellet is resuspended in CAPS buffer as described previously to a concentration of 1 to 2 mg/mℓ. The solution is dialyzed for 18 to 24 hr against two changes of buffer A.

3. Storage

After dialysis, samples of the diffusate from the solution outside of the dialysis bag are saved for use as controls in biological experiments and as a blank for protein determinations. The fibronectin is centrifuged 25,000 × g for 10 min and frozen in polypropylene tubes in liquid nitrogen or (less preferably) in powdered dry ice. It is stored at below −70°C or in liquid nitrogen. Samples not stored in liquid nitrogen tend to lose small amounts of activity after several months, whereas in liquid nitrogen they are stable for at least 2 to 3 years. Samples are defrosted as described for plasma fibronectin; the solubility of cellular fibronectin is much more sensitive to changes in pH, and it is safest to check the pH immediately after defrosting and to adjust it back to pH 11.0 with NaOH if necessary. Exposure to air (e.g., in uncapped tubes that can absorb CO_2) and repeated freezing and thawing should be avoided, since they lead to a loss of activity.

4. Addition of Cellular Fibronectin to Culture Media

If cellular fibronectin is added to tissue cultures, it is added directly to media to a final concentration of 25 to 50 μg/mℓ. It should not be Millipore® filtered, since it is retained on nitrocellulose filters after neutralization. Fibronectin-containing culture media that contain antibiotics have never shown bacterial contamination in our experience, perhaps due to alkaline inactivation of microorganisms. Control cultures receive equal amounts of the diffusate from the same batch of fibronectin.

5. Commercial Sources of Fibronectin

The most reliable and certainly the most economical way to obtain biologically active fibronectin is to prepare it oneself under known purification conditions. However, for small trial experiments or in other situations, it may be desirable to purchase fibronectin. At the moment, plasma fibronectin can be purchased from several companies, including Collaborative Research®, Calbiochem-Behring®, and Bethesda Research Laboratories (BRL)®. Of the preparations we have randomly purchased or obtained and examined by SDS-polyacrylamide gel electrophoresis, the purest were a batch from BRL® and a recent batch from Collaborative®, whereas the worst were several earlier preparations from Collaborative® that showed severe proteolytic degradation. It is important that each and every commercially purchased preparation be examined for purity by SDS gel electrophoresis and (if necessary) for biological activity by the purchaser, since variability is apparently a major hazard. It is not safe to depend on the product descriptions of certain companies which sometimes bear little relationship to the product obtained. Likewise, antifibronectin antisera should be carefully examined for contaminating antibodies, since some commercial preparations are currently contaminated by reactivity to at least one other antigen (probably fibrinogen). Thus, although commercial products reduce the work required to work with fibronectin, hard recent experience indicates that extreme care is required in interpreting any results obtained with these materials unless they have been adequately characterized.

REFERENCES

1. **Yamada, K. M. and Olden, K.,** Fibronectins — adhesive glycoproteins of cell surface and blood, *Nature,* 275, 179, 1978.
2. **Vaheri, A. and Mosher, D. F.,** High molecular weight, cell surface-associated glycoprotein (fibronectin) lost in malignant transformation, *Biochim. Biophys. Acta,* 516, 1, 1978.
3. **Hynes, R. O., Destree, A. T., Perkins, M. E., and Wagner, D. D.,** Cell surface fibronectin and oncogenic transformation, *J. Supramol. Struct.,* 11, 95, 1979.
4. **Pearlstein, E., Gold, L., and Garcia-Pardo, A.,** Fibronectin: a review of its structure and biological activity, *Mol. Cell. Biochem.,* 29, 103, 1980.
5. **Mosher, D. F.,** Fibronectin, *Prog. Hemostasis Thromb.,* 5, 111, 1980.
6. **Mosesson, M. W. and Amrani, D. L.,** The structure and biologic activities of plasma fibronectin, *Blood,* 56, 145, 1980.

7. **Yamada, K. M., Olden, K., and Hahn, L. E.,** Cell surface protein and cell interactions, in *The Cell Surface: Mediator of Developmental Processes*, Subtelny, S. and Wessells, N. K., Eds., Academic Press, New York, 1980, 43.
8. **Ruoslahti, E., Engvall, E., Hayman, E. G., and Spiro, R. G.,** Comparative studies on amniotic fluid and plasma fibronectin, *Biochem. J.*, 193, 295, 1981.
9. **Yamada, K. M. and Kennedy, D. W.,** Fibroblast cellular and plasma fibronectins are similar but not identical, *J. Cell. Biol.*, 80, 492, 1979.
10. **Mosesson, M. W. and Umfleet, R. A.,** The cold-insoluble globulin of human plasma. I. Purification, primary characterization, and relationship to fibrinogen and other cold-insoluble fraction components, *J. Biol. Chem.*, 245, 5728, 1970.
11. **Chen, A. B. and Mosesson, M. W.,** An improved method for purification of the cold-insoluble globulin of human plasma (CIg), *Anal. Biochem.*, 79, 144, 1977.
12. **Chen, A. B., Amrani, D. L., and Mosesson, M. W.,** Heterogeneity of the cold-insoluble globulin of human plasma (CIg), a circulating cell surface protein, *Biochim. Biophys. Acta*, 493, 310, 1977.
13. **Blumenstock, F. A., Saba, T. M., Weber, P., and Laffin, R.,** Biochemical and immunological characterization of human opsonic α_2SB glycoprotein: its identity with cold-insoluble globulin, *J. Biol. Chem.*, 253, 4287, 1978.
14. **Hopper, K. E., Adelmann, B. C., Gentner, G., and Gay, S.,** Recognition by guinea-pig peritoneal exudate cells of conformationally different states of the collagen molecule, *Immunology*, 30, 249, 1976.
15. **Kleinman, H. K., McGoodwin, E. B., and Klebe, R. J.,** Localization of the cell attachment region in types I and II collagens, *Biochem. Biophys. Res. Commun.*, 72, 426, 1976.
16. **Engvall, E. and Ruoslahti, E.,** Binding of soluble form of fibroblast surface protein, fibronectin, to collagen, *Int. J. Cancer*, 20, 1, 1977.
17. **Engvall, E., Ruoslahti, E., and Miller, E. J.,** Affinity of fibronectin to collagen of different genetic types and to fibrinogen, *J. Exp. Med.*, 147, 1584, 1978.
18. **Jilek, F. and Hörmann, H.,** Cold-insoluble globulin (fibronectin). IV. Affinity to soluble collagen of various types, *Hoppe-Seylers Z. Physiol. Chem.*, 359, 247, 1978.
19. **Dessau, W., Adelmann, B. C., Timpl, R., and Martin, G. R.,** Identification of the sites in collagen α-chains that bind serum anti-gelatin factor (cold-insoluble globulin), *Biochem. J.*, 169, 55, 1978.
20. **Vuento, M. and Vaheri, A.,** Dissociation of fibronectin from gelatin-agarose by amino compounds, *Biochem. J.*, 175, 333, 1978.
21. **Klebe, R. J., Bentley, K. L., Sasser, P. J., and Schoen, R. C.,** Elution of fibronectin from collagen with chaotropic agents, *Exp. Cell Res.*, 130, 111, 1980.
22. **Ruoslahti, E., Vuento, M., and Engvall, E.,** Interaction of fibronectin with antibodies and collagen in radioimmunoassay, *Biochim. Biophys. Acta*, 534, 210, 1978.
23. **Zardi, L., Siri, A., Carnemolla, B., Cosulich, E., Viale, G., and Santi, L.,** A simplified procedure for the preparation of antibodies to serum fibronectin, *J. Immunol. Methods*, 34, 155, 1981.
24. **Stathakis, N. E. and Mosesson, M. W.,** Interactions among heparin, cold-insoluble globulin, and fibrinogen in formation of the heparin-precipitable fraction of plasma, *J. Clin. Invest.*, 60, 855, 1977.
25. **Miekka, S. I., Dodd, R., Menache, D., and Ingham, K. C.,** An improved method for isolation of opsonically active plasma fibronectin (Abstract), *Fed. Proc. Fed. Am. Soc. Exp. Biol.*, 39, 1676, 1980.
26. **Jakoby, W. B. and Wilchek, M.,** Affinity techniques, *Methods Enzymol.*, 34, 1, 1974.
27. **Fujikawa, K., Thompson, A. R., Legaz, M. E., Meyer, R. G., and Davie, E. W.,** Isolation and characterization of bovine factor IX (Christmas factor), *Biochemistry*, 12, 4938, 1973.
28. **Lowry, O. H., Rosebrough, N. J., Farr, A. L., and Randall, R. J.,** Protein measurement with the Folin phenol reagent, *J. Biol. Chem.*, 193, 265, 1951.
29. **Alexander, S. S., Colonna, G., and Edelhoch, H.,** The structure and stability of human plasma cold-insoluble globulin, *J. Biol. Chem.*, 254, 1501, 1979.
30. **Yamada, K. M. and Weston, J. A.,** Isolation of a major cell surface glycoprotein from fibroblasts, *Proc. Natl. Acad. Sci. U.S.A.*, 71, 3492, 1974.
31. **Yamada, K. M., Yamada, S. S., and Pastan, I.,** The major cell surface glycoprotein of chick embryo fibroblasts is an agglutinin, *Proc. Natl. Acad. Sci. U.S.A.*, 72, 3158, 1975.
32. **Yamada, K. M., Schlesinger, D. H., Kennedy, D. W., and Pastan, I.,** Characterization of a major fibroblast cell surface glycoprotein, *Biochemistry*, 16, 5552, 1977.
33. **Ali, I. U., Mautner, V., Lanza, R., and Hynes, R. O.,** Restoration of normal morphology, adhesion and cytoskeleton in transformed cells by addition of a transformation-sensitive surface protein, *Cell*, 11, 115, 1977.

34. **Fukuda, M. and Hakomori, S.,** Proteolytic and chemical fragmentation of galactoprotein a, a major transformation-sensitive glycoprotein released from hamster embryo fibroblasts, *J. Biol. Chem.*, 254, 5442, 1979.
35. **Wagner, D. D. and Hynes, R. O.,** Domain structure of fibronectin and its relation to function, *J. Biol. Chem.*, 254, 6746, 1979.
36. **Carter, W. G. and Hakomori, S.,** Isolation and partial characterization of "galactoprotein a" (LETS) and "galactoprotein b" from hamster embryo fibroblasts, *Biochem. Biophys. Res. Commun.*, 76, 299, 1977.
37. **Yamada, K. M., Yamada, S. S., and Pastan, I.,** Quantitation of a transformation-sensitive, adhesive cell surface glycoprotein: decrease on several untransformed permanent cell lines, *J. Cell. Biol.*, 74, 649, 1977.
38. **Vogt, P. K.,** Focus assay of Rous sarcoma virus, in *Fundamental Techniques in Virology*, Habel, K. and Salzman, N. P., Eds., Academic Press, New York, 1969, 198.
39. **McDonald, J.,** Personal communication, 1980.

Chapter 9

LAMININ, A GLYCOPROTEIN FROM BASEMENT MEMBRANES

Jean-Michel Foidart, Rupert Timpl, Heinz Furthmayr, and George R. Martin

TABLE OF CONTENTS

ABSTRACT

Here we describe the methods used in isolating laminin from tissues and cells in culture. Laminin is a large protein concentrated in basement membranes. It mediates the attachment of epithelial and endothelial cells.

I. INTRODUCTION

Basement membranes are extracellular matrices that separate epithelial, endothelial, and muscle cells from underlying connective tissues.[1] They play an important role in cell differentiation, in the establishment of tissue architecture, and as a barrier to the passage of macromolecules. Basement membranes are composed both of collagenous[2,3] and noncollagenous glycoproteins[4] and proteoglycans.[5,6] The biochemical characterization of the noncollagenous proteins has been hampered by not only their limited solubility in aqueous solvents due to their cross-linking, but also by difficulties in obtaining large amounts of basement membranes separate from other tissue elements. A transplantable mouse tumor, the Engelbreth-Holm-Swarm (EHS) sarcoma produces an extracellular matrix of basement membranes.[2] Using this tumor, substantial amounts of noncollagenous proteins can be extracted from the extracellular matrix without resorting to use of proteolytic enzymes or to extraction with chaotropic chemicals commonly used in the isolation of basement membranes components from other sources. A major constituent in these extracts was isolated and found to be a large, noncollagenous glycoprotein termed laminin.[4]

Laminin accounts for approximately half of the total protein in the matrix of the EHS sarcoma. By immunofluorescence, normal tissue basement membranes react with antibody to laminin prepared from the EHS tumor, indicating that laminin or an immunologically related protein is a major component of authentic basement membranes.[7] Laminin is also synthesized by a variety of cultured epithelial, endothelial, smooth muscle, and tumor cells,[7] and milligram quantities have been prepared from such cultures.[8]

II. PURIFICATION OF LAMININ FROM EHS SARCOMA

The EHS tumor is grown either in a s.c. or i.m. site in C57 BL mice and harvested when it reaches a weight of 3 to 6 g. Approximately 1 kg of tumor tissue is homogenized in 3 ℓ of cold 3.4 *M* NaCl, 0.05 *M* Tris-HCl, pH 7.4, containing protease inhibitors (0.01 *M* EDTA, 50 μg/mℓ *p*-hydroxymercuribenzoate; 50 μg/mℓ phenylmethylsulfonyl fluoride) and centrifuged (4°C, 20 min, 10,000 × g) to remove serum and soluble cellular proteins. The tissue is then washed twice. Afterwards the residue is homogenized in 1.5 ℓ of 0.5 *M* NaCl, 0.05 *M* Tris-HCl, pH 7.4, containing the same protease inhibitors as above and extracted overnight by stirring at 4°C (see Figure 1). After centrifugation (4°C, 20 min, 10,000 × g), the residue is extracted two more times each overnight with additional buffer. Since migration of the 200,000- and 400,000-dalton chains of laminin is distinct, the distribution of laminin during purification can be monitored by electrophoresis of the reduced protein. The first extract contains most of the soluble laminin, but is contaminated with serum and other proteins. Subsequent extracts contain greater proportions of laminin, but the laminin in these extracts tends to aggregate and precipitate. The NaCl concentration of the extract is increased to 1.7 *M* to precipitate any Type IV collagen present which is subsequently removed by centrifugation (see Figure 1). The supernatant solution containing laminin is dialyzed against 2 *M* urea, 0.05 *M* Tris-HCl, pH 8.6,

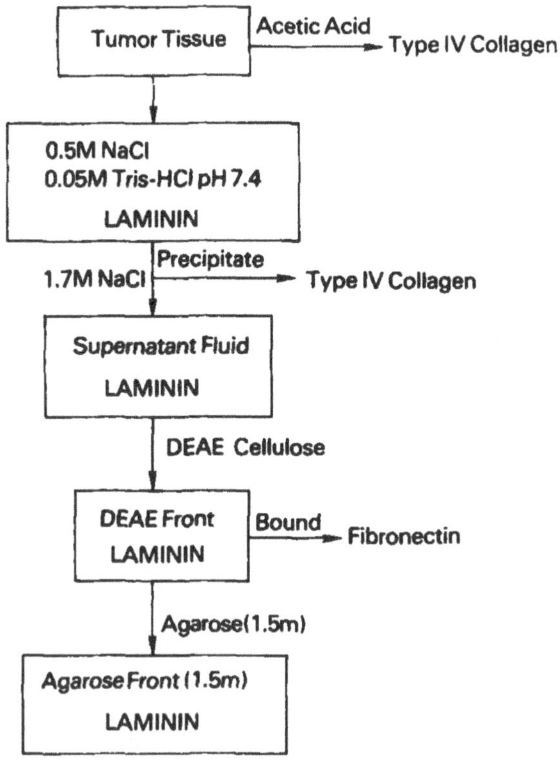

FIGURE 1. Scheme for purification of laminin from the EHS sarcoma.

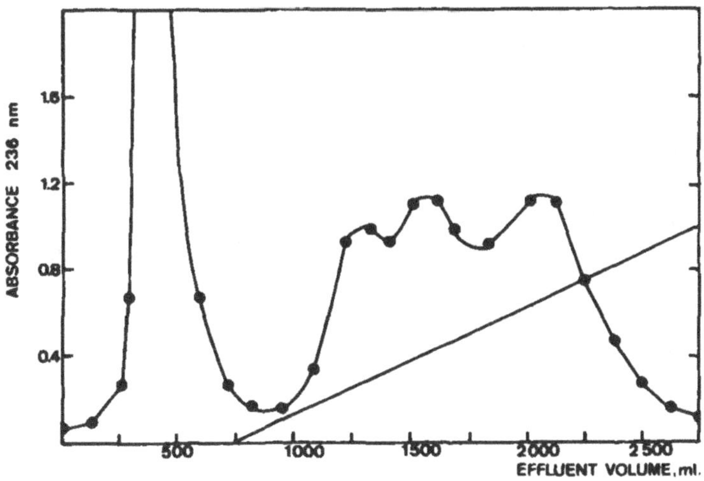

FIGURE 2. Isolation of laminin from a neutral salt extract of the EHS sarcoma by DEAE-cellulose column chromatography. The DEAE-cellulose column (2.5 × 25 cm) is equilibrated at 4° in 0.05 *M* Tris-HCl, pH 8.6, 2 *M* urea. A total of 200 mℓ of a salt extract from 100 g of EHS sarcoma dialyzed against the same buffer was applied to the column. Protein bound to the column was eluted with a linear NaCl gradient from 0 to 0.25 *M* NaCl.

containing 0.25 *M* NaCl and chromatographed on a column of DEAE-cellulose (2.5 × 25 cm) equilibrated with the same solvent at 4°C. The fractions containing laminin, which is unbound, are dialyzed against 2 *M* urea, 0.05 *M* Tris-HCl, pH 8.6, without NaCl and chromatographed on a second column of DEAE-cellulose (2.5 × 25 cm) equilibrated with this solvent at 4°C. Under these conditions, most of the laminin does not bind to the column, while fibronectin, some laminin, and other tumor proteins are bound (see Figure 2). The unbound material is concentrated by ultrafiltration using an Amicon (Diaflo) filter (XM 100 A) to a volume of about 50 mℓ (with only a slight loss of material due to precipitation on the filter) and chromatographed on an agarose (A 1.5 m) column (3.5 × 140 cm) equilibrated in and eluted with 1 *M* $CaCl_2$, 0.05 *M* Tris-HCl, pH 7.4. Laminin elutes in the void volume of the column and is dialyzed against 0.4 *M* NaCl, 0.05 *M* Tris-HCl, pH 7.4, and stored at 4°C. The yield of electrophoretically pure laminin is 0.5 to 1 g/kg of wet tumor. Lyophilization should be avoided since laminin will not redissolve after this treatment. Storage at −20°C tends to generate higher aggregates.

III. PURIFICATION OF LAMININ BY AFFINITY CHROMATOGRAPHY

Since laminin binds to heparin and heparin sulfate,[9] chromatography of laminin on heparin-Sepharose® is useful as an additional step in its purification. In this procedure, laminin (10 to 20 mg) in 0.2 *M* ammonium bicarbonate, pH 7.9, is applied to an heparin-Sepharose® column (2 × 12 cm) (Pharmacia Fine Chemicals,® Uppsala) equilibrated with the same solvent. The column binds 40 to 80% of the laminin which can then be eluted with a linear salt gradient from 0 to 0.5 *M* NaCl or in batch by 0.5 *M* NaCl. Laminin has been isolated in this way from the culture media of epithelial and endothelial cells.[9] While the fibronectin produced by these cells also bind to heparin-Sepharose®, it may be separated from laminin by absorption to gelatin-Sepharose®.[10]

Native laminin interacts only poorly with concanavalin A and wheat germ agglutinin in affinity chromatography.[11] In contrast, peptides generated from laminin by limited proteolysis with pepsin or trypsin show an excellent binding (70 to 80%) to these lectins, indicating that steric hindrance and/or aggregation is responsible for the poor interaction of native laminin.

IV. ISOLATION OF LAMININ FRAGMENT P1 FROM A PROTEOLYTIC DIGEST OF HUMAN PLACENTA[12]

Limited proteolysis of laminin with pepsin produces several unique fragments which are resistant to pepsin and can be separated on agarose A 1.5 m.[11-13] The largest fragment, called P1 (Mr = 290,000) accounts for 25% of the protein and has a high cysteine content (120 to 150 residues per 1000) (see Table 1). This fragment probably arises from the region where the four arms in native laminin intersect.[14]

Human placental tissue which is rich in basement membranes can be used to isolate the P1 fragment from laminin after proteolysis with pepsin.[12] In this procedure, placental tissue is homogenized in 3% Triton® X-100, 0.02 *M* EDTA, pH 7.4 (3.3 ℓ/kg), and extracted for 3 hr at 4°C. Insoluble material is collected by filtration through a nylon net, washed extensively with cold water and then with 0.5 *M* acetic acid, and finally homogenized in 0.5 *M* acetic acid (5 ℓ/kg). The homogenate is

Table 1
COMPARATIVE AMINO ACID COMPOSITION OF TYPE IV COLLAGEN, FIBRONECTIN, AND LAMININ (RESIDUES PER 1000)

	Type IV collagen[a]	Human fibronectin[b]	Laminin[c]	P1[c] Fragment of laminin
Cysteine	7	26	30	135
3-Hydroxyproline	6	0	0	0
4-Hydroxyproline	104	0	0	0
Aspartic acid	53	92	109	113
Threonine	34	97	58	56
Serine	57	68	77	81
Glutamate	102	116	122	106
Proline	65	76	53	63
Glycine	280	80	93	118
Alanine	44	43	76	47
Valine	20	81	48	40
Methionine	13	11	14	6
Isoleucine	25	44	42	20
Leucine	52	57	92	54
Tyrosine	9	45	27	28
Phenylalanine	29	27	31	30
Histidine	11	21	24	33
Hydroxylysine	37	0	2	2
Lysine	11	36	52	30
Arginine	32	52	50	38
Tryptophane	nd	28	nd	nd

Note: nd, not determined.

[a] From Reference 3.
[b] From Reference 16.
[c] From References 4 and 12.

incubated with stirring for 20 hr at 6 to 8°C in the presence of pepsin (0.5 g/kg). Insoluble material is removed by centrifugation and protein is precipitated from the supernatant fluid by adding 1.2 *M* NaCl. The precipitate is dissolved in and then dialyzed against 0.2 *M* NaCl, 0.05 *M* Tris-HCl, pH 7.4, at 4°C for 1 day. After centrifugation, the clarified supernatant fluid is subjected to differential salt precipitation by raising the NaCl concentration to 2 *M* and then to 4 *M*.[15] Material precipitating at 2 *M* NaCl consists of collagenous protein and laminin fragments, including P1. The material precipitated at 4 *M* NaCl contains fragment P1 and several other proteins, but is poor in collagen. P1 from either fraction can be separated on a column of DEAE-cellulose under the conditions described in Figure 2. Collagens elute in the front, while P1, the large laminin fragment, is recovered in the NaCl gradient (0.1 to 0.18 *M* NaCl). The material is then chromatographed on a Sepharose® CL-4B in a column (2.5 × 85 cm) equilibrated in 8 *M* urea, 0.05 *M* Tris-HCl, pH 8.6, from which fragment P1 emerges as a distinct peak at $V_E/V_T = 0.58$. As an additional purification step, fragment P1 may be bound to a column (1 × 5 cm) of concanavalin A-Sepharose® 4B equilibrated with 0.15 *M* NaCl, 0.05 *M* Tris-HCl, pH 7.4, containing 1 m*M* $CaCl_2$. Fragment P1 can be eluted with 0.4 *M* α-methyl mannoside. The purified material is finally dialyzed against water and can be lyophilized.

V. AMINO ACID AND CARBOHYDRATE COMPOSITION

The amino acid composition of laminin is shown in Table 1 and is compared to Type IV collagen[2,3] purified from the EHS tumor and to human plasma fibronectin.[16] In contrast to Type IV collagen, laminin does not have a distinctive composition which could be useful for identification or can be used as criteria for judging purity. The cysteine content is moderately elevated (30 residues per 1000), and there are high levels of acidic amino acids (230 residues per 1000). The amino acid composition is quite different from fibronectin, particularly in threonine, proline, alanine, valine, leucine, and lysine contents.[4]

The chains of laminin separated by electrophoresis show an intense reaction with the periodic acid Schiff procedure[4,13] and react with concanavalin A and wheat germ agglutinin, indicating that laminin is a glycoprotein. Carbohydrate accounts for about 15% of total protein mass[11] and includes 6 to 7 % hexoses (glucose, galatose, and mannose in equal amounts), 5 to 6% sialic acids, 2% glucosamine, and 0.4% galactosamine.

VI. BIOCHEMICAL AND BIOPHYSICAL PROPERTIES

Unreduced laminin, when examined by electrophoresis, migrates as a single high molecular weight protein band. Ultracentrifugal analyses show that, in neutral buffers, laminin consists of monomers (850 K), dimers (1500 K), and larger aggregates, while mainly monomeric components are found in the presence of 6 *M* guanidine.[11,14] Such studies suggest that laminin is a large aggregated macromolecule. Two polypeptide chains with apparent molecular weight of 200,000 and 400,000, respectively, are observed on electrophoresis after reduction of disulfide bonds (see Figure 3). However, labeled laminin chains from vascular and corneal endothelial cells[17] or from other cells[18] have been found to be heterogeneous and to electrophorese as closely spaced doublets.

The presence of chains of 200,000 and 400,000 daltons suggest either a monomer-dimer or precursor-product relationship. However, preliminary data indicated that these chains are related, but not identical,[8] and physical and electronmicroscopic studies suggest that they are present in the same molecule.[14] Electronmicroscopic of laminin shows a unique asymmetric structure with a long arm (75 nm) and three short arms (35 nm) attached together (see Figure 4). The long arm could be formed from a 400,000-dalton chain, and each short arm could be formed from a 200,000-dalton chain. Laminin exhibits an elongated shape with low flexibility and compact domains at the extremities of the molecule.[14,19] Circular dichroism measurements demonstrate a distinct conformation of native laminin with 30% α-helix as well as β-structure and random coil segments. The α-helix elements show a sharp thermal transition around 58°C.[14,19]

VII. IMMUNOLOGICAL PROPERTIES

Rabbits, guinea pigs, and sheep immunized by intradermal injections of 0.1 to 0.25 mg of laminin produce antisera with high antibody titers as assessed by gel diffusion, immunoelectrophoresis, radio-immunoassay, or enzyme-immunoassay.[4,7,20] Antisera or affinity purified antibodies show no reaction with other basement membrane antigens: Types IV and V collagen, fibronectin, bullous pemphigoid antigen(s), or basement membrane proteoglycan.

Disulfide bonds are important for maintaining the characteristic antigenic prop-

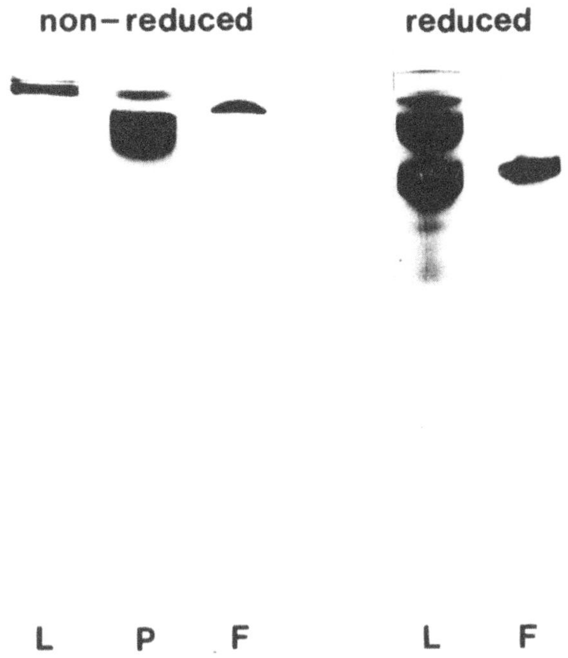

FIGURE 3. Sodium dodecylsulfate polyacrylamide gel electrophoresis of laminin. Electrophoresis was carried out on 5% acrylamide gels in the presence of 1% SDS. The samples were reduced with dithiothreitol (20 m*M*) prior to electrophoresis. L = Laminin; P = Pepsin fragment of laminin; F = Fibronectin.

erties of the protein. Peptide P1 is a highly immunogenic peptide reacting with 70% of the antibodies produced by rabbits against laminin. Minor antigenic determinants shared by the native and reduced laminin have also been identified.[21]

VIII. BIOLOGICAL ACTIVITY

The biological role of laminin is not fully understood. Laminin appears early in embryonic development around the morula stage before the appearance of a recognizable basement membrane.[22] Similarly, laminin appears in embryonic kidney prior to tubule formations.[23] The accumulation of laminin in these sites suggests almost a prematrix template function. Laminin is found mainly in association with basement membrane and has been found by immunohistology in all basement membranes examined.[4,7,21,24,25] In the epidermal-dermal basement membrane[7] and in esophagus basement membranes[7] and glomerular[24] basement, laminin appears by immunoelectron microscopy to be concentrated in the lamina rara zone. In tubular basement membrane, the mesangial matrix, and in Bowmans capsule, laminin appears to be distributed uniformly throughout the basement membrane.[24] Such variations in the distribution of laminin may represent differences in structure and functions among these basement membranes.

Recent studies indicate that laminin promotes the attachment of epithelial and endothelial cells to Type IV collagen,[26,27] the natural substrate for these cells. In these studies, laminin showed no attachment activity for fibroblasts or with other collagens, suggesting that it is specific for these cells and basement membrane collagen.

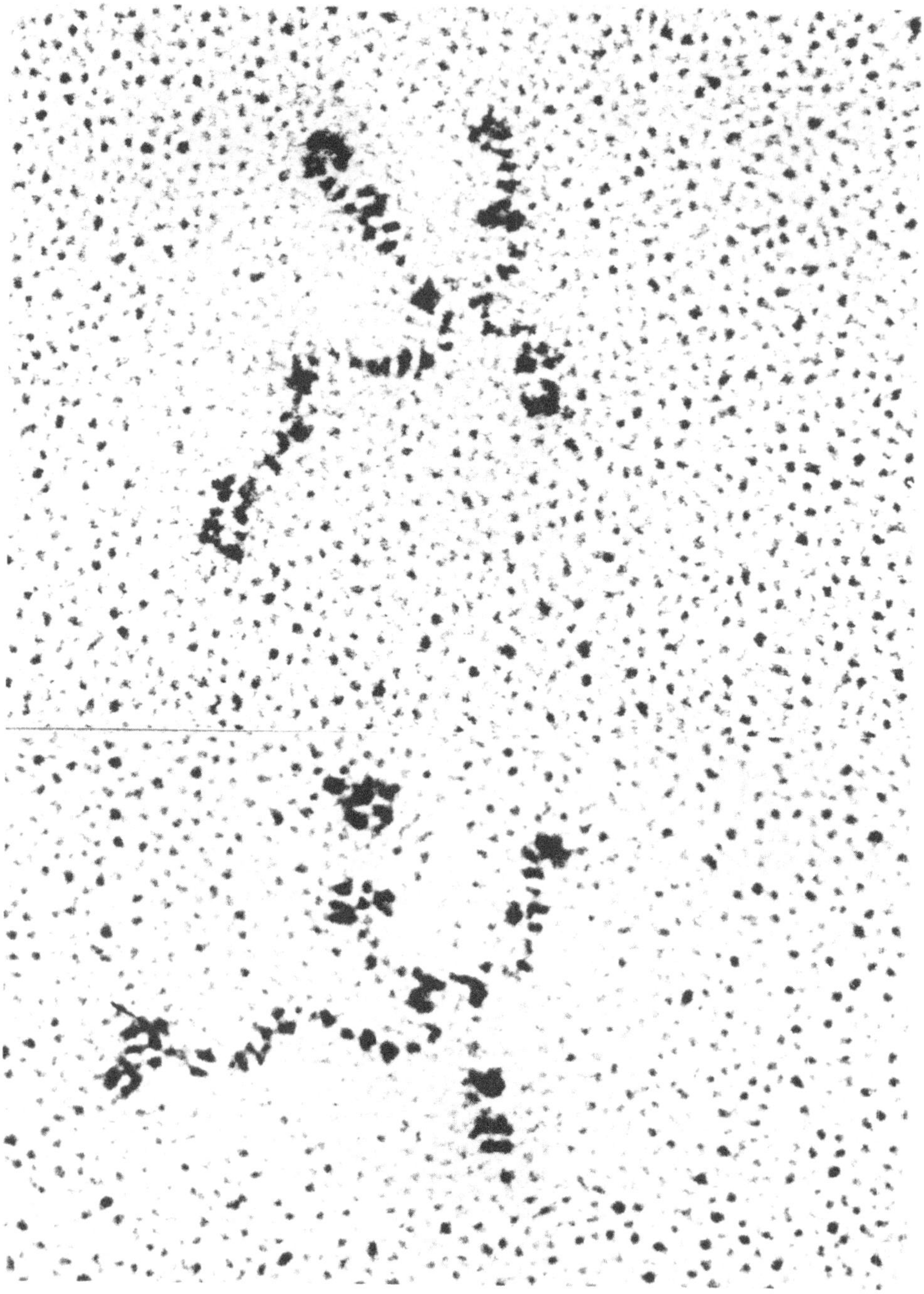

FIGURE 4. Electron micrograph of laminin. A solution of laminin in 70% glycerol was sprayed onto a mica support, and an electron micrograph was obtained after rotary shadowing with platinum-carbon.[14]

Some studies indicate that the synthesis of laminin is restricted to certain cells and varies with cell density and differentiation. For example, laminin is synthesized by epithelial, endothelial, and muscle cells, but not by mesenchymal cells.[7,8,18] The synthesis of laminin by endothelial cells is highest at low cell density,[17] while the synthesis of fibronectin is rather constant at all cell densities. Retinoic acid and cyclic AMP induce the differentiation of teratocarcinoma cells into endodermal cells, and this increases or initiates laminin synthesis by the cells.[28–30] The studies listed here, while preliminary, suggest that laminin is an ubiquitous component of basement membranes with activities affecting the distribution, differentiation, and function of epithelial and endothelial cells.

REFERENCES

1. **Vracko, R.,** Basal lamina scaffold-anatomy and significance for maintenance of orderly tissue structure, *Am J. Pathol.*, 77, 314, 1974.
2. **Orkin, R. W., Gehron, P., McGoodwin, E. B., Martin, G. R., Valentine, T., and Swarm, R. H.,** A murine tumor producing a matrix of basement membrane, *J. Exp. Med.*, 145, 204, 1977.
3. **Timpl, R., Martin, G. R., Bruckner, P., Wick, G., and Wiedemann, H.,** Nature of the collagenous protein in a tumor basement membrane, *Eur. J. Biochem.*, 84, 43, 1978.
4. **Timpl, R., Rohde, H., Gehron Robey, P., Rennard, S. I., Foidart, J. M., and Martin, G. R.,** Laminin — a glycoprotein from basement membranes, *J. Biol. Chem.*, 254, 9933, 1979.
5. **Hassell, J. R., Gehron Robey, P., Barrach, H. J., Wilczek, J., Rennard, S. I., and Martin, G. R.,** Isolation of a heparan sulfate containing proteoglycan from basement membrane, *Proc. Natl. Acad. Sci. U.S.A.*, 77, 4494, 1980.
6. **Kanwar, Y. S. and Farquhar, M. G.,** Presence of heparan sulfate in the glomerular basement membrane, *Proc. Natl. Acad. Sci. U.S.A.*, 76, 1303, 1979.
7. **Foidart, J. M., Bere, E. W., Yaar, M., Rennard, S. I., Gullino, M., Martin, G. R., and Katz, S. I.,** Distribution and immunoelectron microscopic localization of laminin, a noncollagenous basement membrane glycoprotein, *Lab. Invest.*, 42, 336, 1980.
8. **Chung, A. E., Jaffe, R., Freeman, I. L., Vergnes, J. P., Braginski, J. E., and Carlin, B.,** Properties of a basement membrane-related glycoprotein synthesized in culture by a mouse embryonal carcinoma derived cell line, *Cell*, 16, 277, 1979.
9. **Sakashita, S., Engvall, E., and Ruoslahti, E.,** Basement membrane glycoprotein laminin binds to heparin, *FEBS Lett.*, 116, 243, 1980.
10. **Vuento, M. and Vaheri, A.,** Purification of fibronectin from human plasma by affinity chromatography under nondenaturing conditions, *Biochem. J.*, 183, 331, 1979.
11. **Timpl, R., Rohde, H., Ott-Ulbricht, U., Risteli, L., and Bächinger, H. P.,** Chemical characterization of laminin, a major glycoprotein of basement membranes in Glycoconjugates, *Proc. 5th Int. Symp. Kiel*, Schauer, R., Boer, P., Buddecke, E., Kramer, M. F., Vliegenthart, J. F. G., and Wiegandt, H., Eds., Georg Thieme, Stuttgart, 1979, 145.
12. **Risteli, L. and Timpl, R.** Isolation and characterization of pepsin fragments of laminin from human placental and renal basement membranes, *Biochem. J.*, 193, 749, 1981.
13. **Timpl, R., Rohde, H., Wick, G., Gehron Robey, P., Rennard, S. I., Foidart, J. M., and Martin, G. R.,** Characterization of laminin, a major glycoprotein of basement membrane, in *Biochemistry of Normal and Pathological Connective Tissues*, Vol. 2, Robert A. M. and Robert, L., Eds., Centre National de la Recherche Scientifique, Paris, 1980, 225.
14. **Engel, J., Odermatt, E., Engel, A., Madri, J. A., Furthmayr, H., Rohde, H., and Timpl, R.** Domain organizations and flexibility of laminin and fibronectin, two multifunctional proteins of the extracellular matrix, *J. Mol. Biol.*, 150, 97, 1981.
15. **Risteli, J., Bachinger, H. P., Engel, J., Furthmayr, H., and Timpl, R.** 7-S Collagen: characterization of an unusual basement membrane structure, *Eur. J. Biochem.*, 108, 239, 1980.
16. **Mosesson, M. W., Chen, A. B., and Huseby, R. M.,** The cold-insoluble globulin of human plasma: studies of its essential structural features, *Biochem. Biophys. Acta*, 386, 509, 1975.
17. **Gospadorowicz, D., Greenburg, G., Foidart, J. M., and Savion, N.,** The production and localization of laminin in cultured vascular and corneal endothelial cells, *J. Cell. Physiol.*, 107, 171, 1981.
18. **Alitalo, K., Kurkinen, M., Vaheri, A., Krieg, T., and Timpl, R.,** Extracellular matrix components synthesized by human amniotic epithelial cells in culture, *Cell*, 19, 1053, 1980.
19. **Timpl, R., Rohde, H., Risteli, L., Ott, U., Gehron Robey, P., and Martin, G. R.,** Laminin, *Methods Enzymol.*, in press.

20. **Rennard, S. I., Berg, R., Martin, G. R., Foidart, J. M., and Gehron Robey, P.,** Enzyme-linked immunoassay (ELISA) for connective tissue components, *Anal. Biochem.*, 104, 205, 1980.
21. **Rohde, H., Wick, G., and Timpl, R.,** Immunochemical characterization of the basement membrane glycoprotein laminin, *Eur. J. Biochem.*, 102, 195, 1979.
22. **Leivo, I., Vaheri, A., Timpl, R., and Wartiovaara, J.,** Appearance and distribution of collagens and laminin in the early mouse embryo, *Dev. Biol.*, 76, 100, 1980.
23. **Ekblom, P., Alitalo, K., Vaheri, A., Timpl, R., and Saxen, L.,** Induction of a basement membrane glycoprotein in embryonic kidney, *Proc. Natl. Acad. Sci. U.S.A.*, 77, 485, 1980.
24. **Madri, J. A., Roll, J., Furthmayr, H., and Foidart, J. M.,** Ultrastructural localization of fibronectin and laminin in the basement membranes of the murine kidney, *J. Cell Biol.*, 86, 682, 1980.
25. **Scheinman, J. I., Foidart, J. M., Gehron Robey, P., Fish, A. J., and Michael, A. F.,** The immunohistology of glomerular antigens. IV. Laminin, a defined noncollagen basement membrane glycoprotein, *Clin. Immunol. Immunopathol.*, 15, 175, 1980.
26. **Terranova, V. P., Rohrbach, D. H., and Martin, G. R.,** Role of laminin in the attachment of PAM 212 (epithelial) cells to basement membrane collagen, *Cell*, 22, 719, 1980.
27. **Hogan, B. L.,** High molecular weight extracellular proteins synthesized by endoderm cells derived from mouse teratocarcinoma cells and normal extraembryonic membranes, *Dev. Biol.*, 76, 275, 1980.
28. **Howe, C. C. and Solter, D.,** Identification of noncollagenous basement membrane glycopeptides synthesized by mouse parietal endoderm and an endodermal cell line, *Dev. Biol.*, 77, 480, 1980.
29. **Strickland, S., Smith, K. K., and Marotti, K. R.,** Hormonal induction of differentiation in teratocarcinoma stem cells: generation of parietal endoderm by retinoic acid and dibutyryl cyclic AMP, *Cell*, 21, 347, 1980.
30. **Salomon, D., Liotta, L. A., Rennard, S. I., Foidart, J. M., and Yaar, M.,** Synthesis of basement membrane components by mouse embryonal carcinoma derived endoderm cells in serum-free, hormone supplemented medium, *Collagen Relat. Res.*, in press.

Chapter 10

ISOLATION AND PROPERTIES OF CHONDRONECTIN, THE CHONDROCYTE ATTACHMENT FACTOR

A. Tyl Hewitt, Hugh H. Varner, and George R. Martin

TABLE OF CONTENTS

I. INTRODUCTION

Cartilage is a unique tissue containing chondrocytes surrounded by a large amount of matrix composed of Type II (cartilage) collagen, cartilage-specific proteoglycans, link protein, and hyaluronic acid. Recent studies which will be reviewed here indicate that, in addition, cartilage contains a unique glycoprotein, chondronectin. Chondronectin is also present in serum, vitreous humor, chondrocyte-conditioned medium, and extracts of cartilage, but not in extracts of other tissues.[1]

Our studies on chondronectin parallel detailed investigations (reviewed Vol. II, Chapter 6) which showed that fibroblasts and certain other cells require serum for attachment to collagen. Fibronectin, a large glycoprotein, was found to be the factor in serum which is active in promoting the attachment of fibroblasts to collagen.[2,3] Further, these studies showed that serum fibronectin binds to the collagen first and then the cells interact with the collagen-fibronectin complex.[2] Although fibronectin is present in serum and in many tissues, it is not found in mature cartilage.[4,5]

When the attachment of chondrocytes to collagen was studied using in vitro assays analogous to those performed with fibroblasts, it was observed that serum increased the number of attached cells as well as the rate at which they attached in a concentration dependent manner. However, purified fibronectin could not substitute for whole serum. From these observations, a chondrocyte-specific attachment protein, chondronectin, was postulated[1] and subsequently isolated.[6]

II. ASSAY FOR CHONDROCYTE ATTACHMENT

The assays for the chondrocyte attachment factor are based on the procedure developed by Klebe.[2] In these studies, polystyrene petri dishes (35 mm) were coated with collagen by evaporating 1 mℓ of a 10 μg/mℓ solution of collagen in 0.1% acetic acid. Since chondrocytes do not attach to uncoated polystyrene dishes, these were used rather than tissue culture dishes. Type II collagen-coated dishes were routinely used, although experiments on the specificity of attachment were carried out on other collagenous substrates. Type I collagen was purified from lathyritic rat skin,[7] Type II collagen was purified from a rat chondrosarcoma,[8] Type III was purified from fetal calf skin,[9] Type IV was purified from a mouse basement membrane tumor,[10] and Type V was purified from human placenta.[11] Dishes coated with a dried collagen film can be stored at room temperature for several months. Fibronectin and fibronectin-free serum were prepared by adsorbing serum on a collagen-Sepharose® affinity column.[12,13]

Chondrocytes were obtained by digesting chick sternal cartilage from 13- to 17-day-old embryos with 0.4% collagenase (Worthington, CLS II) in Hank's balanced salt solution at 37°C for 3 hr. Chinese hamster ovary (CHO) cells, a fibroblastic cell line requiring fibronectin for attachment to collagen, were used for the comparison of attachment characteristics. These cells were maintained in Dulbeccos Modified Eagles (DME) medium supplemented with 10% fetal bovine serum (FBS) at 37°C in a 5% CO_2-95% air atmosphere and were released from culture flasks with 0.1% trypsin-0.1% EDTA in phosphate buffered saline (PBS).

To measure the attachment of the cells, collagen-coated dishes were first preincubated for 60 min with 1 mℓ of Hams F12 medium containing 200 μg/mℓ of bovine serum albumin plus serum or serum fractions to be tested for attachment activity. Approximately 1.5×10^5 cells in 0.1 mℓ of F12 were then added to each dish. After various times of incubation, unattached cells were removed by rinsing three times with 1 mℓ aliquots of PBS. The attached cells were released from the dish with 0.1%

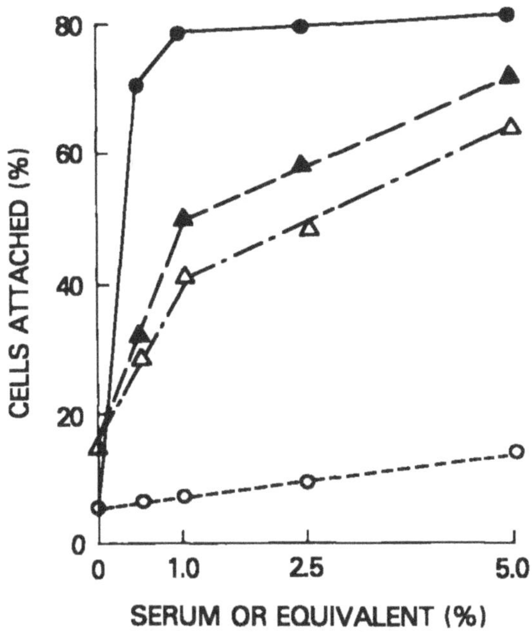

FIGURE 1. Effect of complete serum (filled symbols) and fibronectin-free serum (open symbols) on the attachment of chondrocytes (triangles) and CHO cells (circles) to Type II collagen. Assays were performed as described in the text. Serum stimulates the attachment of both cell types. CHO cells do not attach in the absence of fibronectin, but chondrocyte attachment is unaffected.

trypsin-0.1% EDTA in PBS. Both the number of attached and unattached cells were measured using an electronic particle counter (Coulter Electronics® Inc.), and the ratio of attached to total cells was calculated.

Attachment of chondrocytes to collagen is low in the absence of serum (see Figure 1). Serum stimulates the attachment of chondrocytes in a concentration dependent manner. With 5% serum, approximately 70% of the cells attached. Similar results were obtained with CHO cells except that the rate of attachment of these cells is more rapid[1] and lower levels of serum promote maximal attachment. Fibronectin-free serum stimulates attachment of chondrocytes, but not of CHO cells (see Figure 1), and purified fibronectin does not promote chondrocyte attachment.[1] These data indicate that a serum factor other than fibronectin (i.e., chondronectin) stimulates chondrocyte attachment.

III. ISOLATION OF CHONDRONECTIN

Chondronectin activity is found in chicken, human, and fetal bovine serum. Since preliminary studies indicated that chondronectin did not bind effectively to collagen-Sepharose® affinity columns, alternate methods of purification were used. Isolation of chondronectin from these sera was achieved using the steps outlined in Figure 2. All procedures in the purification were carried out at 4°C in the presence of sodium azide (0.02%) and the protease inhibitors, 6-aminohexanoic acid (25 m*M*) and phenylmethylsulfonyl fluoride (PMSF, 0.2 m*M*).

More than 70% of the chondronectin activity was precipitated from serum with

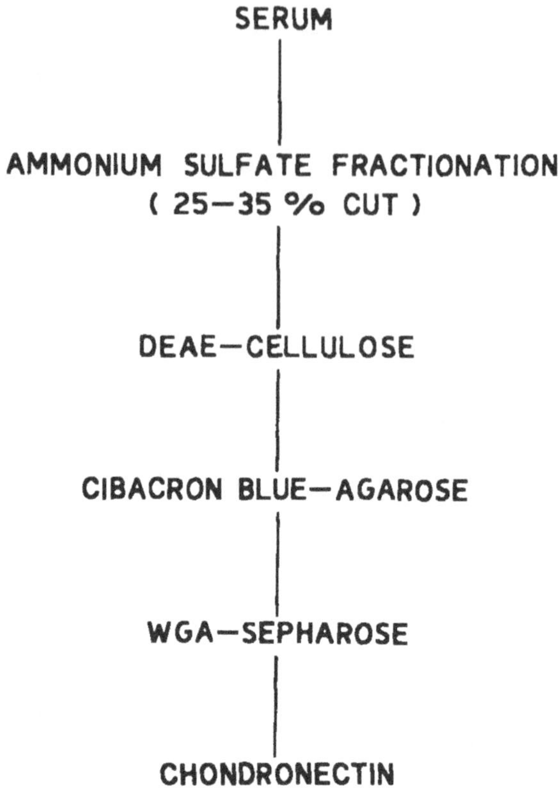

FIGURE 2. Procedure for the isolation of chondronectin from serum. Serum fraction was sequentially chromatographed on columns of DEAE-cellulose, Cibacron® Blue-agarose, and WGA-Sepharose®. See text for a description.

ammonium sulfate between 25 and 35% saturation. The precipitate was dissolved in Tris-buffered saline (TBS; 50 m*M* Tris-HCl, 0.15 *M* NaCl, pH 7.4), dialyzed against 50 m*M* Tris-HCl, pH 7.4, and chromatographed on a column of DEAE-cellulose (5 × 20 cm). Bound protein was eluted with a 2-ℓ linear gradient of 0 to 0.2 *M* NaCl. Aliquots of fractions were dialyzed against F12 medium and assayed for the ability to enhance attachment of either chondrocytes or CHO cells. Figure 3 shows the results of one of the early experiments using a steeper (0 to 1 *M* NaCl) gradient. The procedure described here gives a similar profile, but provides better separation of the two attachment activities.

The fractions containing chondronectin activity were pooled and applied to a column (2.6 × 20 cm) of immobilized Cibacron® blue F3GA (Pierce Chemicals®). Most of the protein was adsorbed, whereas all of the chondrocyte attachment activity passed through the column. Chromatography on a column (1.6 × 10 cm) of wheat germ agglutinin (WGA)-Sepharose® (Pharmacia®) equilibrated in TBS was used as the final step in the purification. Chondronectin which bound to the column was eluted with N-acetyl–D–glucosamine (100 mg/mℓ) in TBS.

Based on its mobility on SDS gels (see Figure 4), its elution position upon gel filtration (not shown), and its adsorption on a WGA-Sepharose® column, chondronectin is a 180,000-dalton glycoprotein. Because of the apparent change in size upon reduction (see Figure 4), the protein contains smaller disulfide-linked chains, suggesting that it is a dimeric molecule.

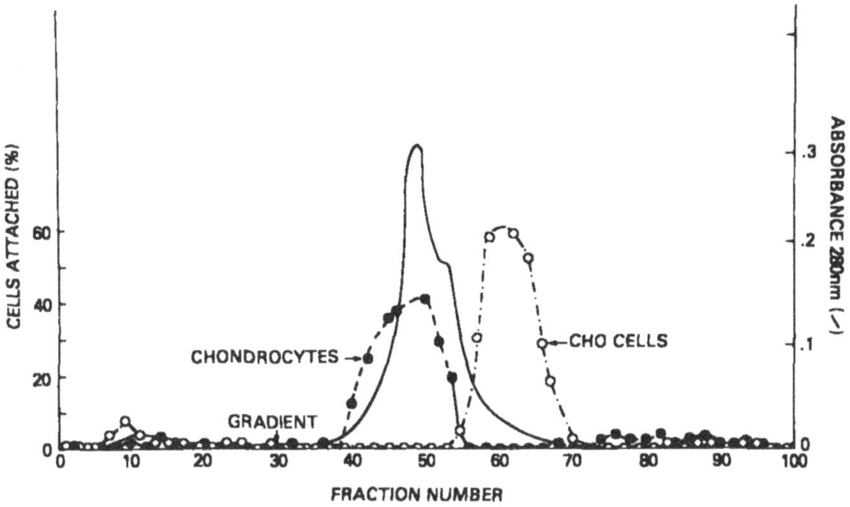

FIGURE 3. Fractionation of the 25 to 35% saturated ammonium sulfate insoluble fraction of fetal bovine serum by DEAE-cellulose ion-exchange chromatography using a 0 to 1 *M* NaCl gradient in 50 m*M* Tris-HCl, pH 7.4, in a total volume of 400 mℓ. Fractions of 5 mℓ were collected, and 1-mℓ aliquots were dialyzed against F12 and tested for attachment activity. Superimposed on the protein elution profile is the pattern of attachment activity for CHO cells and for chondrocytes. The procedure used in this experiment is slightly modified from that presented in Section III, although a similar pattern was obtained using chicken serum and a 0 to 0.2 *M* NaCl gradient. (From Hewitt, A. T., Kleinman, H. K., Pennypacker, J. P., and Martin, G. R., *Proc. Natl. Acad. Sci. U.S.A.*, 77, 385, 1980. With permission.)

Table 1
CHARACTERISTICS OF CHONDRONECTIN AND FIBRONECTIN

Property	Chrondronectin	Fibronectin
Molecular weight (M_r)		
Native[a]	180,000	440,000
Subunit[b]	70,000	220,000
Amount required for attachment activity	5–50 ng/mℓ	1–5 μg/mℓ
Heat lability ($T_{1/2}$)[c]	50–52°C	57–60°C
Concentration in serum	Less than 20 μg/mℓ	300 μg/mℓ
Binds to heparin	Yes	Yes
Lectins bound	Wheat germ agglutinin	Concanavalin A

[a] Determined by gel filtration and SDS-PAGE.
[b] Determined by SDS-PAGE in the presence of dithiothreitol.
[c] Temperature at which attachment factor loses 50% of attachment-enhancing activity upon preincubation for 20 min.

IV. BIOLOGICAL PROPERTIES OF CHONDRONECTIN

In the presence of chondronectin, chondrocytes show a distinct preference for Type II collagen, the cartilage-specific collagen, over Types I, III, IV, and V. Purified chondronectin stimulates chondrocyte attachment to Type II collagen at levels as low as 5 to 10 ng/mℓ.

Chondronectin differs from fibronectin in several respects as outlined in Table 1.

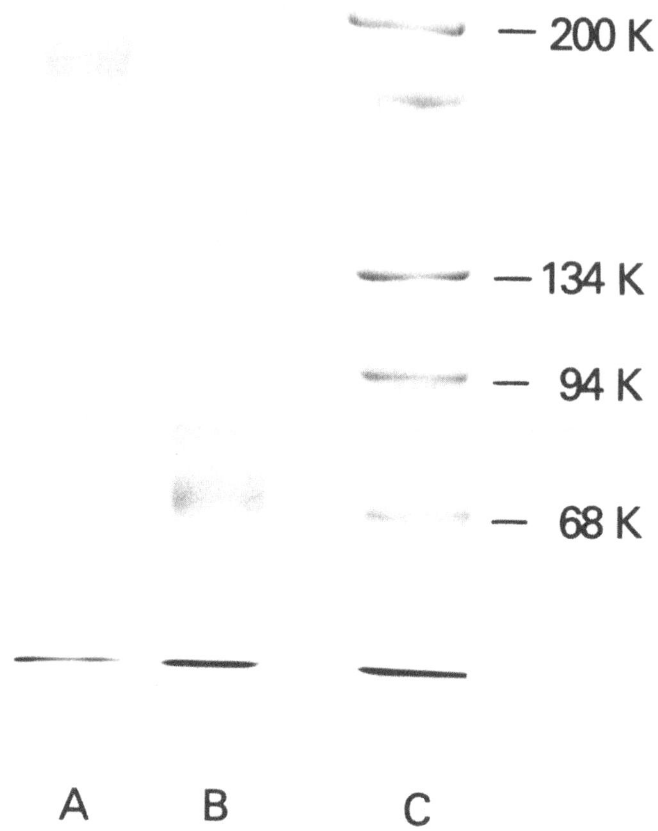

FIGURE 4. SDS gel electrophoresis of chondronectin isolated by affinity chromatography on WGA-Sepharose® (Track A: unreduced; Track B: reduced). Molecular weight standards are in Track C.

These two glycoproteins differ in size, the amount required for enhancing attachment, the quantity found in serum, and heat lability. In addition, fibronectin does not stimulate chondrocyte attachment nor does chondronectin stimulate fibroblast attachment. These proteins are also immunologically distinct[6] as judged by enzyme-linked immunoassay (ELISA).[14]

The events involved in the chondronectin-mediated attachment of chondrocytes to collagen are not clear. Binding of chondronectin to collagen is slow in comparison to that of fibronectin. It is likely that other substances, such as proteoglycans, participate in enhancing or stabilizing the chondronectin-collagen interaction.

Fibronectin causes chondrocytes to lose their characteristic phenotype in vitro. When chondrocytes are cultured in the presence of fibronectin-free serum, they remain differentiated,[15] presumably because chondronectin interacts more efficiently with the cells under these conditions. The isolation of chondronectin should now make it possible to determine what influence it has on the metabolic properties of chondrocytes.

V. SUMMARY

We have described the characteristics of chondrocyte attachment and the isolation of a chondrocyte attachment factor, chondronectin, from serum using DEAE-cellulose ion exchange chromatography followed by chromatography on Cibacron® blue and affinity chromatography on WGA-Sepharose®. Chondronectin is a 180,000-dalton glycoprotein which at low concentrations mediates the attachment of chondrocytes to collagen. As well as differing from fibronectin on the basis of size, specificity, and heat lability, chondronectin and fibronectin are also immunologically distinct.

ACKNOWLEDGMENTS

The authors wish to thank Ms. Charlotte M. Wilkes for the preparation of gels, Ms. Jessica Dowe and Mr. Terry Robbins for their assistance, Ms. Denise Haller for typing the manuscript, and Dr. Hynda K. Kleinman for her critical suggestions during the course of this project. Hugh H. Varner is a National Institutes of Health Postdoctoral Fellow (Award 5F32DE05156-02).

REFERENCES

1. **Hewitt, A. T., Kleinman, H. K., Pennypacker, J. P., and Martin, G. R.,** Identification of an adhesion factor for chondrocytes, *Proc. Natl. Acad. Sci. U.S.A.*, 77, 385, 1980.
2. **Klebe, R. J.,** Isolation of a collagen-dependent cell attachment factor, *Nature*, 250, 248, 1974.
3. **Pearlstein, E.,** Plasma membrane glycoprotein which mediates adhesion of fibroblasts to collagen, *Nature*, 262, 497, 1976.
4. **Linder, E., Vaheri, A., Ruoslahti, E., and Wartiovaara, J.,** Distribution of fibroblast surface antigen in the developing chick embryo, *J. Exp. Med.*, 142, 41, 1975.
5. **Dessau, W., Sasse, J., Timpl, R., Jilek, F., and von der Mark, K.,** Synthesis and extracellular deposition of fibronectin in chondrocyte culture, *J. Cell Biol.*, 79, 342, 1978.
6. **Hewitt, A. T., Varner, H. H., Silver, M. H., Dessau, W., Wilkes, C. M., and Martin, G. R.,** Isolation and partial characterization of chondronectin, an attachment factor for chondrocytes, *J. Biol. Chem.*, in press, 1982.
7. **Bornstein, P. and Piez, K. A.,** The nature of the intramolecular cross-links in collagen. The separation and characterization of peptides from the cross-link region of rat skin collagen, *Biochemistry*, 5, 3460, 1966.
8. **Smith, B. P., Martin, G. R., Miller, E. J., Dorfman, A., and Swarm, R.,** Nature of the collagen synthesized by a transplanted chondrosarcoma, *Arch. Biochem. Biophys.*, 166, 181, 1975.
9. **Epstein, E.,** $[\alpha 1(III)]_3$ human skin collagen released by pepsin digestion and preponderance in fetal life, *J. Biol. Chem.*, 249, 3225, 1974.
10. **Orkin, R. W., Gehron, P., McGoodwin, E., Martin, G. R., Valentine, T., and Swarm, R.,** A unique tumor producing matrix of basement membrane, *J. Exp. Med.*, 145, 209, 1977.
11. **Burgeson, R. E., El Adli, A. F., Kaitila, I. I., and Hollister, D. W.,** Fetal membrane collagens: identification of two new collagen alpha chains, *Proc. Natl. Acad. Sci. U.S.A.*, 73, 2579, 1976.
12. **Hopper, D. E., Adelmann, B. C., Gentner, G., and Gay, S.,** Recognition by guinea-pig peritoneal exudate cells of conformationally different states of the collagen molecule, *Immunology*, 30, 249, 1976.
13. **Engvall, E., and Ruoslahti, E.,** Binding of soluble form of fibroblast surface protein, fibronectin, to collagen, *Int. J. Cancer*, 20, 1, 1977.
14. **Rennard, S. I., Berg, R., Martin, G. R., Foidart, J.-M., and Robey, P. G.,** Enzyme-linked immunoassay (ELISA) for connective tissue components, *Anal. Biochem.*, 104, 205, 1980.
15. **Pennypacker, J. P., Hassell, J. R., Yamada, K. M., and Pratt, R. M.,** The influence of an adhesive cell surface protein on chondrogenic expression *in vitro*, *Exp. Cell Res.*, 121, 411, 1979.

Chapter 11

IMMUNIZATION PROCEDURES, ISOLATION BY AFFINITY CHROMATOGRAPHY, AND SEROLOGICAL AND IMMUNOCHEMICAL CHARACTERIZATION OF COLLAGEN SPECIFIC ANTIBODIES

Heinz Furthmayr

TABLE OF CONTENTS

I. INTRODUCTION

The different types of collagen are the major protein components of the extracellular connective tissue. Undoubtedly some of the collagens fulfill a predominantly structural role to form tendons, ligaments, and the matrix for mineralized bone. Others presumably play a more subtle structural role in concert with noncollageneous glycoproteins in mediating cell-cell interactions at the organ level to build basement membrane barriers at the interface between mesenchyme and epithelia and to form specialized structures, such as the cornea or heart valves. An important role of the extracellular matrix for organ development, cell behavior, maintenance of cellular differentiation, and for pathobiological events in essentially all tissues is widely recognized now, yet there is a lack of understanding of the basic mechanisms underlying these processes.

Immunochemical studies on Type I collagen in the past have provided the necessary knowledge to produce specific antibodies which later has been applied to other collagen types as well. Collagen type-specific antibodies as well as antibodies to particular portions of the native (triple-helical) or denatured (heat denaturation leads to dissociation of the three polypeptide chains) molecule have provided extremely useful tools, particularly for studies related to biological questions. Since different antibody specificities are required for the various applications, in the following section the basic immunochemical features of collagen will be reviewed.[1,2]

A. Basic Immunochemistry of Collagens

Landsteiner in 1936[3] still considered collagen to be nonimmunogenic, and it was not until 1954 that studies by Watson et al.[4] demonstrated that antibodies could be

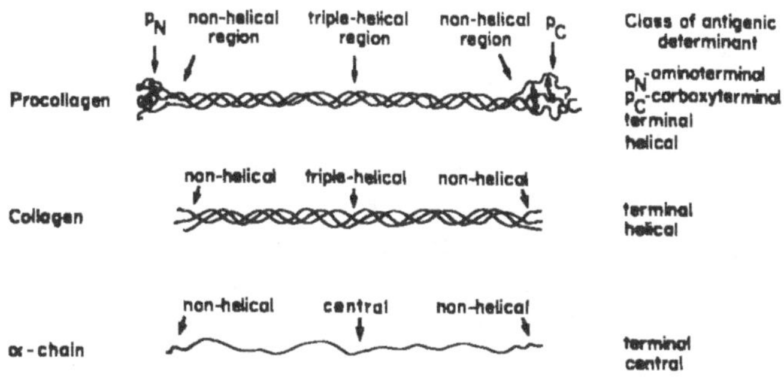

FIGURE 1. Different classes of antigenic determinants of collagen. pN and pC indicate the aminoterminal and carboxyterminal extension peptides of the biosynthetic precursor form of collagen termed procollagen. Removal of the carboxyterminal peptide extension produces the intermediary form p-collagen which carries only antigenic determinants of the pN-class. Two different classes of antigenic sites are associated with the native collagen molecule itself. Terminal determinants are located in the aminoterminal and carboxyterminal nonhelical regions, and helical determinants are found in the large triple-helical body of the molecule. After denaturation the central determinants are uncovered which are located within helical sequences of the denatured α-chains.

raised to native collagen, now known as Type I collagen. However, it was left to later studies to produce immunochemically pure antibody reagents from rabbit antisera by appropriate procedures. Extensive work on the primary structure of Type I collagen in the late 1960s and early 1970s facilitated greatly the immunochemical characterization of the collagen molecule. Based on the structural features of collagen and procollagen (the biosynthetic precursor form of collagen), it was demonstrated that rabbit antisera to procollagen, native, and denatured collagen contain antibodies to many different antigenic sites. This came as no surprise, since the size and particular structure of the molecule allows for a comparable diversity of antibody specificities. Based on these early studies, five general classes of antigenic sites can be distinguished as illustrated in Figure 1. Two of the classes (pN- and pC-determinants) are only found on the procollagen molecule and are located on the so called aminoterminal and carboxyterminal globular domains which are removed by proteolytic cleavage after secretion of the molecule into the extracellular environment. These two classes of antigenic determinants are therefore not detectable on the bulk of collagen molecules isolated from mammalian tissues. These antigenic determinants require that the conformation of the terminal globular domains is preserved, since reduction and alkylation of procollagen destroys antigenic activity. Reduction is also necessary to dissociate the individual polypeptide chains of procollagens during denaturation. Denaturation of native molecules in the absence of reducing agents, however, leads to dissociation of the polypeptide chains of p-collagen isolated from the skin of dermatosparactic calf or sheep, since the carboxyterminal domain containing the interchain disulfide bonds is absent. The pN-extension contains only intrachain disulfide bonds. Denaturation of the p-collagen molecule thus leaves pN-antigenic sites intact even on the dissociated polypeptide, since disulfide bonds do not link the individual domains within the three-chain structure. Current evidence suggests that pN- and pC-antigenic determinants are specific for each segment; a fraction of the antibodies reacts, however, with antigenic sites shared by both the native and reduced form of the pN-peptide, indicating the recognition by such antibodies of sequential antigenic sites in addition to the above mentioned conformation-dependent structures. Further information on these classes of determinants for Types I and III collagen and the characteristic features of the rabbit antibody specificities are discussed in recent reviews[5,87] and are also dealt with in Chapter 7 for molecules isolated from tissue culture and in a section of this chapter for procollagen molecules isolated from tissues.

The three other general classes of antigenic determinants are shared by procollagen and native and/or denatured collagen (see Figure 1). Rabbit antisera or antisera raised in other species to native collagen contain antibodies to so-called terminal, helical, and central determinants. Terminal antigenic site refers to sites located in short peptide regions at the aminoterminal and carboxyterminal ends of the collagen polypeptide chains which are not involved in the formation of the triple-helical body. These determinants are thus equally accessible in the native and denatured molecule and are exquisitely sensitive to proteases. Helical determinants are restricted to the triple-helical body of the collagen molecule and by definition depend on the triple-helical conformation. Heat denaturation destroys this antigenic activity, and various proteases, such as pepsin, trypsin, or pronase, will not affect these antigens. In contrast to bacterial collagenase, these proteolytic enzymes do not digest the triple-helical part of collagen. Central determinants are essentially masked in collagen in the native configuration, but they become exposed after heat denaturation. Since antibodies to such antigenic sites react equally well with polypeptide chains and small fragments, these determinants are considered to depend solely on the amino acid sequence of five or more amino acid residues.

From these earlier studies, it is clear that each of these classes of antigenic sites comprises many individual and in part overlapping determinants. The number of different antibody specificities found in antisera, even after restriction to a particular class, e.g., to helical determinants, by immunoadsorption on a pepsin-treated native collagen adsorbens depends on the selection of the physicochemical form of the immunogen; the species from which it was obtained; the animal species used for immunization; route, dosage, adjuvant, and duration of the immunization course, in addition to genetic influences on the immune response (see Volume II, Chapter 7 and References 2 and 3). Some of this variability is dealt with in later sections in this chapter. It is quite obvious that the rather detailed chemical knowledge of the collagen molecule allows it to design appropriate techniques and procedures not only to develop antibody reagents to class-specific determinants, but furthermore to even more restricted portions of the molecule, i.e., to determinants associated with small peptide regions. It is due to the characterization of such antibodies that reliable reagents became available for a wide variety of applications. From the above discussion, it is also clear that optimal use of collagen antibodies produced by conventional methods in laboratory animals depends on their isolation, enrichment, and immunochemical characterization. Individual antisera from outbred animals such as rabbits will contain a collection of different antibody specifities which will vary considerably from animal to animal. Many of the conflicting data reported in the literature were due to the fact that antisera had been poorly characterized, that antibodies had not been isolated by specific procedures, or that the use of antisera and even isolated antibodies in quite different applications, e.g., serological vs. immunohistological tests, may have given disparate results for various reasons discussed in sections below and Chapters 3 and 4 in Volume II. Collagen is a weak immunogen when compared to other protein antigens, and impurities not easily detectable biochemically can cause severe problems, since antibodies will readily be formed to minute quantities of contaminating material.

Table 1
REACTIVITY OF ANTIBODIES TO HELICAL TYPE I COLLAGEN WITH CHROMATOGRAPHICALLY PURIFIED α-CHAINS

Reactivity				
Strong	Weak	Species immunized	Source of collagen	Ref.
α2	α1(I)	Rabbit	Rat	14
α1(I), α2		Rabbit	Rat	7, 27
α1(I), α2		Chicken	Rat	9
α2		Rabbit	Human	10
α1(I)	α2	Rabbit	Human	11
	α1(I), α2	Rabbit	Calf	13
α2	α1(I)	Rabbit	Guinea pig	14
α1(I), α2		Rabbit	Carp	15

B. Reactivity of Antibodies to Collagen Type I[α1(I)]$_2$α2

The most complete information on the reactivity of collagen-specific antibodies with intact molecules, α-chain components, or smaller fragments is available for Type I collagen from various animal sources. This information is compiled in Table 1 for chromatographically purified α1(I)- and α2-chains which can be used as a source of reference for obtaining α-chain specific antibodies, regardless of the class of determinants recognized. The data in the table indicate that rabbits, chicken, and presumably other species respond to antigen challenge by native collagen with variable amounts of antibodies which react with α-chain components. Because of the considerable sequence homology between α1(I)- and α2-chains, strong interchain cross-reactions are to be expected at least for antigenic sites located within the helical sequences of the polypeptide chains.[9,16] Appropriate immunoadsorption will, however, select for noncross-reactive antibody species, and the distinct antibody specificities for terminal determinants can be isolated as well.

Previous work on collagen antisera raised in various laboratory animals located determinants according to the classes described above. A list of references should be useful in selecting the appropriate system to obtain particular antibodies. The data shown in Table 2 indicate that eventually antibodies can be defined to any region (cyanogen bromide peptide) of the molecule. Procollagen and procollagen peptides are much more potent immunogens than collagen, i.e., a smaller dose is required for immunization, higher titers will be obtained, and most antibodies in rabbit sera will be directed to the globular domains (pN-, pC-determinants). When native Type I collagen is used as the immunogen in rabbits, the majority of the antibodies will be specific for terminal determinants particularly in the carboxyterminal region of the α1-chain. However, antibodies can also be obtained to amino- and carboxy-terminal sites of the α2-chain and the aminoterminal nonhelical region of the α1-chain. Antibodies to terminal determinants are usually species-specific because of the larger variation in amino acid sequence as compared to the central portion of the α-chains having the triplet structure Gly-X-Y. In contrast, although most rabbits produce antibodies to central determinants, these antibodies constitute only a minor fraction of the total antibody activity,[16] and they exhibit considerable cross-reactivity. Rabbits in general respond to native collagen with antibodies to terminal determinants, while most other laboratory animals (chicken, rats, mice, guinea pigs) respond with higher antibody titers to helical determinants. These spe-

Table 2
REACTIVITY TO NATIVE PROCOLLAGEN AND COLLAGEN OF TYPE I $[\alpha1(I)]_2\alpha2$

Determinant class	Active peptide	Species immunized	Source of collagen	Ref.
pN	pα1-CB (0,1) Col 1	Rabbit	Calf	17–21
	pα1-CB 1 Col 1	Rabbit	Chicken	22
	pα1-CB (0,1) Col 1	Mouse	Calf	23
	pα1-CB (0,1) Col 1	Mouse	Sheep	24
Terminal	α1 CB (0,1)	Rabbit	Man	11
	α1 CB (0,1)	Rabbit	Calf	25
	α2 CB 1	Rabbit	Rat	26, 27
	α2 CB 1	Rabbit	Guinea pig	14
	α1 CB 6	Rabbit	Calf	13, 28
	α1 CB 6	Rabbit	Rat	7, 27
	α1 CB 6	Rabbit	Man	11
	α2 CB 5	Rabbit	Calf	13, 28
	α2 CB 5	Rabbit	Rat	7, 27
Helical	$[\alpha1(I)]_2\alpha2$-	Mouse	Calf	29
	fragment	Rat	Calf	30
	$[\alpha1(I)]_3$	Rabbit	Calf	31
	$(\alpha2)_3$	Rabbit	Calf	31
	$[\alpha1(I)]_3$	Rat	Calf	32
	$[\alpha1(I)]_2\alpha2$	Mouse	Calf	29
	$[\alpha1(I)]_2\alpha2$	Rat	Calf	30
	$[\alpha1(I)]_2\alpha2$	Chicken	Calf	9, 32, 33
	$[\alpha1(I)]_2\alpha2$	Rabbit	Chicken	36
	$[\alpha1(I)]_2\alpha2$	Rabbit	Mouse	37
Central	α1 CB 3	Rabbit	Man	11
	α1 CB 8	Chicken	Rat	9, 33
	α1 CB 7	Rabbit	Man	11
		Rabbit	Calf	16
		Chicken	Rat	9, 33
	α1 CB 6	Rabbit	Rat	16
		Chicken	Rat	9, 33
	α2 CB (3,5)	Rabbit	Man	10
		Rabbit	Calf	16
		Rabbit	Rat	16
		Chicken	Rat	9, 33
	α2 CB 4	Chicken	Rat	9, 33
		Rabbit	Man	10

cies are thus of advantage for obtaining collagen type-specific antibodies directed to helical determinants. However, the rabbit is commonly used even for this purpose because of the advantage of the larger amounts of serum and because of their high content of antibodies to terminal, nonhelical determinants which presumably are also type specific.

C. Collagen Type-Specific Antibodies

In comparison to Type I collagen, far less immunochemical work has been done on Types II to V and some of the collagenous structures described most recently. In Table 3, the current information on the various collagens is listed. Because of their insolubility (presumably due to cross-linking), most of these collagens are

Table 3
THE DIFFERENT TYPES OF COLLAGEN

Collagen type	Major tissues[a]	Cells	Chemical characteristics[b]
Type I	Bone, tendon, skin, blood vessels	Fibroblast (tendon, cornea), osteoblast, fibroblast (lung, skin)	33% glycine, 20% imino acids, low hydroxylysine and carbohydrate content
Type II	Cartilage	Chondrocyte	33% glycine, 20% imino acids, high content of hydroxylysine and carbohydrate
Type III	Blood vessels, skin, organs rich in smooth muscle cells	Smooth muscle cell, fibroblast (skin, lung)	33% glycine, 20% imino acids, cysteine, high hydroxyproline and low hydroxylysine content
Type IV	Placenta, lung, basement membranes	Endothelial, smooth muscle and epithelial cells, others?	High carbohydrate, high degree of hydroxylation of proline and lysine residues, low in alanine and arginine, high in hydrophobic residues; interruptions of the typical helical sequence, cysteine; SLS crystallites different from Types I–III
Type V	Placenta, lung, parenchymous organs	Endothelial, smooth muscle and epithelial cells	Amino acid composition similar to Type IV, higher resistance to pepsin; SLS crystallites similar to Types I–III

[a] For a more detailed description of tissue localization, refer to Chapter 4 in Volume 2.
[b] Compare Chapter 1 in Volume 1.

obtained by proteolytic extraction methods of a variety of complex tissues (compare Chapters 4–6 in this volume), and variable portions of the molecules are lost and remain with the insoluble material. Nevertheless, it has been possible to produce antibodies to such pepsin-treated collagens which are specific for the different types of collagen. These antibodies react in serological assays with native, intact molecules in solution or react with molecules in tissues in immunohistochemical procedures (see Chapter 4, Volume II). The major emphasis of the work in recent years has been to develop antibody reagents which can distinguish the various collagens for analysis of biological questions. Since in some instances the chemical nature of the collagen is unknown or under dispute, this makes it even more important to rigorously apply fairly rigid standards for their characterization. Small amounts of contaminants associated with the isolated materials may contribute to immunogenicity and is often difficult to rule out. Despite this drawback, antibodies raised and purified by conventional procedures have been shown successfully to yield useful reagents. All uncertainties and apparent differences in reactivities by various immunological techniques will, however, not be removed until the collagens are better characterized chemically and perhaps with the combination of monoclonal antibody techniques (compare Chapter 12).

In Table 4, a list of references is given, including the various antigens used for immunization. In addition to native collagen and procollagen (Types II, III, IV-EHS, V), fragments (TIX-Type III; 7-S-Type IV), α-chains (αB-Type V; αC-Type IV), and pepsinized preparations of basement membranes (Type IV) have been used. In most instances, it could be shown that the antibodies react with various structures

Table 4
COLLAGEN TYPE-SPECIFIC ANTIBODIES

Type of collagen	Species immunized	Source of collagen	Cross-reactivity	Ref.
$\alpha 1(II)_3$	Rabbit	Calf		30, 35
	Rat	Calf		30, 35
	Mouse	Calf		29
	Goat, guinea pig	Chicken		36
$\alpha 1(III)_3$	Rabbit	Calf		20
	Rabbit	Chicken		37
	Rabbit	Mouse		38
	Rabbit	Rat		38
T1X (Type III)	Rabbit	Calf		39
Procollagen (Type III)	Rabbit	Calf	Human Sheep	10, 19, 40
	Rabbit	Sheep	Calf Human	19
	Rabbit	Rat		38
	Rabbit	Human		51
Type IV				42, 43
EHS	Rabbit	Mouse	All species tested	44, 45
placenta (pepsin)	Rabbit	Human		41
Lung (pepsin)	Rabbit	Calf		46
7-S	Rabbit	Human, mouse		47
Col 1(IV)	Rabbit	Mouse		
Type V(AB_2)				
$[\alpha 1(V)]_2\alpha 2(V)$	Rabbit	Human	All species tested	48, 50
	Rabbit	Chicken		59
$\alpha 1(V)$-chains (B-chains)	Rabbit	Human		45

in tissues by immunohistochemical methods (see Chapters 3 and 4, Volume II) in a tissue-specific fashion. For instance, Type II antibodies will react with cartilagenous structures wherever cartilage is found; type IV antibodies will react with basement membranes or the basal lamina around smooth muscle cells. The reactivity of some of the type-specific antibodies, however, cannot always be related to particular structures: Type IV and Type V antibodies will localize to the sinusoids in the mammalian liver, although basement membranes cannot be demonstrated by light and electron microscopy.[52] The reliability of such results will thus depend largely on the proven specificity of the antibodies and will furthermore require confirmation by independent methods.[50]

II. METHODS

A. Immunization

It is impossible to describe a simple immunization schedule valid for all collagen antigens. Collagen in general is a weak immunogen requiring larger amounts and multiple injections of antigen to obtain reasonable quantities of antibodies. However, this is not true for every type of collagen, and procollagen is certainly the exception requiring only small doses of antigen. There is ample evidence in the literature that the antibody response in particularly outbred animals such as the rabbit will vary qualitatively and quantitatively. This effect is seen when antibodies of certain specificities are desired. Some animals will give high antibody titers, and others will show little or no response. Remarks are added to help the reader to choose the most appropriate methods proven to yield antibodies, but these are by far not exclusive, since in many instances comparative studies were not performed.

1. Antigen

Some of the collagens are quite soluble in native triple-helical form in neutral solvents at physiological pH, and in most instances, native collagen will be more immunogenic than the isolated α-chain in random configuration. Neutral salt-soluble collagens have been obtained in sufficient yields from the skin of small laboratory animals after treatment with lathyrogens added to their diet or from the skin of larger animals at young age (e.g., calf). From human tissues, the yield of neutral salt-soluble and even acid-soluble collagens is low, and some of the collagen types can only be obtained at present by proteolytic extraction. Intact native collagens isolated from tissue culture may provide an alternative source in the future (see Chapter 7). Although in many early studies collagen solutions in 0.05 *N* acetic acid mixed with complete Freund's adjuvant have been used for injection, acetic acid has been replaced by a solvent less painful for the animal. Collagen antigens which are insoluble in neutral salts are first dissolved in 0.05 *N* acetic acid at approximate concentrations of 5 mg/mℓ and then dialyzed against 0.1 *M* calcium acetate (or against 0.1 *M* phosphate, pH 8, containing 0.4 *M* NaCl) before mixing with the adjuvant. The following information should serve as a guideline for the dose of antigen that will be required to obtain good antisera in rabbits (though not every rabbit will give the same response quantitatively or qualitatively; the reader is referred also to other chapters in this volume):

Antigen	Dosage	Minimum number of injections
Native Type I Collagen	5 mg	3–4
Native Type II Collagen	5 mg	3
Native Type III Collagen	5 mg	2–4
Native Type IV Collagen		
EHS sarcoma, acid soluble	250 μg	3
Human placenta, pepsin	250–500 μg	3
Native Type V	5 mg	4
Type I procollagen	0.5–1 mg	2–3
Type III procollagen	0.5–1 mg	2–3
Peptides		
T1X (Type III)	1 mg	4
Col 1	100 μg	4

2. Adjuvants

Various adjuvants have been employed in previous studies, including incomplete and complete Freunds adjuvant, aluminum hydroxide, and others. Insufficient data are available to allow comparisons for every collagen antigen, and it is thus recommended to use complete Freunds adjuvant (such as supplied by Difco Laboratories, Detroit, catalogue #0638-60 Bacto Adjuvant complete Freund). No appreciable quantities of antibodies are obtained when antigen is injected without adjuvant. A water-in-oil emulsion is prepared by recycling equal volumes of the antigen solution and the well mixed adjuvant through an 18-gauge needle and using a 2- to 5-mℓ glass syringe until the emulsion becomes rather viscous. Alternatively, small volumes can be prepared by pushing the mixture back and forth through a double-hubbed needle connected to two syringes. Larger volumes require more time, and the use of a rotating shaft homogenizer (Tissumizer; Tekman Co., Cincinnati, Ohio) is recommended. The latter method will involve, however, the loss of finished emulsion which coats the surfaces and which cannot be retrieved. Before use, the water-in-oil emulsion is tested simply by allowing a few drops to fall onto cold water in a small beaker. The emulsion should remain as white drops on the surface.[53]

3. Route of Inoculation

Complete Freund's adjuvant when injected into footpads produces severe swelling, ulceration, and possibly abscesses. The procedure is painful and does not allow the animal to move freely. It has been our experience that antibody titers are not different when other sites for injections are chosen. We routinely have used for a long period now the intradermal route by injecting the antigen with the needle positioned flat into the dorsal skin on the rear part of the rabbit. After shaving or removing hair with scissors, the skin is cleaned with 70% ethanol, and approximately 0.5 mℓ of the adjuvant-antigen emulsion is injected per site using a glass syringe and a 19- to 21-gauge needle. Booster injections are given the same way, but alternatively the antigen has been given i.p. in solvent alone.

Rats and guinea pigs are injected in a similar fashion. For the inoculation of chicken and mice, the adjuvant-antigen mixture is deposited s.c. or is injected into the musculature of the thigh. Highly immune animals will respond to the booster injection with an Arthus reaction at the site of the injection and may develop an ulcer or a sterile abscess. Ulcers are treated with antibiotic ointments.

4. Immunization Schedule

Many different immunization schemes have been used by other groups and in our own laboratories. We routinely use one schedule now: the first booster injection is given 2 weeks after the initial inoculum, and further booster doses are given in biweekly intervals until desired antibody levels have been achieved. It is obvious that some thought is given to the test system prior to initiating immunization which will be used to monitor antibody levels by comparison with preimmune sera. Blood samples are taken from the animal 12 to 14 days after the last booster. A lucid and rather useful account of the various bleeding techniques for small laboratory animals can be found elsewhere and will not be included here.[53]

B. Serological Assays

A number of serological assays have been adapted specifically to the immunochemical study of the collagens. The assays most commonly used now are based on the principle of passive hemagglutination, on the complex formation of radiolabeled antigen in solution, and on immunoenzymatic techniques by binding of antibodies to immobilized antigen (ELISA). Complement-fixation and immunodiffusion methods are not widely used, the latter being not very sensitive and of restricted value largely because of the size of the collagen antigen and the comparatively low precipitating antibody titers in rabbit antisera. Since the radioimmunoassay and ELISA methods are discussed elsewhere in this volume, we will describe and give a critical account of only the passive hemagglutination tests here.

1. Passive Hemagglutination

The agglutination of particles containing antigens on their surface is a simple and rather sensitive method for the detection of antibodies. Much less antibodies are required for this test than to precipitate antigens in solution. Some soluble antigens can be adsorbed onto inert materials, such as latex beads or bentonite, but such antigen-coated particles are difficult to standardize and have not found very widespread use. Red blood cells, on the other hand, have traditionally been used in agglutination reactions to monitor for blood group antigens expressed on their surface, and these studies using direct hemagglutination have revealed an elaborate antigenic surface which is mainly determined by complex carbohydrate structures linked to proteins and lipids. After washing red cells free of serum components, some antigens, particularly polysaccharides, can coat the red cell surface simply by incubation for a short while. While proteins can absorb to unmodified red cells, the amounts are usually too small to provide enough antigenic sites for agglutination of the cells when antibodies are added. This is due to negative repulsion of the negatively charged (caused by sialic acid) red cells. Although the more avid IgM antibodies may still overcome these repulsive forces to give visible cell agglutinates, it is generally more useful to create more antigenic sites after modification of the cell surface prior to antigen exposure. This will greatly improve agglutinability and thus the sensitivity of the hemagglutination assay. Passive hemagglutination provides thus a means to visualize antibody activity to proteins or antigens in general which are foreign to the red cell surface by coating the cell surface with a particular antigen. Boyden discovered that various proteins can be adsorbed effectively onto tannic acid treated red cells,[54] and Stavitsky[55] improved the method to the extent that the method has become useful for a wide variety of different antigens. This technique of coating tanned erythrocyte is somewhat different from coupling protein antigens to the red cell surface by chemical means. To achieve the latter, red cells are commonly treated with bifunctional reagents, such as *bis*-diazo-benzidine, carbodiimides, or aldehydes, before exposure to antigenic substances.

All the methods of passive hemagglutination share one possible pitfall, namely the agglutination of red cells due to the presence in the sera of most species of natural heterophile antibodies. Although this agglutination is specific in a serological sense, it is nonspecific for the passive hemagglutination assay using a foreign antigen. This pitfall can easily be avoided by removal of such heterophile antibodies by absorption with washed red cells. To prevent the cells from complement-mediated lysis during this step, the antiserum or normal serum is heat-inactivated for 30 min at 56°C. Absorption of the natural antibodies is then done by mixing equal volumes of packed, washed red cells and serum and incubation for 1 hr at room temperature prior to centrifugation.

Two methods of passive hemagglutination have been used widely in studies of antibodies to collagen: the tanned cell technique and the glutaraldehyde cell technique. Both methods have advantages and drawbacks which will be discussed below.

2. *Hemagglutination with the Tanned Cell Technique*

Because they are readily available, most laboratories use either human blood group O or sheep red cells. Human cells are less fragile than animal cells and may be of advantage also because of greater sensitivity.

Human blood is drawn from a blood group O donor on citrate dextrose solution USP Formula A by mixing 15 mℓ of the sterile anticoagulent with 100 mℓ of blood.

Dextrose · H_2O	1.65 g
Citric acid · H_2O	0.54 g
Sodium citrate · $2H_2O$	1.49 g
Distilled water	67.5 mℓ

The blood should be stored for 2 to 3 days before use, and the cells will be useful up to 2 weeks. Alternatively, outdated blood from the blood bank can be tried.

Whether coated or not, fresh cells can thus be used only for a limited period of time before they become fragile and easily hemolyzed or clumped during the tannic acid treatment. Preservation techniques, such as fixation and formalin, have not found widespread uses for collagen antibody work. Fixation potentially affects coating if done prior to addition of collagen or affects collagen on the cells if preservation is carried out after coating.

Tanned and sensitized red cells without preservation are stable only for about 8 to 10 hr and have thus to be prepared freshly for every working day.

a. Materials for the Tanned Cell Test

i. Tannic acid

A 1% stock solution is prepared by dissolving reagent grade tannic acid in saline and keeping it in a dark bottle at 4°C. The stock solution is discarded after 1 month, or earlier if it becomes turbid or discolored. In general a dilution of 1 part stock with 199 parts of saline (0.005%) is prepared before use. Although there apparently is little difference in effectiveness between the reagents of different manufacturers, it is still advisable to determine the optimal concentration of tannic acid, i.e., the concentration which will yield strong aglutination in addition to giving the most sensitive assay for a given antigen. The result of such a titration may appear as in the example given below:

Concentration of tannic acid (dilution from stock)	Dilution of antiserum (−log 2)									
0.02% (1:50)	−	+	+	++	++	−	−	−	−	
0.01% (1:100)	+	+	++	++	+++	++	++	+	−	−
0.005% (1:200)	++	++	+++	+++	+++	++	++	+	−	−
0.0025% (1:400)	−	±	+	+	+	±	−	−	−	−

ii. Buffers

Phosphate buffered saline (0.075 *M*) solutions as described originally by Boyden[54] are used at pH values of 7.2 and 6.4. They are prepared according to the table:

Solution A: 0.15 *M* KH_2PO_4 (20.4 g/ℓ)
Solution B: 0.15 *M* Na_2HPO_4anhydrous (21.3 g/ℓ)

pH	Solution A mℓ/ℓ	Solution B ml/ℓ	Sodium chloride g/ℓ
7.2[a]	119.5	380	4.5
6.4	338.5	161	4.4

[a] After preparation the pH should be adjusted as necessary.

iii. Diluents and Antisera

Antisera or antibody solutions are diluted in 1% normal rabbit serum (1 mℓ/100 mℓ of phosphate buffered saline, pH 7.2). The same diluent is used also during the final washes for the preparation of the antigen-coated "sensitized" cells. The protein-additive is required to prevent the tannic acid-treated and antigen-coated cells from agglutinating spontaneously, since not all sites on the surface of the cells are occupied by the antigen. It is evident that the proteins in the additive become attached to the cells, the amount being related to the concentration used. The usual assumption is that the availability of the antigen (used for sensitization) for interaction with antibodies is not affected. Since rabbit normal serum as well as serum from other species contains antibodies reacting with heterologous red cells, the normal serum and the antisera as well (except when affinity-purified antibodies are used) have to be absorbed. The sera are heated at 56°C for 30 min to inactivate hemolytic complement. The heat-inactivated sera are then mixed with washed, packed erythrocytes of human blood group O donors (or sheep cells if these are to be used for testing) at equal volumes, then incubated at room temperature for 30 min, and after centrifugation, the absorbed serum is removed with a Pasteur pipette. Absorbed reagents are stored at −20 to −30°C in appropriate aliquots.

iv. Erythrocytes

Human blood group O or sheep erythrocytes are taken into sterile ACD solution and are stored at 4°C. Allow 2 to 3 or more days for stabilization of the cells before first use. On the day of the testing, remove blood and wash the cells several times free of serum with excess volumes of phosphate-buffered saline, pH 7.2. After prolonged storage (3 to 4 weeks) or if infection had occurred, the cells will exhibit lysis, and the first washing fluid will show excess hemoglobin. Most of the time, these cells cannot be washed free of hemoglobin, and they should be discarded.

v. Antigen

An important consideration is the form in which collagen antigens are used to coat the cells. The protein has to be soluble in physiological buffers, and since low concentrations are used to sensitize the cells, this normally does not present a problem. Collagen can be obtained in native triple-helical form or as gelatine, its heat-denatured form. Denaturation can also be achieved by chaotropic agents such as 2 *M* KSCN. After separation and purification of the constituent α-chains, these can be used to coat cells as well as cyanogen bromide-derived peptides. It has been found empirically that antibodies to the denatured form of collagen react with stronger agglutination and higher titers in the tanned cell technique than in the glutaraldehyde cell technique. This is the case with cells coated with denatured collagen, α-chains, or smaller peptides. On the other hand, antibodies to native collagen ("type-specific" antibodies) will react better with the glutaraldehyde technique described below.

The acid-soluble collagens are dissolved in 0.1 *M* citrate pH 3.6 (10.5 g citric acid. H_2O, 50 mℓ 1 *N* HCl, 100 mℓ 1 *N* NaOH for 1 ℓiter) or in 0.05 to 0.1% acetic acid at a concentration of 1 mg/mℓ, and the stock solutions are stored at −20°C in small aliquots. Prior to sensitization, an appropriate volume is removed and neutralized by addition of 0.2 volumes of 0.9 *M* NaCl in 0.188 *M* Na_2HPO_4, and the collagen solution is diluted to 0.01 mg/mℓ with phosphate buffered saline pH 6.4.

To prepare denatured collagen, a solution of 10 mg/mℓ in acid is heated at 50°C for 30 min prior to sensitization. Alternatively, lyophilized collagen is dissolved at 2 mg/mℓ in 2 *M* KSCN and stored at 4°C. Dilutions of the denatured collagens are prepared by addition of phosphate buffered saline, pH 6.4 to a concentration of 0.01 mg/mℓ. Isolated α-chains or peptides are dissolved at concentrations of 5 to 10 mg/mℓ and are diluted 1 to 100 to 1 to 200 to give a final concentration of 0.05 mg/mℓ. It is appropriate when deciding on the optimal antigen concentration for sensitizations to test different concentrations. As shown in the example given below, high concentrations of the antigen will give strong agglutination, but nonspecific agglutination will also be observed with normal serum and in the absence of antibodies. Low concentrations will result in lower titers and a loss of sensitivity of the assay.

Concentration of native collagen for sensitization (mg/100 mℓ(w/v)	Titer with antiserum to native collagen (−log₂)	Nonspecific agglutination with normal serum
10	2560	+
3	2560	±
1[a]	2560	−
0.1	640	−

[a] Optimal sensitizing antigen concentration.

Concentration of denatured collagen for sensitization (mg %)	Titer with antiserum to native collagen (−log₂)	Nonspecific agglutination with normal serum
10	1024	+
3	1024	±
1[a]	512	−
0.1	32	−

[a] Optimal sensitizing antigen concentration.

Other materials for the test Centrifuge tubes, Pasteur pipettes, 100-mℓ cylindrical measure, and spot test plates (white porcelain, 180 × 140 mm, 30 wells; Thomas Co. 7812–C52).

b. Sensitization of Erythrocytes

The method described here is for the preparation of 20 mℓ of a 2.5% suspension of antigen-coated cells.

1. Tannic acid treatment — Cells are washed three times with physiological saline or phosphate-buffered saline pH 7.2 (PBS 7.2) by sedimentation at about 750 × g for 5 to 10 min in a refrigerated centrifuge, removing the supernate and resuspending in fresh saline. It is convenient to have a Pasteur pipette attached to a filter pump for removal of supernate from the cell pellet. 0.6 mℓ of the packed cells are pipetted into each of two centrifuge tubes (50 mℓ) or screw-cap bottles and suspended in 10 mℓ of PBS 7.2. A starting volume of 0.6 mℓ rather than 0.5 mℓ is used to allow for loss of cells during the many washing steps.
2. Ten millilitres of tannic acid solution (~1 to 200 from a 1% stock solution in saline) are added to the cell suspension in each tube, and the contents are well mixed. The tubes are incubated for 15 min at 37°C in a water bath. Both tubes are then centrifuged for 5 min at 750 × g, the supernate is removed, and the cells are resuspended and washed once with 20 mℓ of PBS 7.2.
3. Antigen coating — The cell pellets in both tubes are then resuspended well in 10 mℓ of PBS 6.4, and to one tube an equal volume of appropriately diluted antigen solution in PBS 6.4 is added, and the mixture is incubated for 30 min at room temperature. The other tube is set aside to provide uncoated cells for controls.
4. The cells are spun out and the supernate is removed. The supernate in some instances may be saved for further sensitization, since only a small fraction of

the antigen coats the cells. After two washings in 20 mℓ of PBS 6.4 containing 1% heat-inactivated and absorbed rabbit normal serum (or serum from the species used to obtain specific antibodies), the cells are finally suspended in 20 mℓ of PBS 7.2 containing 1% normal serum to give a 2.5% suspension. During these last washes, the cells have to be suspended well each time by shaking to dissociate small clumps of red cells. If considerable lysis occurs after the tannic acid treatment or after sensitization, fresh cells should be obtained or the buffer solutions checked. The cell suspension is kept at 4°C and remains stable for hemagglutination for about a working day before lysis occurs.

There is good evidence that only a very small fraction, less than 1%, of the antigen available is taken up by the cell. However, the quantity absorbed will depend on the concentration of the antigen present during the sensitization step. The following figures may serve as a guide to assist in chosing optimal antigen concentrations. These figures are for the antigen concentration in PBS 6.4 before addition to the cells:

Antigen	Optimal Concentration (mg/mℓ)	Ref.
Native collagen	0.05–0.01	56,57
Denatured collagen	0.05–0.01	57,58
α-Chains	0.1–0.05	7,11
CNBr-peptides	0.1–0.05	7,16

c. *Assay Method*

i. Porcelain plates

Prior to assay, all sera are inactivated by heating at 56°C to prevent hemolysis of the protein-coated red cells. They are also absorbed with uncoated red cells to remove heteroagglutinins. The hemagglutination test is carried out on 30 well porcelain spot test plates at room temperature. Serial two-fold dilutions of the antiserum are made in 7 × 10 mm round bottom glass tubes with final volumes of 0.1 to 0.5 mℓ depending on how many different sensitized cells are being tested. Initial dilutions are 1 to 10 or 1 to 100 in the case of strong antisera. As the diluent, PBS 7.2 is used containing 1% heat-inactivated and absorbed normal serum. The reagents are dropped into the wells of the test plate with a Pasteur pipette (approximately 25 to 50 μℓ per drop): one drop of diluent, one drop of dilute serum (use the same pipette and start pipetting with the highest dilution), and two drops of sensitized erythrocytes. The plates are gently agitated at room temperature, and readings are taken after 10 min. Agglutination is indicated by larger to smaller clumps of cells which can easily be distinguished from the suspended homogeneously appearing red cells in a negative well. Grading is done as ++++, +++, ++, +, − and eventually with half-step intervals. Some experience is necessary to read the plates consistently, but this can be gained readily after observing a number of titrations and agglutination patterns. With some high titered sera, prozones may be seen indicated by a weaker agglutination at the higher serum concentrations. There are differences in sensitivity of the assay, depending on the antibody class being tested. As with other agglutination systems, IgM is several hundred-fold more efficient to agglutinating cells than IgG.

The test may be carried out in the wells of plastic agglutination plates using Takatsy-type loops for making the serum dilutions directly in the wells (see below). However, conditions have not been established with the antigen-sensitized cells described above.

ii. Microtiter plates

Michaeli et al.[14,62] have applied the methods described by Herbert[63] using formalinized, tanned sheep cells for antigen coating. Since fresh cells can only be used for a limited period of time, their useful life can be extended by preservation or fixation before tannic acid treatment and sensitization.

d. Preparation of Formalinized Cells

1. Freshly collected sheep red blood cells are washed five times with physiological saline and packed after the final wash (750 × g for 15 min). Then 25 mℓ of the packed cells are suspended in 200 mℓ PBS 7.2 (36.0 g NaCl, 7.4 g Na_2HPO_4 anhydrous, 2.15 g KH_2PO_4 anhydrous in 5 ℓ of distilled water) and placed in a 500-mℓ flask.
2. Fifty millilitres of commercial formalin (40% formaldehyde) are filled to a length of two thirds into a dialysis bag, leaving no air in the bag, and the tied-off bag is submerged into the red cell suspension. The suspension is gently agitated for 3 hr at room temperature, and at that time, the dialysis bag is punctured to allow the formalin to freely mix with the cell suspension. After a further 12 hr of gentle mixing, the dark brown cells are filtered through gauze and are washed five times with physiological saline. To resuspend the cells after each centrifugation, a spatula may be of help. It is of importance though that resuspension is complete before each wash.
3. Finally the cells are suspended to 25 or 50% in saline and are stored at 4°C. To prevent bacterial growth, 0.2% formalin or 1 to 10,000 thiomersal can be added as a preservative.

i. Tannic Acid Treatment of Formalinized Cells

Tanning of the formalinized cells is done by using the tannic acid concentration described above for fresh cells. After 15 min of incubation at room temperature, the cells are centrifuged and washed twice in PBS 7.2, making sure to resuspend the cells carefully and completely each time.

ii. Sensitization of the Formalinized and Tanned Cells

The antigen solution is added to the 5% suspension of the cells (0.5 mℓ packed cells per 10 mℓ of PBS 7.2) at concentrations of 5 μg/mℓ for native or denatured collagen and of 5 μg/mℓ for isolated α-chains. After 45 min of gentle agitation at room temperature, the cells are centrifuged and washed twice with PBS 7.2 containing 1% normal rabbit serum (heat-inactivated, absorbed). They are made up to 50 mℓ of 1% suspension and are stored at 4°C. A test can be carried out to determine the amount of normal serum required to prevent nonspecific agglutination: a double dilution series of heat-inactivated and absorbed normal serum is mixed with the coated red cells, and if agglutination is observed at a given serum concentration, sufficient serum is added to bring the serum concentration to this amount.

iii. Assay Method

Doubling dilution series of 25 μℓ antisera and normal serum are made with Takatsy loops: 25 μℓ of diluent is added to each well with a standardized pipette. The first well receives 25 μℓ of antiserum diluted 1 to 10 to 1 to 100, depending on the antibody titers. With the loops, 25 μℓ of diluted serum is transferred, and the last 25 μℓ are discarded. This step is followed by addition of 25 μℓ of diluent to each well in microtiter plates (U-shaped wells). Then 50μℓ of the 1% sensitized red cells are added, and the plates are well agitated to obtain mixing. After incubation at 4°C overnight (12 hr), the agglutination results are obtained by reading the plates: agglutination is indicated by an even carpet of cells covering the bottom of the well. Agglutination-negative wells are seen as small round buttons at the bottom with sharp margins.

As for all the hemagglutination assays, controls include appropriate normal serum dilutions with coated cells and antiserum with uncoated cells to check for nonspecific agglutination. A further and important test for antigenic specificity is provided by inhibition of the hemagglutination with various proteins or peptides related or unrelated to the antigen being tested.

3. *Inhibition of Hemagglutination*

The inhibition of passive hemagglutination is used in order to assess the specificity and activity of antigenic substances in a semiquantitative way. The reaction measures indirectly the antigen-antibody interaction which takes place in solution, prior to addition of antigen-coated red cells. Its principle is therefore based on the competition for antibody combining sites between soluble antigen and fixed antigen (on the red cell surface). If the inhibiting compound has a structure closely similar or identical to the specific antigenic determinant to which the antibodies are directed, interaction will occur, and this is observed as a reduction in hemagglutination titer in comparison to a control dilution series from which the inhibitor has been omitted. The extent of inhibition is related not only to the amount of inhibitor, but also to its affinity for the antibody combining sites.

The sensitivity of passive hemagglutination-inhibition is high, since the primary antigen-antibody interaction occurs independently of the monitoring system (the antigen-coated red cell) and requires only small amounts of inhibitors. The procedure can thus be used to great advantage for the determination of antigenic activity associated with small peptides which may be too small or available in insufficient quantities to directly coat the tanned erythrocyte for direct agglutination.

The limitations of the passive hemagglutination-inhibition tests lie in their lack of precision. The nature of the agglutination assay implies an inherent error of at least ± one titer step (±50% antibody concentration). This nonprecise data collection does not allow for absolute determination of affinity constants, but can yield nevertheless relative comparisons. Since only reductions in 50% of the antibody concentration or titer steps are measured, the procedure will be of most value when dealing with monospecific antibodies (specific for a single antigenic determinant). With complex mixtures of different antibody specificities and antigens with multiple antigenic sites, the presence of a single determinant on a small peptide inhibitor may not be detectable even if the specific antibody amounts to 50% of the total antibody content of the serum. Furthermore, the reactions may become nonspecific beyond a certain maximal concentration of the inhibitor, i.e., the cells are agglutinated at a high dose of the inhibitor. Nonspecific effects should also be suspected if the inhibitory activity of an antigen cannot be "diluted out," and inhibition is observed over a wide range of concentrations. Sometimes this may be due to bacterial contamination, wrong pH adjustments, or due to high viscosity.

The inhibition procedure can be set up, however, in a semiquantitative way, and the activity of various inhibitors can be expressed in terms of weight or moles per liter in comparison. This procedure, the inhibition of four or eight agglutinating units, is described below.

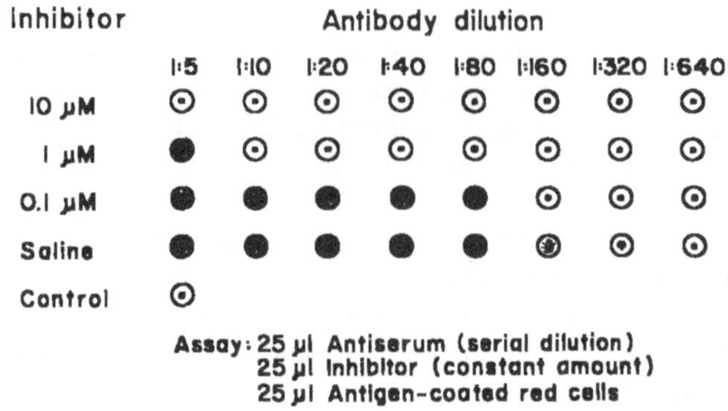

FIGURE 2. Passive hemagglutination-inhibition (Method A). To double serial dilutions of antibodies, an equal volume of constant amounts of antigen (inhibitor) is added before the antigen (inhibitor)-antibody reaction is monitored by addition of antigen-coated red cells. In this figure, the procedure is shown according to Method A (see text). The inhibition pattern is obtained for three concentrations of the inhibitor. The saline control series does not contain the inhibitor (antibody control), and the single control well contains the highest concentration of the inhibitor, but no antibody (inhibitor control). The agglutination pattern shown here is obtained with microtiter plates: the small center button indicates absence of agglutination (the red cells sediment to the bottom of the U-shaped wells); the filled wells indicate agglutination (the red cells are prevented by aggregation to slide down the walls of the wells). On porcelain plates (see text), the reverse pattern is obtained: absence of agglutination is indicated by a homogeneous suspension of red cells, while agglutination is seen as clumps in different sizes of red cells. The data are expressed as the reduction in number of titer steps in which agglutination is observed, as compared to the saline control, and can be illustrated as shown in Figure 3.

a. Inhibitors

The test substances are standardized in terms of weight or moles of peptide or protein. This is usually done by preparing stock solutions in PBS 7.2 which contain a preweighed amount of lyophilized material. If greater accuracy is desired or needed, the protein content of the stock solution is determined by amino acid analysis and using norleucine as an internal standard. Dilutions of the inhibitor are prepared by appropriate dilutions.

b. Inhibition Method

The test can be carried out for both the plate and microtiter well method as described above and consists of preincubation of serial doubling dilutions of antibodies with several concentrations of the inhibitor (Method A) or by preincubation of four or eight agglutinating units of antibodies with serial doubling dilutions of the antigen (see Method B) before addition of antigen coated cells.

i. Method A: inhibition with constant antigen concentration

For this procedure, several serial doubling dilutions of the antiserum in small test tubes or in the wells of the microtiter plate are prepared (see Figure 2). An equal volume of the inhibitor dissolved in PBS 7.2 is added at concentrations (one concentration per serial dilution) spanning two to three log units (e.g., 10, 1, 0.1 μmol/

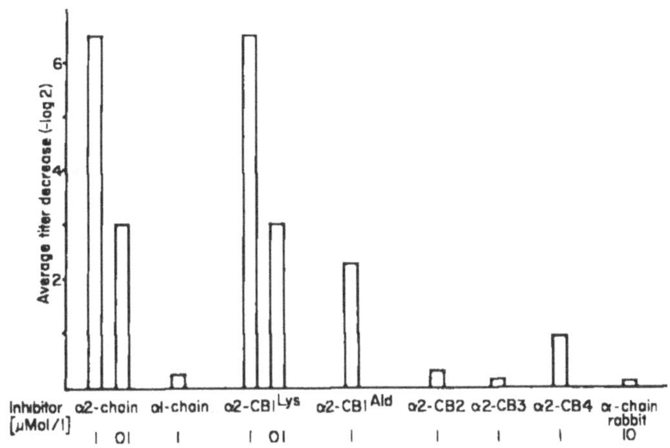

FIGURE 3. Passive hemagglutination-inhibition (Method A): comparative analysis of different peptide fragments for antigenic activity with rabbit antibodies to rat collagen Type I. A single antigenic determinant (terminal determinant) is located within the nonhelical region of the rat α2-chain. A comparable decrease in titer is observed for the α2-chain and the small CNBr-derived peptide α2-CB1^{Lys}. Modification of the ε-amino group of the single lysine residue to the aldehyde form affects antigenicity, as is indicated by the more than tenfold higher concentration of α2-CB1^{Ald} required to achieve the same decrease in titer as α2-CB1^{Lys}. None of the other CNBr-fragments of the α2-chain is inhibitory nor is the ten- to hundred-fold higher concentration of rabbit collagen α-chains.

ℓ). To another serial dilution, physiological saline or diluent is added to serve as a noninhibited control and to determine the titer of the antiserum against which the agglutination patterns of the other series are read. Controls also include normal serum in the presence or absence of inhibitor to check for nonspecific agglutination of the coated red cells or the inhibitor. After 10 to 15 min, the antigen-coated red cells are added, and the agglutination is read as usual. The data are expressed as decrease in titer ($-\log_2$ units) of the inhibited series vs. the saline control. One example is given in Figure 3 taken from Reference 26. The data describe antibodies to a terminal determinant in the aminoterminal telopeptide region of rat α2-chains present on the small fragment α2-CB1. None of the other cyanogen bromide derived fragments carry this antigenic determinant. Furthermore, modification of the lysyl side chain in α2-CB1 to convert the ε-amino group to an aldehyde apparently affects binding affinity.

ii. Method B: inhibition with variable antigen concentration

For this test, the titer of the antiserum or antibody solution is first determined by setting up the usual passive hemagglutination assay using the porcelain plate or microtiter plate. Four and eight agglutinating units[64] are determined as shown in Figure 4; these are the dilutions of the antiserum which remain constant in the following inhibition assay, and appropriate volumes of these dilutions are prepared. The second step consists of preparing serial doubling dilutions of the antigen in PBS 7.2 or diluent. To every one of the antigen-containing wells, an equal volume of the antiserum dilution (four or eight agglutinating units) is added, and the plates are incubated after mixing for 15 min to 1 hr at room temperature. After this period, two volumes of the antigen-coated red cells are added, and the agglutination is read

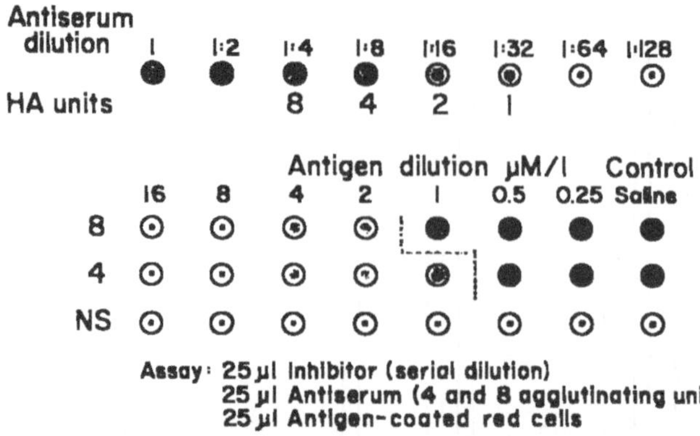

FIGURE 4. Passive hemagglutination-inhibition (Method B): inhibition of four to eight hemagglutinating units. The titer of the collagen antiserum is first determined by preparing a doubling-dilution series of the antiserum or antibody solution, addition of diluent (25 µℓ) and addition of antigen-coated cells. This is shown at the top of the figure for the agglutination pattern observed in the wells of microtiter plates; the small button in the center of the wells indicates absence of agglutination (the red cells sediment); the filled wells indicate agglutination (the red cells do not sediment). From the agglutination pattern, the titer of the antiserum is translated into the dilutions of the antiserum containing four or eight agglutinating units. These are the dilutions three or four titer steps more concentrated than the antibody dilution in the last well, clearly showing agglutination. To determine semiquantitatively the serological or antigenic activity of an inhibitor, twofold serial dilutions of the inhibitor are prepared, and 25 µℓ of four or eight agglutinating units of the antiserum is added before addition of the antigen-coated red cells. NS indicates the inhibitor control (no antibody), and saline controls give the strength of agglutination of four or eight agglutinating units, respectively, of the antibodies in the absence of the inhibitor. The inhibitory activity is measured as the lowest concentration of the antigen (inhibitor) which completely (as indicated by the filled wells) inhibits the agglutination of the antigen-coated red cells.

as usual. The end point of inhibition is between the last tube showing no agglutination (complete inhibition) and the tube showing slight agglutination.

Each antigen titration series includes controls consisting of: (1) the antiserum standards (four and eight agglutinating units) to which diluent has been added instead of inhibitor; and (2) a normal serum control — antigen dilutions to which diluted normal serum has been added. If different inhibitors are compared, they should be run in parallel in the same tests and using the same batch of cells.

The activity is expressed on a weight or molar basis as shown in the example in Table 5.

4. Passive Hemagglutination: the Glutaraldehyde-Coupling Technique

In contrast to coating of tanned red cells, there are a number of techniques to covalently link protein antigens to the surface of red cells taking advantage of bifunctional reagents. Among the various compounds, glutaraldehyde has been successfully used in immunochemical studies on the various collagen types. The method described here was adapted for collagens from a procedure described earlier for other protein antigens[65] and was first used for the characterization of chicken antisera to native collagen.[32,33] The test uses microtiter plates requiring overnight incubation

Table 5
HEMAGGLUTINATION-INHIBITION OF RABBIT ANTIBODIES SPECIFIC FOR CALF α1-CB(0.1)[a]

Peptide[b]	Position in α1-CB(0.1)[b]	Minimum concentration of peptide to inhibit agglutination of red cells[c] (μM) Anticalf collagen 344	Antihuman collagen 313
α1-CB(0.1)	1–19	1	0.5
α1-CB(0.1)[Ald]	1–19	0.5	0.25
α1-CB(0.1) T1[a]	1–9	80	0.5
α1-CB(0.1) T2	10–19	80	80
α1-CB(0.1)[Ald] Th1	1–12	40	1
α1-CB(0.1)[Ald] Th1[a]	1–5	nd[d]	1
α1-CB(0.1)[Ald] Th2	13–19	40	80
α1-CB(0.1)[Ald] C1	1–4	40	40
α1-CB(0.1)[Ald] C2	5–19	8	20
α1-CB(0.1)[Ald] E1	1–13	1	nd

[a] Adapted from Reference 25.

[b]

```
                    C   Th                       T              Th         E
                    ↓ 5 ↓                        ↓ 10           ↓          ↓    15
calf,   pGlu-Leu-Ser-Tyr-Gly-Tyr-Asp-Glu-Lys-Ser-Thr-Gly-    -Ile-Ser-Val-Pro-
        Gly-Pro-Hse
human,  pGlu-Leu-Ser-Tyr-Gly-Tyr-Asp-Glu-Lys-Ser-Thr-Gly-Gly-Ile-Ser-Val-Pro-
        Gly-Pro-Hse
```

calf, ____________________________________, Maximum size of antigenic determinant;
human, ____________

[c] Red cells were coated with calf α1-chains; eight agglutinating units of calf or human collagen antibodies were used.

[d] nd, not tested.

until the results are obtained and cannot be adapted for the faster hemagglutination technique on porcelain plates. The test is simple and reliable, and it is the method of choice in characterizing collagen type-specific antisera by hemagglutination and hemagglutination-inhibition.

a. Materials

1. Glutaraldehyde (Sigma®, Grade I, 25% aqueous solution; stored frozen at −70°C)
2. 0.1 *M* calcium acetate
3. 0.15 *M* NaCl
4. 3.8% sodium citrate (stored at 4°C)
5. Phosphate-buffered saline, pH 7.2 (see p. 16)
6. Normal serum (e.g., rabbit, heated at 56°C for 30 min absorbed with red cells)
7. Human blood (freshly drawn in sodium citrate, one part to four parts of blood)

b. Antigens

Lyophilized collagens of the various types are dissolved at a concentration of 3 mg/mℓ in 0.05 *N* acetic acid at 4°C on a magnetic stirring motor. The solution is then dialyzed against 0.1 *M* calcium acetate to provide the native collagen preparation for coupling to the red cells.

Denatured collagen is prepared from the native antigen by heating for 30 min at 50°C prior to addition to red cells. To avoid effects caused by renaturation of the antigen upon storage of the antigen-coupled cells at 4°C, prior to the agglutination test, the heat treatment is repeated with the respective cell suspension.

c. Preparation of Antigen-Coupled Erythrocytes

We find it convenient to draw the small amounts of blood needed for the assay from laboratory personnel. A 3.8% solution of sodium citrate (0.6 mℓ is drawn into a disposable syringe, and blood is taken to a total volume of 2.5 to 3 mℓ.

1. The blood is washed three times in a refrigerated centrifuge with excess volumes of cold PBS 7.2. Aliquots of 0.5 mℓ packed cells are made to a 10% suspension by addition of 4.5 mℓ of PBS 7.2. To each aliquot, 1.5 mℓ of a freshly prepared 2.5% solution of glutaraldehyde is added, and the mixture is stirred for 15 min at room temperature. The cells are washed three times with 0.15 *M* saline and are then suspended in 5 mℓ of 0.15 *M* saline. These washes with saline are obviously important, since coupling of the collagens is done in calcium acetate. Phosphate buffers during the following step will lead to severe clumping of the red cells due to calcium phosphate.
2. To the glutaraldehyde-treated red cells, 5 mℓ of the collagen solution in 0.1 *M* calcium acetate is added, and the mixture is stirred for 90 min at room temperature.
3. After the coupling step, the cells are washed twice with 0.1 *M* calcium acetate, followed by two washes in 0.15 *M* saline, taking care of complete dispersion and resuspension of the cells prior to each centrifugation. After the final sedimentation, the cells are suspended in 5 mℓ of PBS 7.2 containing 0.1% heat-inactivated and absorbed normal serum and 0.02% sodium azide. When stored at 4°C, the cells can be used for several weeks to months. Storage of the cells for 1 to 2 days prior to first use decreases nonspecific agglutination. For the assay, the cells are diluted to a 0.5% cell suspension in PBS 7.2 containing 0.1% normal serum.

d. Assay

The assay is performed on microtiter plates with U-shaped wells. A doubling dilution series of the antiserum or antibody solution is made by prepipetting 25 μℓ of diluent (PBS 7.2, 0.1% normal serum) into each well. To the first well, 25 μℓ of (diluted) serum is added, and 25 μℓ are transferred with Takatsy loops from well to well. A control series contains normal serum. To each well, 25 μℓ of diluent is then added (which could be replaced by the inhibitor for the hemagglutination inhibition assay described above), followed by 50 μℓ of the antigen-coupled red cells. After mixing the plates are stored at 4°C for 16 hr, and the results are recorded as + to + + + + agglutination. The absence of agglutination is indicated by a small button of cells at the bottom of the well. Agglutination-negative wells sometimes do not form this small, tight button, but rather a ring. The reading remains, however, unambiguous by comparing the agglutination pattern with the normal serum control.

e. Remarks

In the case of nonspecific clumping or aggregation of the cells, different antigen concentrations should be tried. Particularly with Type III collagen preparations we

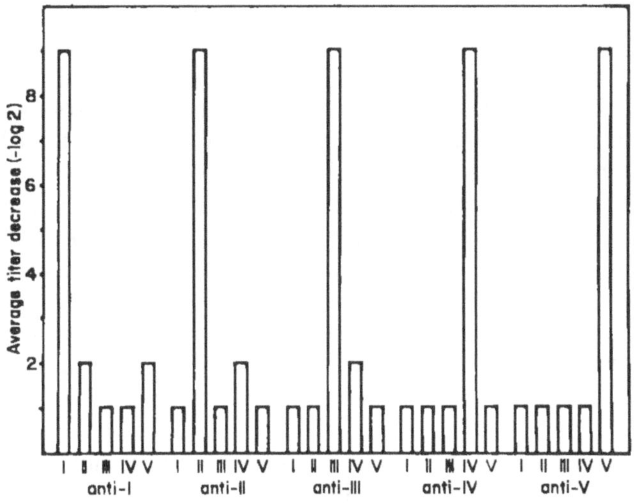

FIGURE 5. Serological characterization of type-specific antibodies to various collagens by passive hemagglutination-inhibition (Method A, see Figure 2). Human red cells were coated with either type I, II, III, IV, or V collagen. Antibodies to the respective types of collagen were purified by cross-adsorbtion on various collagen immunoadsorbentia and were finally eluted from the specific collagen type immunoadsorbens (see Figure 7). After concentration of the eluted type-specific antibodies, serial doubling dilutions from 1:100 to 1:25,600 were made. The different collagen types were added as inhibitors at a concentration of 2.5 μg per well. As can be seen in the figure, the specific collagen type to which the antibodies had been raised showed complete inhibition (they inhibit the agglutination of the antibodies to at least a dilution of 1 to 100), while the other collagens exhibit no or only slight inhibition.

have observed that, possibly due to the high viscosity of our preparations, much lower amounts of antigen had to be used to obtain useful cells. The concentrations ranged from 0.15 mg to 1 mg/mℓ. In order to distinguish between specific and nonspecific agglutination, the cells coupled with native collagen can be heat-treated to denature the collagen, or they can be treated with bacterial collagenase (CLSPA, Worthington) by incubation of 5 mℓ of a 1% suspension with 0.05 mg enzyme for 2 hr at 37°C. Care should also be taken to use comparable amounts of red cells coated with different collagen antigens for the assay to facilitate reading. The concentrations of the red cells are either determined by counting and appropriate adjustments of the dilution factor or by empirically making suspensions at different concentrations and observing the size of the cell button in diluent.

f. Inhibition Assay

Specificity of the agglutination reaction is usually determined by an inhibition assay particularly when antibodies are desired specific for a given type. Low titers of cross-reactive antibodies may not be detected by direct agglutination. The assay is done similarly as described above by substituting 25 μℓ of the inhibitor at appropriate concentrations for diluent. In the example given below (see Figure 5), type-specific antibodies to collagen Types I to V were checked for cross-reactivity using this assay.[43]

C. Isolation of Collagen Antibodies by Immunoadsorption

Specific immunoadsorption procedures are crucial for the purification of collagen antibodies with defined specificity due to the generally weak nature of collagen antigens as immunogens. Among the various methods for isolation of antibodies, the immunoadsorbent techniques are the simplest and have thus been used most commonly. They provide a convenient way of removing antibodies to contaminating substances which invariably are present in collagen antisera. These impurities give rise to antibodies of largely unknown specificities, although antibodies to serum proteins (albumin, fibrinogen) are frequently present. Although such antibodies may not interfere with immunochemical studies in hemagglutination (assays, radioimmunoassays, or other assays, they pose severe problems when collagen antisera characterized solely by serological methods are used particularly in immunohistological studies which unfortunately are not always done with appropriate specificity controls.

This lack of careful characterization of collagen antibodies has led to a considerable accumulation of reports in the literature with conflicting results. The information provided by many of the studies has to be viewed with scepticism unless additional and independent data can be provided. Although the ways described here to obtain specific antibodies can safeguard against the most obvious pitfalls, the application of immunoadsorption procedures per se to collagen antisera cannot completely remove this critical attitude toward data obtained with "specific" collagen antibodies in tissue and organ sections, since the collagen preparation used to prepare specific immunoadsorbentia may contain the same contaminants as the one injected into the animal. Nevertheless, there are good reasons in favor of immunoadsorbentia:

1. Immunoadsorbentia can be used to select against antibodies to contaminants, since the amount of the contaminants in the collagen preparation linked covalently to the adsorbens is usually below 1% and remains of the total protein undetectable biochemically. A heterogeneous mixture of antibodies passed through and eluted from an adsorbens column will at least be "operationally" specific for a particular collagen, since antibodies to contaminants in these preparations are usually below the level of sensitivity of the assay systems. Such antibodies to contaminants are, however, easily detectable in the nonbound fraction.[66]
2. Immunoadsorbentia can be used to isolate antibodies of defined specificity to small antigenic fragments, native or denatured collagen only, or to single antigenic determinants.
3. Immunoadsorbentia can be used to concentrate antibodies which may sometimes be present only in minute quantities in an antiserum.

Considerable information has been obtained with collagen antibodies obtained by these conventional means, but careful work and sometimes rather elaborate work is required to obtain antibodies of defined specificities for meaningful studies. It remains still to be determined whether this conventional approach to produce specific and useful collagen antibody reagents can be abandoned in favor of monoclonal antibodies, obtained by somatic cell hybridization techniques. This alternative would certainly circumvent some of the ambiguities expressed above which cannot be completely avoided with antibodies raised in rabbits or other species short of preparing antibodies to single determinants. Although the latter approach allows one to be reasonably confident in the quality of the antibody reagent, this may not be possible for every collagen type, and this approach would be extremely difficult at the present time for some of the more recently discovered molecules for which the chemical background does not exist to make such attempts even feasible.

What is an immunosorbent? It may be defined as an insoluble material to which antigen is covalently attached in such a fashion that the antigen retains antibodies. In order to be able to recover only the specific antibodies from immunosorbents, the insoluble matrix should be chemically inert — it should not allow interaction of other serum proteins including serum IgG. Provided this is the case, such adsorbents can be used also as an analytical tool for the detection and estimation of antibody protein present in antisera. The other important property of the adsorbent is its capacity. This depends on the number of reactive groups on the adsorbent, the molecular size of the polymer unit structure, the molecular size and shape of the antigen to be attached to the adsorbens, and for some adsorbentia, the pore size of the matrix. The latter property, for instance, may allow binding of a small antigenic peptide inside the matrix, but it could nevertheless be lost for antibody interaction because the much larger antibody molecule may be excluded from the interior of the matrix due to its size. The capacity in this case solely depends on the number of reactive antigen molecules present on the matrix surface. It is therefore not always possible to predict the capacity of an adsorbens from chemical information and the procedures used are mostly empirical in nature.

Among the many different types of insoluble supporting matrices, investigators have been commonly using only two — *p*-aminobenzyl cellulose (PAB-cellulose) and Sepharose®.

1. Preparation of PAB-Cellulose Immunoadsorbens (Native and Denatured Collagen)

p-Aminobenzyl cellulose is commercially available (Serva, Heidelberg, Germany) and is made by reaction of *p*-nitrobenzylchloride and cellulose, followed by reduction of the nitrogroup. The resulting aminoaryl cellulose can be diazotized and coupled to proteins that react with diazonium salts.[67] For coupling of collagen to diazotized PAB-cellulose, the procedure of Webb and Lapresle[68] was followed:[69,70]

1. Six grams of the *p*-aminobenzyl cellulose is suspended in 160 mℓ 0.2 *N* HCl at 4°C, and with constant stirring 1.2 g of sodium nitrite in 5 mℓ distilled water is added dropwise. The temperature is kept between 3 to 4°C for 30 min, and then the material is sedimented in a refrigerated centrifuge and washed once with 200 mℓ of 5% sodium acetate, once with 200 mℓ of 1 *M* urea, and three times with 200 mℓ of distilled water.
2. The diazotized cellulose derivative is added to a solution of native collagen (100 mg/100 mℓ 0.1 *N* acetic acid and dialyzed against phosphate buffer pH 7.4, I = 0.15), and the mixture is stirred at 4°C for 3 days. After this coupling step, the material is sedimented by centrifugation in the cold and washed three times with ice cold 0.15 *M* saline.
3. To block the unreacted diazonium groups, the sediment after the last centrifugation is stirred with 1 ℓ 0.1% α-naphtol, pH 8.0 (1 g α-naphtol dissolved in 4 to 6 mℓ 2 *N* NaOH, diluted to approximately 1 ℓ with distilled water and the pH adjusted to 8.0 with acetic acid), for 4 hours at 4°C. This is followed by two washes with 0.15 *M* saline.

For the coupling of PAB-cellulose with denatured collagen, the identical procedure is followed from Steps 1 to 3, except that all manipulations are done at room temperature. One hundred mg of collagen in 100 mℓ of 0.1 *N* acetic acid or sodium

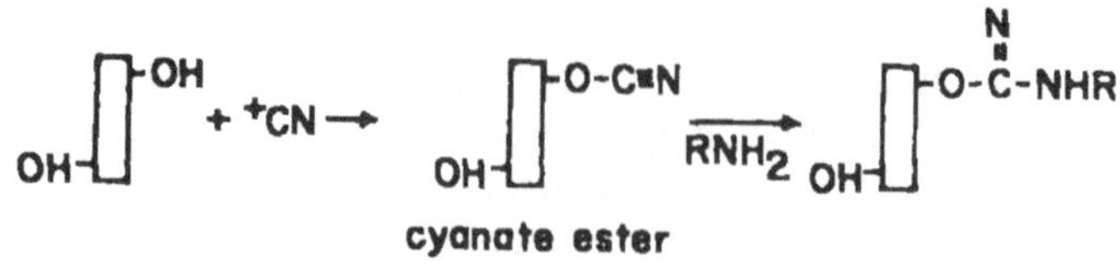

FIGURE 6. Chemistry of the coupling of proteins to cyanogen bromide-activated Sepharose®.

citrate, pH 3.7 (31.52 g citric acid. H_2O, 22.02 g sodium citrate. 2 H_2O in 1 ℓ of distilled water), is heated at 40°C for 90 min and then dialyzed at room temperature against 0.2 *M* sodium borate, pH 8.5. After Step 3, the sediment is suspended in 200 mℓ 0.15 *M* saline and is stirred overnight at 37°C to extract any noncovalently bound collagen. After two additional washes in saline, the immunosorbent is suspended in 200 mℓ of PBS 7.2 and is kept frozen at −20°C. Hydroxyproline determination of different immunoadsorbents showed that 1 g contains generally between 5 to 8 mg covalently attached collagen.

2. Preparation of Sepharose® 4B Immunoadsorbens

Agarose beads have been one of the most useful and versatile polysaccharide supports for affinity chromatography. In order to covalently attach proteins to this solid support, the agarose must first be activated. The most widely used method for activation is the reaction with cyanogen bromide (CNBr) introduced in 1976.[71] During the reaction of CNBr with agarose, reactive cyanate ester groups are formed which will readily react with amino groups in proteins (see Figure 6).[71,72] Although a number of procedures have been described using solid CNBr[73] or CNBr dissolved in organic solvents such as acetonitrile,[74] studies on the activation product of Sepharose® by a colorimetric method suggest that maximal activation occurs with solid CNBr and within a few minutes.[72] The method described here for coupling of collagen to cyanogen bromide-activated agarose follows the published procedures closely and has been adapted with only some minor modification:

1. Sepharose® 4B (Pharmacia) is washed thoroughly with 0.1 *M* NaCl and distilled water and is finally suspended with a volume of water equal to the volume of the beads. The volume of the beads is determined in milliliters of packed gel after centrifugation.
2. The slurry in a beaker is placed on a magnetic stirrer in an ice bath and is placed into a fume hood. A balance and pH meter are required in addition in the hood area. The CNBr in small crystals is weighed (50 mg/mℓ of beads) and is added at once to the beads, and the pH is titrated and kept between 11 and 11.5 by addition of 4 *N* NaOH with gentle stirring. The initial consumption of the base will be high as CNBr dissolves, but less base will have to be added after a few minutes to keep the pH within range. After 8 to 10 min, the CNBr is dissolved, and the slurry is poured into a chilled sintered-glass funnel containing chunks of ice, and the gel is rapidly filtered with suction and washed with ten volumes of ice-cold distilled water (within 2 to 3 min) and five to ten volumes of 0.1 *M* calcium acetate pH 8.8.
3. The packed activated Sepharose® is spooned out into the collagen, dissolved or dialyzed into 0.1 *M* calcium acetate pH 8.8 (5 mg/mℓ of solvent for 1 mℓ of gel), and the solution is shaken gently for 24 hr at 4°C on a shaker. Stirring of

the agarose suspension by using a magnetic stirring bar should be avoided. The broken agarose beads will cause slower flow rates and increased nonspecific adsorption.

4. After this coupling step, the suspension is centrifuged, and the supernate is saved (for determination of nonbound protein). After brief washes, an equal volume of 0.2 *M* calcium acetate, pH 8.8, containing 1% ethanolamine (or 1 *M* glycine), is added to promote masking of excess activated groups which have not been utilized for coupling. After shaking the suspension for 6 hr at 4°C, the gel is washed sequentially with 0.1 *M* calcium acetate, pH 8.0; 1 *M* acetic acid containing 0.15 *M* NaCl (the pH of the gel suspension should be 3); 1 *M* Tris-(hydroxymethyl)-aminomethane (to pH of 7.0); and finally with PBS 7.2 containing 0.02% sodium azide. The gel is stored at 4°C.

a. Modification of the Conditions for Activation and Coupling

The amount of ligand coupled to the gel depends on the amount of CNBr added. Although the binding capacity of Sepharose® increases with its cyanate ester content, it is not possible to establish a simple direct correlation between the concentration of cyanate ester groups and the coupling capacity for a specific protein. Therefore, if low capacity of the adsorbens presents a problem, higher amounts of CNBr can be used. Typically this varies between 50 and 300 mg of CNBr per milliliter of gel. Larger amounts of proteins have also been used for coupling, but higher substitution does not necessarily lead to an absorbens with greater capacity. For collagen, α-chains, or smaller peptides, 5 mg/mℓ of beads produces typically an effective immunoadsorbens.

3. Isolation of Antibodies by Immunoadsorption

The usual methods for the isolation of antibodies employ either a batch or column procedure. We find it convenient to continuously monitor column effluents with a spectrophotometer and have thus used the column technique exclusively. The gel slurry is filled into an appropriately sized column. In general 20 to 30 mℓ of immunoadsorbens is used for the isolation of antibodies from 20 mℓ of antiserum. The gel is washed with PBS 7.2, and the column effluent is monitored at 280 nm. The flow rate is maintained at about 60 mℓ/hr. Ten to 20 mℓ of serum is applied and washing of the column is continued with PBS until the absorbance reaches the buffer-baseline again. To dissociate and elute the bound antibodies from the adsorbens, the washing buffer is substituted for a 1 *M* acetic acid solution containing 0.15 *M* NaCl until the pH of the effluent is around 3, followed by 0.05 *M* HCL, 0.15 *M* NaCl. Then the buffer is switched to 1 *M* Tris-(hydroxymethyl)-aminomethane until the effluent reaches pH 7.0, and finally the column is washed with PBS 7.2 to prepare the adsorbens for the next cycle. Antibodies usually begin to elute around pH 5, and this is seen as a rise in adsorbance at 280 nm. The effluent is collected on a fraction collector. The material from the nonbound protein-containing peak and the eluted antibodies are pooled immediately. The pH of the eluate is adjusted to near neutrality by addition of 2 *M* Tris-(hydroxymethyl)-aminomethane, and this is followed by extensive dialysis in the cold room against PBS 7.2. Antibodies are concentrated by negative pressure dialysis or by ultrafiltration on Amicon filters to a concentration of about 0.5 to 1 mg/mℓ (OD_{280} 1-2).

Both the nonbound serum fraction and the eluted antibody-containing fraction are tested by passive hemagglutination in comparison with the unseparated antiserum to analyze for complete absorption and for recovery.

The specificity of the eluted antibodies is determined by appropriate procedures — by passive hemagglutination-inhibition to obtain semiquantitative data, or quantitatively by radioimmunoassay as described elsewhere.

The antibody solutions are stored frozen at −20°C in aliquots of 0.5 mℓ or less. Multiple freezing and thawing will result in loss of antibody activity due to precipitation or for other reasons. Long-term storage of antibodies did not usually affect antibody activity. The preparations can also be lyophilized and stored at 4°C. Before use the sample is reconstituted with an appropriate volume of distilled water.

a. Peptide Elution Technique

In the original work on the isolation of antibodies to Type I collagen, tryptic peptides were used to dissociate antibodies with particular affinity from the immunoadsorbens.[69,70] This method of elution has proven to be highly efficient and less severe than the acidic conditions described above. Although not all the peptide material could be removed from the eluted antibodies by dialysis, this did not interfere with subsequent testing. The method is useful for antibodies to Type I collagen, but may not be widely used for other collagens because of the amount of peptides needed which limits its application. Furthermore, since peptides are used for elution, antibodies to central and perhaps terminal determinants (see this chapter) will be dissociated,[70,33] but antibodies to the conformation dependent helical sites will not be recovered. The method has, however, been used to separate and purify various antibody specificities.

b. Preparation of Peptides

Twelve grams of a freeze-dried calf skin split (dermis) is cut into small pieces with scissors and is suspended in 500 mℓ of distilled water. Swelling of the material is achieved by incubation at 65°C for 40 min. The pH is adjusted to 8 with 2 *N* NaOH, and after centrifugation at 10,000 r/min for 30 min, the residue is suspended by homogenization in 0.2 *M* $NaHCO_3$, pH 8. Trypsin is added (1 mg/g wet weight) and digestion is done for 3 to 4 hr at 37°C. After this period, the solution is heated to 60°C for 20 min, sodium azide is added to a concentration of 0.02%, and the insoluble material is sedimented by centrifugation at 10,000 r/min for 1 hr. The supernate is filtered through a 0.45 μm filter and then is concentrated threefold by rotary evaporation. The concentration of the peptides are in the order of 2.5 to 4 mg protein *N* per milliliter.

In a typical experiment, 6 g collagen-immunoadsorbens (PAB-cellulose) is filled into a column (1.6 × 10 cm), and 10 mℓ of the antiserum is loaded. After extensive washing with PBS 7.2, 5 to 10 mℓ of the peptide solution is passed through the column with monitoring of the column effluent at 230 nm. This is followed by washing with PBS 7.2. Since not all the antibodies can be eluted with the peptide solution, the adsorbens can be completely regenerated by elution of the remaining antibody protein with 1 *M* acetic acid containing 0.15 *M* NaCl, followed by neutralization with 1 *M* Tris-(hydroxymethyl)-aminomethane and PBS 7.2.

The antibodies in the peptide eluate are concentrated by addition of an equal volume of saturated ammonium sulfate (720 g in 1 ℓ of distilled water). The precipitate is recovered after 12 hr at 4°C by centrifugation and is dissolved in 1 mℓ of PBS 7.2 or of normal rabbit serum (heat treated at 56°C for 30 min and absorbed with human red cells). After dialysis against PBS 7.2 in the cold, the antibody solution is ready for testing.

III. PREPARATION OF COLLAGEN TYPE-SPECIFIC ANTIBODIES

Most of the genetically distinct types of collagen are quite similar to one another in their macromolecular structure, yet they are sufficiently different in amino acid sequence to permit the production of specific antibodies, specific for each type of collagen in a given species. Thus, antibodies to Type I, for instance, will not cross-react with Type II collagen or will do so only weakly. Cross-reactions between Types I and III collagen on the other hand may be stronger when compared to other collagens.[20] Despite the homology in primary structure of the α-chains of a particular collagen type from different species, the cross-reactivity of antibodies raised in rabbits or other species with the same collagen type from another species is quite restricted and for some types is generally not observed. Because of this species specificity, antibodies to a particular type and for a particular experimental system have to be developed and characterized separately, and only in some instances will more universally reactive antibodies be obtained. Table 6 gives a selected list of type-specific antibodies and their observed cross-reactivities. In compiling this list, the major emphasis has been placed on antibodies obtained from species in which reasonable amounts of antibodies can be expected. Some data are available in the literature on mouse antibodies to collagen (see Volume II, Chapter 7), but unless these are isolated from peritoneal fluid after stimulation, the amounts of antibodies which can be recovered are small.

Monoclonal antibodies to collagen types are very likely even more restricted in their cross-reactivity and species specificity and may react only with collagen of the same type and species, even in the case in which rabbit antibodies exhibit considerable reactivity across species.

A. Antibodies to Native Collagens

This section will deal only with the preparation of type-specific antibodies by conventional procedures. Because antisera to any type of collagen frequently contain variable amounts of antibodies that react with other collagen types in addition to noncollagenous impurities, the antisera have to be purified by appropriate cross-absorptions on immunoadsorbentia. These cross-absorptions should precede the final binding to and elution of the antibodies from the specific immunoadsorbens. Since cross-reactive antigenic sites are not necessarily shared by the various collagen types, it is recommended that several collagens are used to remove such antibodies, e.g., removal of antibodies to Type II collagen does not eliminate the cross-reaction with Type III collagen of antisera to Type I collagen.[76] For this reason, antisera routinely are absorbed on immunoabsorbentia prepared with several or all the other collagens except for the one used for immunization. Previous work has shown that antisera even after extensive cross-absorptions still retain high titers for the collagen to which the antibodies had been raised, and these antibodies can be recovered from the antiserum as almost pure antibody protein from a specific immunoadsorbens. In Figure 7, the purification of antibodies to Type V collagen is shown as an example. The scheme uses a series of immunoadsorption steps, with the final step being the elution of the desired type-specific antibodies from the appropriate collagen-immunoadsorbens. Not shown in this scheme is the cross-adsorption on Type IV collagen or laminin. Since Type V (AB_2) collagen usually is obtained from tissues rich in basement membranes, the potential contamination with these proteins should not be overlooked. It is convenient to run the immunoadsorbens columns in series and regenerate the adsorbentia separately with acidic eluants. It is also possible to

Table 6
SPECIES-SPECIFICITY OF COLLAGEN-TYPE SPECIFIC ANTIBODIES[a]

Collagen	Species immunized	Source of collagen	Species cross-reactive	Ref.
Type I	Rabbit	Calf	Human	5
	Rabbit	Mouse	Rat, hamster, guinea pig	37
	Guinea pig	Calf		61
	Guinea pig	Chicken	Human	36, 59
	Mouse	Chicken		49[b]
	Rabbit	Chicken		36
pN	Rabbit	Calf	Human	5
	Rabbit	Sheep	Human, calf	19
	Rabbit	Rat	Mouse	38
pC	Rabbit	Chicken	Rat, human	84
Pn	Mouse	Calf		23
Type II	Rat	Calf	Human, calf	30, 36
	Rabbit	Rat		30
	Rabbit	Chicken	Rat, calf	36
	Rabbit	Calf	Chicken, human	36, 76
	Mouse	Calf		29
pC	Rabbit	Chicken		84, 94
Type III	Rabbit	Calf	Human	5, 20
	Goat	Chicken		86
	Rabbit	Human	Calf	43, 50, 60
	Guinea pig	Chicken		86
pN	Rabbit	Rat	Mouse	38
	Rabbit	Calf	Chicken	5, 20, 39, 85, 86
Type IV	Rabbit	Human	Calf, mouse, rat, chicken	44
	Rabbit	Mouse	Calf, mouse, rat, chicken	43, 44, 50, 60
	Mouse	Calf		41[b]
	Guinea pig	Mouse		44
	Mouse	Human		41[b]
Type V	Rabbit	Human	Calf, mouse, rat, chicken	43, 48, 50
	Mouse	Human	Calf, human	93

[a] Compare also Chapters 3 and 4 in Volume 2.
[b] Monoclonal antibodies.

mix the various adsorbentia for the removal of the cross-reactive antibodies. The disadvantage of the latter procedure may be the loss of activity of one of the components. Since the level of cross-reaction is usually quite small, this will possibly remain obscured. Because of the low amounts of cross-reactive antibodies in collagen antisera, much larger volumes of antisera can be applied to a given amount of adsorbens without exhausting its capacity.

As outlined in one of the sections of this chapter, collagen type-specific antibodies isolated from an immunoadsorbens of native collagen will be specific for and react with helical or terminal determinants. If antibodies to the triple-helical body of collagen of a particular type are desired, these two antibody specificities can be distinguished and separated by immunoadsorption on a denatured collagen adsorbens prior to the native collagen adsorbens. It is evident that the immunoadsorption procedure provides a powerful tool to concentrate antibodies which may occur only

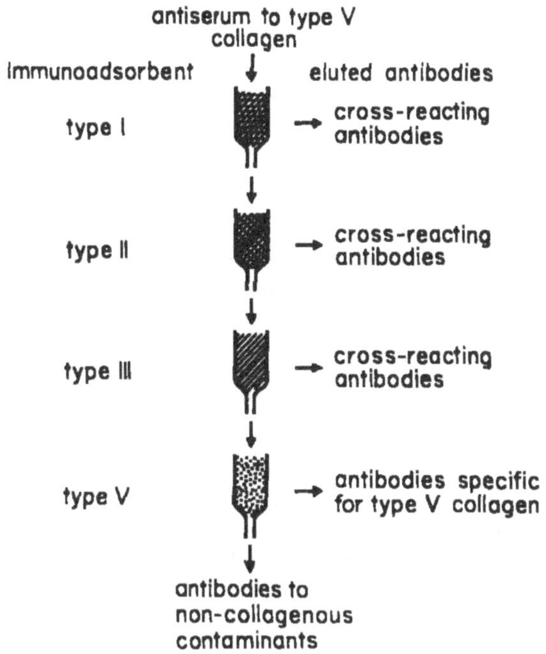

FIGURE 7. Purification and isolation of collagen type-specific antibodies. The crude antiserum is passed over a series of different immunoadsorbentia (only adsorbentia of collagen Types I, II, and III are shown) to remove cross-reactive antibodies, and finally the type-specific antibodies are adsorbed onto and eluted from the specific immunoadsorbens (in this case, Type V collagen).

with low titers in the crude antiserum. Rabbit antisera, for instance, contain only low levels of antibodies reacting with helical determinants. These antibodies are only useful, if they can be enriched.

B. Antibodies to Procollagen as Collagen Type-Specific Reagents

The biosynthetic precursor form of the collagen molecule is termed procollagen. This molecule contains aminoterminal and carboxyterminal peptide extensions on each collagen α-chain — aminopropeptide (pN) and carboxypropeptide (pC) — which account for 10 and 20% of the total mass (M_r = 450,000 daltons for Type I or Type III collagen). Although most studies to date have dealt with Type I and Type III procollagen, similar precursor forms containing the typical features of extension peptides apparently exist for Type II,[77,78] Type IV (Chapter 1), and Type V collagen.[79] Since the procollagen has consistently been found to be a better immunogen than collagen and since cross-reactions of specific antibodies to procollagen are usually not observed either between the amino- and carboxyterminal peptide, nor between extension peptides on different collagen types, these antibodies thus by definition are type-specific and can provide very useful reagents which exhibit high affinity constants. One has to keep in mind though in using such antibody reagents for various applications that they recognize structures on precursor molecules only. It is not clear at the moment what the kinetics of the conversion of procollagen into collagen are for the various types in vivo. Although numerous studies have shown that procollagens are detectable by the immunofluorescence technique in various tissues,[5,52] only small and negligible amounts can be isolated.

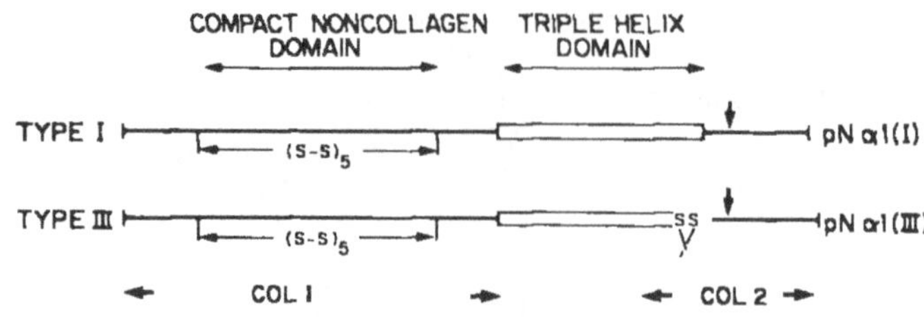

FIGURE 8. The domains of the pN aminoterminal extension peptide of procollagen. The α1(I)- and α1(III)-chains contain a compact noncollagenous domain that is stabilized by five intrachain disulfide bridges. It is flanked by flexible peptide regions on both sides. At the carboxyterminal side of this domain a triple-helical region connects the pN-peptide to the nonhelical region of the collagen molecule (see Figure 1). The vertical arrows indicate the cleavage site at which procollagen N-protease clips the polypeptide chains of procollagen after secretion from the cell. Bacterial collagenase releases the two fragments Col 1 and Col 2 which mainly consist of noncollagenous sequences. Antigenic sites are predominantly found associated with the compact domain of peptide Col 1.

The major source of aminopropeptides has been the skin of dermatosparactic calf or sheep which contains considerable amounts of pN-collagen. In distinction to procollagen, pN-collagen α-chains contain only the aminopropeptide. However, procollagen can also be obtained in sufficient quantities from fetal skin,[80–82] the skin of colchicine-treated rats,[83] and from organ culture systems.[84] The latter study clearly demonstrates that potentially all procollagens can be obtained from tissue culture systems in quantities sufficient for immunochemical work. In Chapter 7 of this volume, the experimental procedures are given in more detail to particularly obtain antibodies to the pC-peptide.

The discussion of procollagen antibodies will be restricted thus to antibody specificities related to pN-collagen. Antibodies to bovine Type I and Type III pN-collagen, although specific for and not cross-reactive with each procollagen antigen, show better interspecies cross-reactivity (see Table 6). Antibodies to bovine and sheep Type I pN-peptide react with human, but not rat or chick antigen, while antibodies to bovine Type III pN-peptide bind the chicken antigen.[85,86] The major antigenic sites are dependent on the complex conformation of the pN-peptide, since cleavage with proteases within the peptide alters the affinity of the antibodies for the peptide and since reduction of the intrachain disulfide bridges and alkylation abolishes antigenicity.[87] The antigenic sites are found mostly associated with the noncollagenous domain Col 1 (see Figure 8) which can be obtained by collagenase digestion.

Antibodies with specificity for the aminoterminal extension peptide can be isolated from antisera raised in rabbits by multiple injections of 500 μg quantities of pN-collagen or 100 μg Col 1 peptides. The sera are isolated on affinity systems as described above: on a native collagen immunoadsorbens to remove antibodies to collagen and by binding to and elution from a pN collagen, pN-α1(I)- or pN-α1(III)-chain, or Col 1 peptide immunoadsorbens.

Various studies have demonstrated that such antibodies provide useful tools to analyze tissue culture cells,[85,86,90–92] the extracellular matrix in vivo,[39,52] or biological fluids[19] for the presence of the pN-peptide component. Because of the limited amounts of peptide material which are normally available, radioimmunoassay procedures as described in Chapter 13 are the method of choice to analyze for antibody activity and specificity.

ACKNOWLEDGMENTS

Work in the author's laboratory was supported by a contract from the National Cancer Institute, NO1-CB-84225, grants from the Fairfield Chapter of the American Heart Association, the Anna Fuller Fund, and NIH/DRR Basic Research Support Grants, and a Research Award from the American Cancer Society. I would like to thank R. Timpl and R. Glanville for a preview of their article on the aminopropeptide of collagen.

REFERENCES

1. **Furthmayr, H. and Timpl, R.,** Immunochemistry of collagens and procollagens, *Int. Rev. Connect. Tissue Res.*, 7, 61, 1976.
2. **Timpl, R.,** Immunochemistry of collagens and procollagens, in *Biochemistry of Collagen*, Ramachandran, G. N. and Reddi, A. H., Eds., Plenum Press, New York, 1976, 310.
3. **Landsteiner, K.,** *The Specificity of Serological Reactions*, Bailliere Tindall, London, 1936, 28.
4. **Watson, R. F., Rothbard, S., and Vanamee, P.,** The antigenicity of rat collagen, *J. Exp. Med.*, 99, 535, 1954.
5. **Timpl, R., Wick, G., and Gay, S.,** Antibodies to distinct types of collagens and procollagens and their application to immunohistology, *J. Immunol. Methods*, 18, 165, 1977.
6. **Michaeli, D., Senyk, G., Maoz, A., and Fuchs, S.,** Ascaris cuticle collagen and mammalian collagens — cell mediated and humoral immunity relationships, *J. Immunol.*, 109, 103, 1972.
7. **Timpl, R., Fietzek, P. P., Furthmayr, H., Meigel, W., and Kühn, K.,** Evidence for two antigenic determinants in the C-terminal region of rat skin collagen, *FEBS Lett.*, 9, 11, 1970.
8. **Lindsley, H. B., Mannik, M., and Bornstein, P.,** Immunogenicity of rat and human collagen polypeptide chains, *Arthritis Rheum.*, 12 (Abstr.), 676, 1969.
9. **Furthmayr, H., Stoltz, M., Becker, U., Beil, W., and Timpl, R.,** Chicken antibodies to soluble rat collagen. II. Specificity of the reactions with individual polypeptide chains and cyanogen bromide peptides of collagen, *Immunochemistry*, 9, 789, 1972.
10. **Michaeli, D. and Epstein, E. H.,** Isolation and identification of the antigenic determinant of human collagen, *Isr. J. Med. Sci.*, 7, 462, 1971.
11. **Furthmayr, H., Beil, W., and Timpl, R.,** Different antigenic determinants in the polypeptide chains of human collagen, *FEBS Lett.*, 12, 341, 1971.
12. **LeRoy, C. E.,** Precipitating and complement-fixing antibodies to collagen with species and collagen subunit specificity, *Proc. Soc. Exp. Biol. Med.*, 128, 341, 1968.
13. **Pontz, B., Meigel, W., Rauterberg, J., and Kühn, K.,** Localization of two species specific antigenic determinants on peptide chains of calf skin collagen, *Eur. J. Biochem.*, 16, 50, 1970.
14. **Michaeli, D., Martin, G. R., Kettman, J., Benjamini, E., Leung, D. Y. K., and Blatt B. A.,** Localization of antigenic determinants in polypeptide chains of collagen, *Science*, 166, 1522, 1969.
15. **Meigel, W., Pontz, B., Timpl, R., Hieber, E., Nordwig, A., Steffen, C., and Kühn, K.,** Comparative immunological studies on fish and invertebrate collagens, *J. Immunol.*, 107, 1146, 1971.
16. **Timpl, R., Beil, W., Furthmayr, H., Meigel, W., and Pontz, B.,** Characterization of conformation independent antigenic determinants in the triple-helical part of calf and rat collagen, *Immunology*, 21, 1017, 1971.
17. **Timpl, R., Wick, G., Furthmayr, H., Lapière, C. M., and Kühn, K.,** Immunochemical properties of procollagen from dermatosparactic calves, *Eur. J. Biochem.*, 32, 584, 1973.
18. **Becker, U., Timpl, R., Helle, O., and Prockop, D. J.,** NH_2-terminal extensions on skin collagen from sheep with a genetic defect in conversion of procollagen into collagen, *Biochemistry*, 15, 2853, 1976.

19. **Rohde, H., Nowack, H., Becker, U., and Timpl, R.,** Radio immunoassay for the aminoterminal peptide of procollagen p α1(I)-chain, *J. Immunol. Methods*, 11, 135, 1976.
20. **Nowack, H., Gay, S., Wick, G., Becker, U., and Timpl, R.,** Specific antibodies against type III collagen and procollagen, *J. Immunol. Methods*, 12, 117, 1976.
21. **Rohde, H., Becker, U., Nowack, H., and Timpl, R.,** Antigenic structure of the aminoterminal region in type I procollagen. Characterization of sequential and conformational determinants, *Immunochemistry*, 13, 967, 1976.
22. **von der Mark, K., Click, E. M., and Bornstein, D.,** Immunology of procollagen. I. Development of antibodies to determinants unique to pro-α1 chains, *Arch. Biochem. Biophys.*, 156, 356, 1973.
23. **Nowack, H., Rohde, H., Götze, D., and Timpl, R.,** Genetic control and carrier and suppressor effects in antibody response of mice to procollagen, *Immunogenetics*, 4, 117, 1977.
24. **Rhode, H., Nowack, H., and Timpl, R.,** Localization of antigenic activity and immunogenic capacity in different conformational domains of procollagen peptide, *Eur. J. Immunol.*, 8, 141, 1978.
25. **Rauterberg, J., Timpl, R., and Furthmayr, H.,** Structural characterization of N-terminal antigenic determinants in calf and human collagen, *Eur. J. Biochem.*, 27, 231, 1972.
26. **Timpl, R., Furthmayr, H., and Beil, W.,** Maturation of the immune response to soluble rat collagen: late appearance of antibodies directed to N-terminal sites of the α2-chain, *J. Immunol.*, 108, 119, 1972.
27. **Lindsley, H., Mannik, M., and Bornstein, P.,** The distribution of antigenic determinants in rat skin collagen, *J. Exp. Med.*, 133, 1309, 1971.
28. **Becker, U., Timpl, R., and Kühn, K.,** Carboxyterminal antigenic determinants of collagen from calf skin — localization within discrete regions of the nonhelical sequence, *Eur. J. Biochem.*, 28, 221, 1972.
29. **Nowack, H., Hahn, E., and Timpl, R.,** Specificity of antibody response in inbred mice to bovine type-I and type-II collagen, *Immunology*, 29, 621, 1972.
30. **Hahn, E., Timpl, R., and Miller, E. J.,** Production of specific antibodies to native collagens with chain compositions $[\alpha1(I)]_3$, $[\alpha1(II)]_3$ and $[\alpha1(I)]_2\alpha2$, *J. Immunol.*, 113, 421, 1974.
31. **Hahn, E. Timpl, R.,** Involvement of more than a single polypeptide chain in helical antigenic determinants of collagen, *Eur. J. Immunol.*, 3, 442, 1973.
32. **Beil, W., Furthmayr, H., and Timpl, R.,** Chicken antibodies to soluble rat collagen. I. Characterization of the immune response by precipitation and agglutination methods, *Immunochemistry*, 9, 779, 1972.
33. **Beil, W., Timpl, R., and Furthmayr, H.,** Conformation dependence of antigenic determinants on the collagen molecule, *Immunology*, 24, 13, 1973.
34. **von der Mark, K., von der Mark, H., and Gay, S.,** Study of differential collagen synthesis during development of the chick embryo by immunofluorescence. II. Localization of type I and type II collagen during long bone development, *Dev. Biol.*, 53, 153, 1976.
35. **Hahn, E., Timpl, R., and Miller, E. J.,** Demonstration of a unique antigenic specificity for collagen α1(II) chains from cartilagenous tissue, *Immunology*, 28, 561, 1975.
36. **von der Mark, H., von der Mark, K., and Gay, S.,** Study of differential collagen synthesis during development of the chick embryo by immunofluorescence. I. Preparation of collagen type I and type II specific antibodies and their application to early stages of the chick embryo, *Dev. Biol.*, 48, 237, 1976.
37. **Roll, J. F., Madri, J. A., and Furthmayr, H.,** Preparation of mouse collagen type I, III, IV and V specific antibodies, unpublished, 1981.
38. **Roll, J. F. and Furthmayr, H.,** Preparation and reactivity of rat procollagen type I and type III specific antibodies, unpublished, 1981.
39. **Becker, U., Nowack, H., Gay, S., and Timpl, R.,** Production and specificity of antibodies against the aminoterminal region in type III collagen, *Immunology*, 31, 57, 1976.
40. **Nowack, H., Olsen, B. R., and Timpl, R.,** Characterization of the aminoterminal sequence in type III procollagen, *Eur. J. Biochem.*, 70, 205, 1976.
41. **Foellmer, H., Madri, J. A., and Furthmayr, H.,** Preparation of antibodies specific for type IV collagen isolated from calf lung, unpublished, 1981.
42. **Timpl, R., Martin, G. R., Bruckner, P., Wick, G., and Wiedemann, H.,** Nature of the collagenous protein in a tumor basement membrane, *Eur. J. Biochem.*, 84, 43, 1978.
43. **Madri, J. A. and Furthmayr, H.,** Collagen polymorphism in the lung: an immunochemical study of pulmonary fibrosis, *Hum. Pathol.*, 11, 353, 1980.
44. **Timpl, R., Glanville, R. W., Wick, G., and Martin, G. R.,** Immunochemical study on basement membrane (type IV) collagens, *Immunology*, 38, 109, 1979.
45. **Sherman, M. I., Gay, R., Gay, S., and Miller, E. J.,** Association of collagen with preimplantation and peri-implantation mouse embryos, *Dev. Biol.*, 74, 470, 1980.

46. **Risteli, J., Bächinger, H. P., Engel, J., Furthmayr, H., and Timpl, R.,** 7S collagen: characterization of an unusual basement membrane structure, *Eur. J. Biochem.*, 108, 239, 1980.
47. **Timpl, R., Risteli, H., and Bächinger, H. P.,** Identification of a new basement membrane collagen by the aid of a large fragment resistant to bacterial collagenase, *FEBS Lett.*, 101, 265, 1979.
48. **Madri, J. A. and Furthmayr, H.,** Isolation and tissue localization of type AB_2 collagen from normal lung parenchyma, *Am. J. Pathol.*, 94, 323, 1979.
49. **Linsenmayer, T. F., Hendrix, M. J. C., and Little, C. D.,** Production and characterization of a monoclonal antibody to chicken type I collagen, *Proc. Natl. Acad. Sci. U.S.A.*, 76, 3703, 1979.
50. **Madri, J., Dreyer, B., Pitlick, F. A., and Furthmayr, H.,** The collagenous components of the subendothelium. Correlation of structure and function, *Lab. Invest.*, 43, 303, 1980.
51. **Sherr, C. J., Taubman, M. B., and Goldberg, B.,** Isolation of a disulfide-stabilized three-chain polypeptide fragment unique to the precursor of human collagen, *J. Biol. Chem.*, 248, 7035, 1973.
52. **Wick, G., Brunner, H., Penner, E., and Timpl, R.,** Diagnostic application of specific anti-procollagen sera. II. Analysis of liver biopsies, *Int. Arch. Allergy App. Immunol.*, 56, 316, 1978.
53. **Herbert, W. J.,** in *Handbook of Experimental Immunology*, Vol. 3, Weir, D. M., Ed., Blackwell Scientific, Oxford, 1978, A3.1, A4.1.
54. **Boyden, S. V.,** The adsorption of proteins on erythrocytes treated with tannic acid and subsequent hemagglutination by antiprotein sera, *J. Exp. Med.*, 93, 107, 1951.
55. **Stavitzky, A.,** Micromethods for the study of proteins and antibodies. I. Procedure and general applications of hemagglutination and hemagglutination-inhibition reactions with tannic acid and protein-treated red blood cells, *J. Immunol.*, 72, 360, 1951.
56. **Steffen, C., Timpl, R., and Wolff, I.,** Immunogenicity and specificity of collagen. III. Induction and demonstration of anti-collagen antibodies and serological comparison of native and denatured soluble collagen, *Z. Immunitaetsforsch. Allergy Klin. Immunol.*, 134, 91, 1967.
57. **Steffen, C., Timpl, R., and Wolff, I.,** Immunogenicity and specificity of collagen. V. Demonstration of three different antigenic determinants of calf collagen, *Immunology*, 15, 135, 1968.
58. **Timpl, R., Wolff, I., Wick, G., Furthmayr, H., and Steffen, C.,** Immunogenicity and specificity of collagen. VII. Differences between various collagens demonstrated by cross-reaction studies, *J. Immunol.*, 101, 725, 1968.
59. **von der Mark, K.,** Localization of collagen types in tissues, *Int. Rev. Connect. Tissue Res.*, in press, 1981.
60. **Roll, J. F., Madri, J. A., Albert, J., and Furthmayr, H.,** Codistribution of collagen types IV and AB_2 in basement membranes and mesangium of the kidney, *J. Cell. Biol.*, 85, 597, 1980.
61. **Gentner, G. J. and Adelmann, B. C.,** The relation between the cell-mediated immunological response and the induction of circulating antibodies to collagen in guinea pigs, *Immunology*, 31, 95, 1976.
62. **Michaeli, D., Benjamini, E., and Leung, D. Y. K.,** Immunochemical studies on collagen. II. Antigenic differences between guinea pig skin collagen and gelatin, *Immunochemistry*, 8, 1, 1971.
63. **Herbert, W. J.,** in *Handbook of Experimental Immunology*, Vol. 1, Weir, D. M., Ed., Blackwell Scientific, Oxford, 1978, 20.1.
64. **Springer, G. F. and Williamson, P.,** Immunochemical significance of L- and D-fucose derivatives, *Biochem. J.*, 85, 282, 1962.
65. **Avrameas, S., Taudon, B., and Chuilon, S.,** Glutaraldehyde, cyanuric chloride and tetra azotized *O*-dianisidine as coupling reagents in the passive hemagglutination test, *Immunochemistry*, 6, 67, 1969.
66. **Wick, G., Furthmayr, H., and Timpl, R.,** Purified antibodies to collagen: an immunofluorescence study of their reaction with tissue collagen, *Int. Arch. Allergy Appl. Immunol.*, 48, 664, 1975.
67. **Campbell, D. H., Luescher, E., and Lerman, L. S.,** Immunologic adsorbents. I. Isolation of antibody by means of a cellulose-protein antigen, *Proc. Natl. Acad. Sci. U.S.A.*, 37, 575, 1951.
68. **Webb, T. and Lapresle, C.,** Isolation and study of rabbit antibodies specific for certain of the antigenic groups of human serum albumin, *Biochem. J.*, 91, 24, 1964.
69. **Timpl, R., Furthmayr, H., Steffen, C., and Doleschel, W.,** Isolation of pure anti-collagen antibodies using a specific immunoadsorbent technique, *Z. Immunitaetsforsch. Allerg. Klin. Immunol.*, 134, 391, 1967.
70. **Timpl, R., Wolff, I., Furthmayr, H., and Steffen, C.,** Immunogenicity and specificity of collagen. VI. Separation of antibody fractions with restricted specificity from anti-collagen sera using an immunoadsorbent technique, *Immunology*, 15, 145, 1968.
71. **Axen, R., Porath, J., and Ernbäck, S.,** Chemical coupling of peptides and proteins to polysaccharides by means of cyanogen halides, *Nature*, 214, 1302, 1967.
72. **Kohn, J. and Wilchek, M.,** Colorimetric method for monitoring activation of Sepharose by cyanogen bromide, *Biochem. Biophys. Res. Commun.*, 84, 7, 1978.

73. **Cuatrecasas, P., Wilchek, M., and Anfinsen, C. B.,** Selective enzyme purification by affinity chromatography, *Proc. Natl. Acad. Sci. U.S.A.*, 61, 636, 1968.
74. **March, S. C., Parikh, I., and Cuatrecasas, P.,** Simplified method for cyanogen bromide activation of agarose for affinity chromatography, *Anal. Biochem.*, 60, 149, 1974.
75. **Parikh, I., March, S., and Cuatrecasas, P.,** Topics in the methodology of substitution reactions with agarose, *Methods Enzymol.*, 34, 77, 1974.
76. **Wick, G., Nowack, H., Hahn, E., Timpl, R., and Miller, E. J.,** Visualization of type I and type II collagens in tissue sections by immunohistologic techniques, *J. Immunol.*, 117, 298, 1976.
77. **Merry, A. H., Harwood, R., Woolley, D. E., Grant, M. E., and Jackson, D. S.,** Identification and partial characterization of the non-collagenous amino- and carboxyterminal extension peptides of cartilage procollagen, *Biochem. Biophys. Res. Commun.*, 71, 83, 1976.
78. **Uitto, J.,** Biosynthesis of type II collagen. Removal of amino- and carboxyterminal extensions from procollagen synthesized by chick embryo cartilage cells, *Biochemistry*, 16, 3421, 1977.
79. **Kumamoto, C. and Fessler, J. H.,** Biosynthesis of A,B procollagen, *Proc. Natl. Acad. Sci. U.S.A.*, 77, 6434, 1980.
80. **Blackwell, B. A. and Bensusan, H. B.,** Purification and characterization of proteins, including procollagen type III, in salt extracts of fetal bovine skin, *Biochem. Biophys. Res. Commun.*, 75, 94, 1977.
81. **Byers, P. H., McKenney, K. H., Lichtenstein, J. R., and Martin, G. R.,** Preparation of type III procollagen and collagen from rat skin, *Biochemistry*, 13, 5243, 1974.
82. **Nowack, H., Rhode, H., and Timpl, R.,** Isolation of a procollagen peptide from amniotic fluid, *Hoppe-Seylers Z. Physiol. Chem.*, 357, 601, 1976.
83. **Smith, B. D., McKenney, K. H., and Lustberg, T. C.,** Characterization of collagen precursors found in rat skin and rat bone, *Biochemistry*, 16, 2980, 1977.
84. **Pesciotta, D. M., Silkowitz, M. H., Fietzek, P. P., Garies, P. N., Berg, R. A., and Olsen, B. R.,** Purification and characterization of the aminoterminal propeptide of pro $\alpha 1(I)$ chains from embryonic chick tendon procollagen, *Biochemistry*, 19, 2447, 1980.
85. **Fessler, L. I., Timpl, R., and Fessler, J.,** Assembly and processing of procollagen type III in chick embryo blood vessels, *J. Biol. Chem.*, 256, 2531, 1981.
86. **Herrmann, H., Dessau, W., Fessler, L. I., and von der Mark, K.,** Synthesis of types I, III and AB_2 collagens by chicken tendon fibroblasts in vitro, *Eur. J. Biochem.*, 105, 63, 1980.
87. **Timpl, R. and Glanville, R. W.,** The aminopropeptide of collagen, *Clin. Orthop.*, in press.
88. **Bruckner, P., Bächinger, H.-P., Timpl, R., and Engel, J.,** Three conformationally distinct domains in the aminoterminal sequence of type III procollagen and its rapid triple helix $\rightleftharpoons$ coil transition, *Eur. J. Biochem.*, 90, 595, 1978.
89. **Nowack, H., Olsen, B. R., and Timpl, R.,** Characterization of the aminoterminal sequence in type III procollagen, *Eur. J. Biochem.*, 70, 205, 1976.
90. **Gay, S., Martin, G. R., Müller, P. K., Timpl, R., and Kühn, K.,** Simultaneous synthesis of types I and III collagen by fibroblasts in culture, *Proc. Natl. Acad. Sci. U.S.A.*, 73, 4073, 1976.
91. **Vaheri, A., Kurkinen, M., Lehto, W. P., Linder, E., and Timpl, R.,** Codistribution of the pericellular matrix proteins in cultured fibroblasts and loss in transformation, *Proc. Natl. Acad. Sci. U.S.A.*, 75, 4944, 1978.
92. **Alitalo, K., Kurkinen, M., Vaheri, A., Krieg, T., and Timpl, R.,** Extracellular matrix components synthesized by human amniotic epithelial cells in culture, *Cell*, 19, 1053, 1980.
93. **Madri, J. A. and Kemp, J. D.,** Immunogenetics of the collagens, in *Immunochemistry of the Extracellular Matrix*, Vol. 2, Furthmayr, H., Ed., CRC Press, Boca Raton, Fla., 1981.
94. **Pesciotta, D. M., Curran, S., and Olsen, B. R.,** Preparation of antisera specific for the amino and carboxyl propeptides of type I and type II procollagens, *Immunochemistry of the Extracellular Matrix*, Vol. 1, Furthmayr, H., Ed., CRC Press, Boca Raton, Fla., 1981.

Chapter 12

PRODUCTION OF MONOCLONAL ANTIBODIES TO COLLAGENS AND THEIR IMMUNOFLUORESCENCE LOCALIZATION IN EMBRYONIC CORNEA AND CARTILAGE

Thomas F. Linsenmayer and Mary J. C. Hendrix

TABLE OF CONTENTS

I. INTRODUCTION

During recent years, antibodies specific for the different genetic types of collagens have become frequently used reagents, especially for immunocytochemical localizations of collagens in developing embryos, pathological specimens, and cultured cells.[1–3] Compared to most other protein antigens, however, collagen is a very poor immunogen, and, in addition, there is a high degree of cross-reactivity among the different genetic types of collagens. Thus, to obtain antibodies with a specificity suitable for most purposes, antisera must be purified by affinity chromatography on immunoabsorbants for both heterologous and homologous collagen types (see Furthmayer, Chapter 11; Timpl et al.[4]). In addition, there are other complications, such as obtaining sufficient quantities of antibodies to perform certain experiments and the unavoidable variations in antibodies which can occur due to differences within individual animals or groups of animals. We felt that some of these difficulties, at least, could be overcome by applying the recently devised cell hybridization technology for the creation of lymphocyte-myeloma hybrid cells, "hybridomas", capable of producing monoclonal antibodies against the different genetic types of collagen.

II. BACKGROUND

The basis for the specificity of hybridoma-produced antibodies lies in the clonal selection theory of antibody production which proposes that each individual lymphocyte and its progeny, once committed to antibody synthesis, produces a single type of antibody molecule with a single antigenic specificity. The spectrum of antibody diversity within an individual, then, reflects the diversity of individual lymphocytic cells within the organism rather than the production by each lymphocyte of more than one type of antibody. Thus, if one were able to clone individual lymphocytes, these clones would provide a source of "monoclonal" antibodies with absolute specificity against a single antigenic site. Thus far at least, the direct cloning of lymphocytes has not been possible, since they seem to have a limited division capacity in culture.

As initially demonstrated by Köhler and Milstein,[5] however, this goal has been achieved indirectly by fusing lymphocytes with myeloma cells to form hybrid cells called "hybridomas". Hybridomas are capable of both specific antibody production and indefinite propagation in vitro. Myeloma cells (see Table 1) are derived from plasmacytomas and probably represent tumor cell lines. Most are derived from BALB/c mice, and being tumorigenic, they are capable of indefinite propagation both in vivo and in vitro. In addition, most myeloma cell lines produce and secrete large amounts of their own immunoglobulin molecules of a single subclass, but with no known antigenic specificity. For example, one myeloma cell line, MPC-11, 45.6TG1.7 (abbreviated MPC-11), produces IgG composed of γ 2b heavy chains and K-light chains (see Table 1).[6] Another line, P3-X63Ag8 (abbreviated P3), produces IgG composed of γ 1 heavy chains and K-light chains.[5]

In initial experiments, Cotton and Milstein[7] were able to demonstrate that when two different antibody-producing cells were fused, the hybrid cells synthesized immunoglobulin molecules of both parental cells. This initial fusion experiment was done between cells of a mouse myeloma cell line and a rat tumor cell line. When analyzed through the use of antimouse and antirat sera plus isoelectric focusing, hybrid clones were found to synthesize immunoglobulin molecules containing chains from both parental cell types. The evidence showed that heavy and light chains of both parental lines could be joined into hybrid molecules. Similar interspecies fusions

Table 1
MYELOMA LINES AVAILABLE

Cell line	Parental line	Immunoglobulin synthesized	Immunoglobulin secreted	Ref.
(abbreviated name)				
MPC-11 — 45.6TG1.7 (MPC-11)	Same	IgG (γ 2b,K)	Same	6
P3-X63-Ag8 (P3)	Same	IgG (γ 1,K)	Same	5
P3-NS1/1 — Ag4-1 (NS1)	P3-X63-Ag8	(—,K)	None	17
Sp2/0-Ag14 (Sp2)	P3-X63-Ag8	None	None	18
X-63-Ag8.653 (653)	P3-X63-Ag8	None	None	19

between mouse myeloma cells and human peripheral lymphocytes yielded cells capable of producing both human and mouse immunoglobulin chains.[8] Again precipitation of immunoglobulin molecules with species-specific antisera, followed by separation of the chains and subsequent reprecipitation of the separated chains with the antisera against the alternate species, showed that interspecies hybrid molecules had been produced.

Importantly, in these studies and subsequent ones[5,6] involving hybridization between cells of two different myeloma lines, no evidence could be detected for hybrid immunoglobulin chains containing the variable regions from one parental chain and the constant regions of the other parent. Köhler and Milstein[5] fused cells from two different myeloma lines of BALB/c origin, one parent secreting IgG with γ 2a heavy chains and the other secreting IgG with γ 1 chains. When the proteins secreted by the hydrid cells were analyzed by isoelectric focusing in the unreduced state, several new bands were observed in addition to those bands which corresponded to the proteins secreted by each parental cell type. When the hybrid proteins were focused under reducing conditions, however, no new bands were observed and the pattern obtained was simply the sum of the patterns obtained from each parental cell line. Thus, the process of cell fusion did not result in any detectable difference in the primary structure of the synthesized chains as would be expected if the variable and constant regions from the different parental cell types had combined within the same heavy or light chain. These results were both important and encouraging, since the synthesis of scrambled individual chains with various segments from each parent would probably preclude the production of antibodies with useful affinities, if they had any affinity at all.

The definitive experiment which demonstrated the feasability of the hybridoma approach to antibody production was also performed in this same study by Köhler and Milstein.[5] A fusion was performed between cells of the P3 myeloma line with splenic lymphocytes from a mouse that had been immunized with sheep red blood cells. This fusion resulted in clonable hybrid cells which had a stable karyotype, suggesting the presence of both myeloma and lymphocyte genomes. The hybrids secreted immunoglobulins which, in addition to having chains characteristic of the parental myeloma cell line, also contained new components presumably of lymphocyte origin. Most importantly, the immunoglobulins produced by the hybrid cells, in the presence of complement, were able to lyse sheep red blood cells, thus attesting to the production of molecules with a specific antibody activity which could be attributed to the lymphocyte parent.

In these experiments, as is the case for the production of all hybridoma antibodies now produced, the information for producing antibody molecules with a defined antigenic specificity is supplied by the lymphocyte parent, and the ability to be propagated indefinitely and continue immunoglobulin synthesis is supplied by the myeloma parent. Thus far, monoclonal antibodies have been produced against numerous cell surface antigens, such as on T-lymphocytes[9] and B-lymphocytes,[10] more specific antigens, such as influenza virus[11] and rabies virus,[12] and other antigens (see Melchers et al.),[13] including a few purified proteins such as the acetylcholine receptor[14] and collagen Types I[15] and II.[16]

In evaluating the information which will now be presented on the production and characterizations of monoclonal antibodies against purified collagens, it must be remembered that only two positive hybridoma clones have thus far been isolated — one each against chicken Type I and Type II collagens. The data which have been obtained demonstrating the considerable advantages of this technique, plus some potential disadvantages, are from this extremely small, and thus potentially unrepresentative, sample. We think, however, that they can serve as guidelines for the types of results that one can expect to obtain with the use of monoclonal antibodies. We also want to emphasize that it is not impossible for a monoclonal antibody made against one genetic type of collagen to cross-react with some other type of collagen — either known or unknown. The only definite statement that can be made is that the antibody will not cross-react with those collagen types against which it has already been tested and shown to be negative.

In addition, for the methods presented here, we have included some updated information and procedures that were not available when we began our original work. We are now using most of these in attempts to produce additional monoclonal antibodies to extracellular matrix molecules such as an animal collagenase and several other collagens.

III. HYBRIDOMA FORMATION

A. Myeloma Cell Lines, Choice of a Line and Its Propogation

For the production of anticollagen monoclonal antibodies, we have used the parental myeloma cell line MPC-11 which as stated earlier produces its own IgG composed of γ 2b heavy chains and K-light chains. For work in progress we are using nonsecretor and nonproducer lines (see Table 1) which are now readily available. The nonsecretor cell line P3-NSI/1-Ag4-1 (abbreviated NS1) produces no heavy chain, but produces some K-light chain which is not secreted.[17] The nonproducer cell lines Sp2/0-Ag1 (abbreviated SP2)[18] and x63-Ag8.653 (abbreviated 653)[19] neither produce nor secrete any heavy or light chains. All of these cell lines have been shown to be capable of participating in hybridoma formation with various efficiencies, and although not yet used for anticollagen antibody production, they may offer definite advantages over the secretor myeloma lines.

The disadvantage of using a parental myeloma cell line which makes its own nonspecific immunoglobulin molecules is that the hybridoma cells that are constructed produce hybrid molecules containing all combinations and permutations of heavy and light chains from both the lymphocyte and myeloma cell parents. Thus, for the antibodies produced by the hybridomas, many of the individual binding sites and even whole antibody molecules are inactive. A hypothetical example would be an IgG secreting hybridoma in which both parental genomes remained equally active. Since each binding site within an antibody molecule is composed of one heavy and one light chain and an IgG molecule is a dimer containing two antigenic sites, there are 16 possible combinations and permutations of the parental heavy and light chains. In this case, 7 of the 16 possible IgG molecules produced would have one active

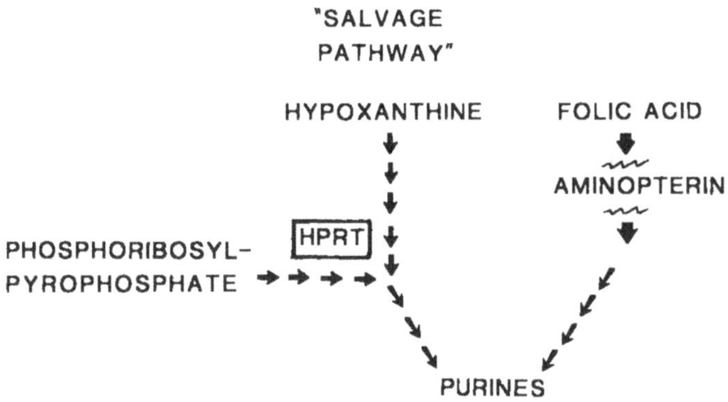

FIGURE 1. HAT (hypoxanthanine: aminopterin: thymidine) method for selection of hybridoma cells. HPRT, hypoxanthine phosphoribosyl transferase.

antigenic site consisting of both light and heavy chains of lymphocyte origin, but in only one of these molecules would both antigenic sites consist solely of specific chains of lymphocyte origin. Thus, although completely good antibody molecules are produced, a large amount of inactive antibody is also produced by such hybrids. In addition, some evidence exists that, in general, monoclonal antibodies produced from secretor myeloma parents have a lower affinity than those produced by the nonsecretor lines.[35] Possibly the presence of one or more inactive chains within an IgG molecule interferes with the overall binding of the molecule.

The isolation of hybridoma cells requires the use of a biochemical selective technique. Thus, all of these myeloma cell lines have also been selected for resistance to 8-azaguanine (in the case of P3) or 6-thioguanine (in the case of MPC-11). Cells resistant to these compounds are deficient in the enzyme hypoxanthine phosphoribosyl transferase ($HPRT^-$). Such cells are killed in a selective medium[20] containing hypoxanthine, aminopterin, and thymidine (HAT) in which normal cells being $HPRT^+$ can survive.

This selective mechanism, which was initially developed for use in somatic cell genetic analyses,[20] depends on the poisoning of the metabolic pathways normally requiring folic acid by including the folic acid antagonist, aminopterin, in the selective medium (see Figure 1). Folic acid is normally required for reactions leading to the synthesis of purines and thymidilate. The block to thymidilate synthesis is overcome by including exogenous thymidine as a component of the HAT medium. The block to purine synthesis can also be circumvented by normal cells which contain a functional HPRT enzyme and thus are able to effectively utilize the "salvage pathway" for purine synthesis. In these cells, the exogenously added hypoxanthine in the HAT medium is condensed with phosphoribosylpyrophosphate by the enzyme HPRT, ultimately resulting in purine formation. The $HPRT^-$ myeloma parental cells cannot utilize the exogenously added hypoxanthanine and thus are killed by HAT selective medium — unless, they have been hybridized to a normal cell, such as a lymphocyte, which can supply the hybrid with the information necessary to synthesize an active $HPRT^+$ enzyme. Hybridomas live in HAT medium.

Most parental myeloma cell lines grow well in Dulbeccos Modified Eagles Medium (DMEM) with high glucose (4.5 g/ℓ) or in RPMI 1640. Complete medium is also supplemented with 10% fetal calf serum plus penicillin (100 units per milliliter) and streptomycin (100 μg/mℓ). The fetal calf serum should come from a lot that has been tested for its ability to promote the growth of mass cultures of both the parental myeloma cells and a hybridoma line, if available. More importantly, it should also be tested for ability to promote the growth of cells at clonal densities. Once an adequate lot of serum is obtained, a large quantity should be purchased and stored frozen at −85°C in a low-temperature freezer.

The myeloma cells are routinely grown in petri plates or flasks in a humidified tissue culture incubator with the CO_2 concentration adjusted to about 6 to 7%. The cells are allowed to grow up to a maximum density of about 5×10^5 cells per milliliter before being passaged. Every 2 to 4 days the cultures are split approximately 1:5 or 1:10 by either removing about 90% of the cells and medium and then refeeding or by transfering about 10% of the cells to a fresh plate before feeding. When the cells are healthy, they appear as round, smooth-surfaced spheres. If the cells are allowed to reach too high a density or are not passaged frequently enough, their cytoplasm rapidly develops a granular or vacuolated appearance. Such cultures, even if refed, frequently die, and are best discarded. Alternatively, if the cells are passaged at too low a density, again they may die or at best will grow very slowly. Each cell line will differ slightly in its growth characteristics, density dependence, and maximum obtainable density. These parameters will become obvious as one becomes familiar with handling the cells.

To remove any $HPRT^+$ revertant cells that arise in the parental myeloma population, every 3 to 4 weeks cells are routinely passaged through either 6-thioguanine (10 μg/mℓ) in the case of MPC-11 cells or 8-azaguinine (20 μg/mℓ) in the case of the NS1 cells. This procedure insures a population composed of only $HPRT^-$ cells.

To insure adequate availability of parental myeloma cells and hybridomas once they are isolated, adequate numbers of cells should be frozen and stored in liquid N_2. Cells that are to be frozen should be healthy appearing and in the log growth phase. About 5 to 10×10^6 cells are harvested by centrifugation (400 × g for 5 min in a clinical centrifuge). The complete growth medium is aspirated off and replaced with 1 mℓ of freezing medium. Freezing medium consists of complete medium with 10% dimethylsulfoxide (DMSO). The cells are gently resuspended with a pasteur pipette and are quickly tranferred into a freezing vial. The cells should be frozen slowly. A good way to do this is to place the vials upright into a small styrofoam container packed with gauze as a support. The styrofoam top is taped onto the container, and it is placed into a −85°C freezer overnight. The next day, or later, the vials are transferred to a liquid N_2 freezer where they can be stored indefinitely. One vial from each freezing should be thawed and tested for cell viability.

To obtain a high percentage of viable cells from a frozen sample, the vials should be rapidly thawed in a 37°C water bath. As soon as the freezing medium has melted, the cells are transferred to 10 mℓ of complete medium and then are centrifuged. The medium is aspirated to remove DMSO, and the cells are resuspended in several milliliters of complete medium and transferred to a 60-mm petri dish which is then put into the incubator. (After several hours, we usually change the medium again, but this is probably unnecessary.) By the next day, a definite increase in cell number should be noted, and from this point the cells are handled as any mass culture.

B. Source of Lymphocytes

To produce hybridomas which will synthesize anticollagen antibodies, the source of lymphocytes must be an animal which has been immunized with the type of collagen to which an antibody is desired. Since at present the only myeloma cell lines readily available are of mouse origin, the lymphocytes should also be of mouse origin. Interspecies hybrids such as mouse-rat have recently been formed, but these

Table 2
ANTIBODY RESPONSE TO COLLAGEN TYPES I AND II[24,25]

Mouse strain	Response to Type I collagen	Response to Type II collagen
SJL/J	High	High
DBA/1J	Low	High
C57B1/10	High	Medium

may be less stable than intraspecies ones. The lymphocyte parent supplies all the hybridomas which grow in HAT medium with the information to make a functional $HPRT^+$ enzyme. In addition, and most importantly, they supply a percentage of the hybridomas, small though it may be, with the information to synthesize anti-collagen antibodies. Unhybridized, parental lymphocytes, being normal cells, are not killed in HAT medium. However, since they do not divide *in vitro*, they are also selectively removed with time in culture and only the hybridomas survive.

It is known that different strains of mice can have varying antigenic responses to the different types of collagens[21,22] and to Type I procollagen.[23] Thus, it seems reasonable to assume that the strain of mouse used for immunization could affect the probability of obtaining hybridoma clones producing antibodies. We used a Jackson Laboratory strain, ASW/SN, for our work in producing antibodies against collagen Types I and II. Some other strains that have been reported as being high and low responders to collagens are given in Table 2, and to maximize the probability of obtaining positive clones, lymphocytes from one of the high responders should probably be used. BALB/c, the strain from which most myeloma cells are derived falls into the low-responder class[24] and probably should not be used as a lymphocyte donor. Thus, the hybridomas will be produced with myeloma cells from one strain of mouse and lymphocytes from another. Unfortunately, this complicates the production of antibody by the ascites method of growing the hybridomas in the peritoneal cavity of a mouse, since the host will undoubtedly recognize determinants of the opposite strain on the hybridoma cells as foreign and reject them. This potential problem, as discussed later, can be overcome.

The regimen used for immunization of the lymphocyte donor undoubtedly affects the number of lymphocytes available which are carrying information to produce antibodies against a specific type of collagen. It is probably best to hyperimmunize with several injections of native collagen given at 2-week intervals. The first injection is emulsified in complete Freunds adjuvant and given s.c., and subsequent ones are given either without adjuvant or with incomplete adjuvant and administered i.p. We have used 200 μg of collagen for each injection, but it may be possible to use less (see Hahn et al.),[22] About 2 weeks after the last injection, the mice should be bled and the serum should be tested for the presence of antibody activity and titer, using a method such as passive hemagglutination (discussed later). Mice with a high antibody titer should then be allowed to rest for an additional 4 weeks or more without any immunizations. Then 3 or 4 days before a fusion is to be performed, the donor mouse should receive a final injection. The best results are generally thought to be obtained if this final injection is given intravenously in the tail vein, but we obtained a satisfactory result with an i.p. injection.

C. Hybridoma Formation

The general method of hybridoma formation that we have used will first be described, and then a detailed description of each step will be given. An outline of the

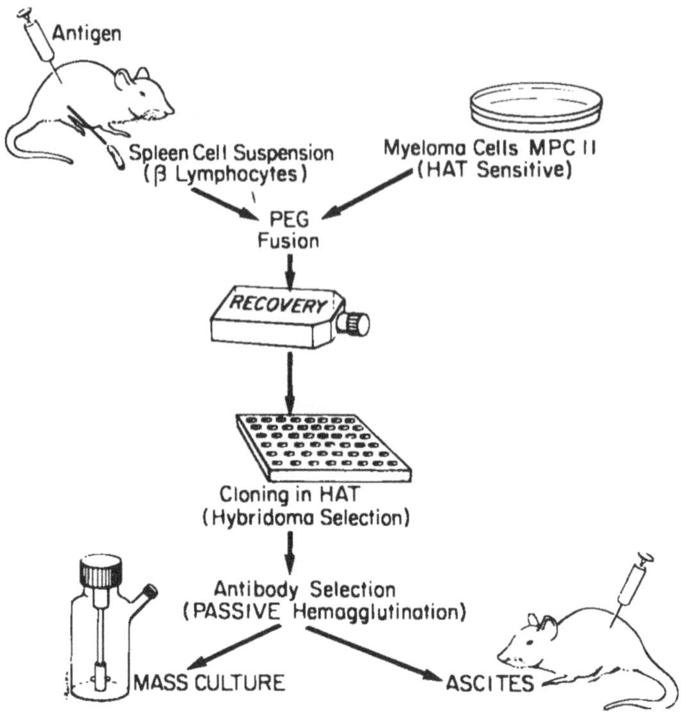

FIGURE 2. Diagram of a method for producing monoclonal antibodies; HAT = hypoxanthine: aminopterin: thymidine; PEG = polyethylene glycol.

methods we have used to produce hybridomas synthesizing anticollagen monoclonal antibodies is diagramatically shown in Figure 2. MPC-11 myeloma cells were fused with lymphocytes removed from a mouse immunized with highly purified collagen. The agent used to promote cell fusion was polyethylene glycol (PEG) 1000. Subsequent to fusion, cells were allowed to recover in normal medium and then hybridomas were selected by plating the entire cell suspension into numerous tissue culture wells containing HAT medium. Only the hybridoma cells can both survive and grow in HAT, so the initial selection simply involves examining wells for cell growth. Then to select for those hybridomas synthesizing anticollagen antibodies, medium from hybridoma-containing culture wells was tested for the presence of antibody activity by passive hemagglutination. Cells from antibody positive wells were cloned by limiting dilution in microtiter wells and retested for continued antibody production. The clones were grown up to large numbers, and antibody was harvested either directly from the spent culture medium in which the hybridomas had been growing or from the ascites fluid of mice which had received intraperitoneal injections of hybridoma. Cloned cells were stored frozen for future use.

D. Cell Fusion Procedure

The myeloma cells used for fusion should be from a healthy culture in log phase growth. Depending on the number of splenic lymphocytes obtained and the ratio of spleen cells to myeloma cells used, about 1 to 3 $\times$ 10^7 myeloma cells will be needed.

Lymphocytes are obtained from the spleen of an immunized mouse. The spleen is sterilly removed into a 100-mm petri dish containing serum-free medium. Originally we released the lymphocytes by mincing the spleen and then agitating the pieces using a vortex mixer. Recently, we have turned to a more gentle method for flushing out the lymphocytes. Two hypodermic syringes equipped with 25-gauge needles are filled with medium, and while one is used to immobilize the spleen against the bottom of the petri plate, the other is used to repeatedly puncture the organ while gently squirting a stream of medium into the spleen to flush out the cells. Once most of the cells are removed, as judged by loss of the bright red color of the spleen, the cells are counted. To eliminate erythrocytes from the cell count, a 25-μℓ portion of the cell suspension is diluted up to 250 μℓ with 2% acetic acid, and the suspension is transferred to a hemocytometer for counting. The 2% acetic acid lyses the erythrocytes so that only the white cells are included in the count.

Although the technique for producing hybridomas is only 4-years-old and has only been in general use for the past 2 years, the number of modifications of fusion procedures is probably about as great as the number of workers performing fusions. I will present two fusion procedures, one according to Geftner et al.[25] which we used for our original production of monoclonal antibodies to collagens and a second[26] which we have been using for the past several months in efforts to produce monoclonal antibodies to other extracellular matrix molecules. The latter method[26] appears to produce a larger number of hybrids.

All procedures currently in use involve PEG as a fusion agent. Various studies have used either PEG 1000 or 1500 at concentrations ranging from 30 to 50%. We routinely use PEG 1000 at a concentration of 37.5%. A good source of PEG is Koch-Lite (Colnbrook Bucks, England), whose product does not dramatically lower the pH of the medium when it is added. The PEG we have used from several other sources does lower the pH. To make up a sterile PEG solution, we autoclave the solid, waxy PEG, and after a period of cooling, but before it resolidifies, we add the desired volume of PEG to serum-free medium with rapid pipetting.

1. Fusion Procedure 1[25]

All operations are performed at room temperature. Myeloma cells and lymphocytes are washed and resuspended in serum-free medium. Then myeloma cells (10^6) and lymphocytes are mixed together in a ratio of 1:3 to 1:10 in a 50-mℓ conical bottom centrifuge tube. The cells are centrifuged at 250 × g for 5 min, and the supernatant fluid is aspirated off. Then 0.2 mℓ of 37.5% PEG in serum-free medium is added, and the tube is gently agitated by hand to resuspend the cells. Next, the cells are centrifuged for 3 min at 500 × g and then are allowed to sit undisturbed for an additional 5 min. Then 5 mℓ of serum-free medium are added to the tube; the cells are resuspended gently with a Pasteur pipette and then are repelleted by centrifugation at 250 × g for 5 min. The supernatant is removed, and the cells are dispersed in 10 to 20 mℓ of complete medium in preparation for dispensing into tissue culture wells.

2. Fusion Procedure 2[26,27]

Myeloma cells (10^7) and spleen cells (3×10^7 to 1×10^8) are washed several times in serum-free medium and then are mixed together in a 50-mℓ, round-bottomed centrifuge tube. Then they are pelleted at 400 × g for 10 min at room temperature. The supernatant fluid is aspirated off, and the tube is transferred to a beaker of 37°C H_2O, in which all further manipulations are performed. With a pipette, 1 mℓ of PEG solution at 37°C is slowly added over a 1-min period, using the tip of the pipette to gently stir the cell pellet. The pellet is gently stirred for another minute, resulting in a suspension of cell clumps. During a second 2-minute period, an additional 2 mℓ of serum-free medium is slowly added, and finally an additional 7 mℓ of serum-free medium is stirred in during a final 2 to 3-min period. The suspension is centrifuged at 400 × g for 10 min at room temperature. The supernatant is removed, and 10 to 20 mℓ of complete medium is pipetted onto the cell pellet. At this point, the suspension can be further diluted to the proper density for being dispersed into culture wells.

A simplified modification of the above procedure has been devised (courtesy of Dr. Jack Wands, Massachusetts General Hospital). In Wands procedure, the myeloma cells and lymphocytes are mixed together and centrifuged at 400 × g for 10 min into a pellet at the bottom of a 50-mℓ round-bottomed centrifuged tube. Then 2 mℓ of PEG solution, prewarmed to 37°C, is gently pipetted down the side of the tube so as to disturb the cell pellet as little as possible. The tube is gently rotated for 2 min in a 37°C water bath; 10 mℓ of serum-free medium is added, and the cells are suspended by very gentle pipetting. The cells are recentrifuged, the medium is aspirated off, and the cells are resuspended in 10 to 20 mℓ of complete medium.

It has been reported[28] that when cells are fused in Ca^{++}-free medium and then are allowed to recover for a short period of time in the same medium, a higher percentage of hybridoma survival is noted. Thus, for any of the above fusion procedures, it may be desirable to make the diluted PEG solution with a culture medium which is devoid of Ca^{++}. If we use the Ca^{++}-free procedure, the cells are initially washed in Ca^{++}-free medium, and after the fusion procedure is completed, the cells are suspended in 10 mℓ of Ca^{++}-free, serum-free medium instead of complete medium. They are incubated in this medium for 15 to 20 min at 37°C. Then 10 mℓ of complete DMEM-containing serum is added and the cells are pelleted by centrifugation and finally resuspended in 10 to 20 mℓ of complete medium with serum.

E. Hybridoma Selection and Testing for Antibody

After fusion, the cells are dispersed into multiple tissue culture wells. Previously we performed this initial plating in 24-well plates (Costar #3524). Recently, however, we have been dividing the cells into smaller aliquots in 96-well plates (Costar #3596). The cells, now in complete medium, are adjusted to a concentration of 2×10^4 myeloma cells per 100 μℓ, and 100-μℓ aliquots of suspension are dispersed into wells of 96-well culture plates. The next day, 100 μℓ of HT medium (HAT medium minus the aminopterin) is added, and on the third day, 100 μℓ of medium containing a 2× concentration of HAT supplement is added. Subsequently, every 3 days about one third of the medium in each well is replaced with fresh (1×) HAT medium.

HAT medium[20] consists of complete medium supplemented with 0.1 m*M* hypoxanthine, 0.4 μ*M* aminopterin, 16 μ*M* thymidine, and 3 μ*M* glycine. The HAT supplement can be made up as 50× concentrate, sterilized and stored frozen in 10-mℓ aliquots suitable for making up 500-mℓ batches of HAT medium. In making up the concentrate, the hypoxanthine goes into solution best if the pH is made slightly acidic, and the aminopterin dissolves best at a basic pH. The pH can then be subsequently readjusted to neutrality.

After 2 to 3 days in HAT medium, almost all the myeloma cells will have developed a shrunken, granular appearance due to killing by the aminopterin. Seven to ten days later, if the fusion has been successful, many of the wells will have one or more foci of cell growth, as evidenced by nodules of large, round cells with the appearance of the original myeloma parents. These nodules should be allowed to grow until they cover about one tenth of the bottom of the well or more. At this point, those wells with cells should be tested for antibody production.

For this we have routinely used the passive hemagglutination method of Beil et al.,[29] but other assays such as the recently devised enzyme-linked immunoabsorbant assay (ELISA) should be equally suitable (see Rennard, Chapter 14). For passive hemagglutination, collagens are coupled to human type O erythrocytes (RBCs). Then 10 mℓ of RBCs are washed three times with a physiological saline (such as Hanks), and the final concentration of cells is adjusted to 10%. Then, 1.5 mℓ of 6% gluteraldehyde is added to 15 mℓ of the RBC suspension, and the mixture is stirred for about 15 min at room temperature. The cells are washed three times in 0.1 *M* calcium acetate and are suspended in 10 mℓ of the same solution. They are mixed with 10 mℓ of a 0.3% collagen solution in the same solvent for 2 hours at room temperature. The cells are washed twice with calcium acetate, twice with 0.15 *M* NaCl, and then are brought up to 10 mℓ with 0.1% bovine serum albumin in physiological saline (BSA saline). They are stored at 4°C and can be used for about 2 weeks to 1 month.

To test for antibody producing hybridomas, spent medium (100 μℓ) from each culture well is transferred to a well of a round (Linbro #76-111-05) or conical (Linbro #76-121-05) bottom microtiter plate (Linbro/Titertek). Then, a 100-μℓ aliquot of the derivatized RBCs diluted 1:20 with BSA saline is added to each well. Then the plates are stored overnight at 4°C before being scored for agglutination.

F. Cloning

Once the positive wells have been identified, the cells must be cloned. Then they are grown to populations large enough for producing useful quantities of antibody and for freezing for future use. The process of cloning the anticollagen antibody-secreting hybridomas eliminates the possibility of the production of other noncollagen antibodies, since each hybridoma contains the genome of a single lymphocyte parent. In addition, repeated cloning ensures that the desired hybridoma will not be overgrown by other hybridoma cells which may be in the population or by nonsecreting revertants of the anticollagen producing hybrid.

We clone by the method of limiting dilution. In this method, cells are distributed in wells so that statistically each well receives one or less than one viable cell (in a 96-well plate, 25 or fewer wells have cell growth). The number of wells exhibiting growth depends upon two parameters — the number of cells plated in each well and the percentage of cells which are viable at clonal densities (cloning efficiency). The cloning efficiency itself will also depend primarily upon two parameters — the myeloma cell line used for fusion and the lot of serum used in the medium. This is why, as stated earlier, each lot of serum should be screened before purchase. Cloning efficiencies can be greatly improved through the use of a "feeder" cell layer which we now routinely employ.

Cells that can serve as feeder layers range from established tissue culture cell lines to freshly liberated thymocytes from mice. We have successfully used Mytomycin C-treated 3T3 fibroblasts (suggested to us by Dr. Jack Wands). To produce a feeder layer, cultures of 3T3 cells are treated for 2 hr with freshly dissolved Mytomycin C at a concentration of 5 μg/mℓ. The cell layers are then washed 2 times with saline and released with 0.25% trypsin. The trypsin is inactivated by adding an equal volume of complete medium; the cells are collected by centrifugation and are resuspended in complete medium at a concentration of 2×10^5 cells per milliliter. Then 100-μℓ portions of cell suspension are added to each well of a 96-well culture dish. The cells are returned to the incubator to attach for 24 hr, and the medium is replaced with 50 μℓ of HAT medium at which time they are ready to be seeded with the hybridoma cells.

Hybridoma cells in the antibody-positive wells of the 96-well plate are suspended by pipetting and then are counted in a hemocytometer. A portion of the cells is diluted to a concentration of 10 cells per milliliter and 100 cells per milliliter with

HAT medium, and 50 μℓ of cell suspension are added to each well of the plates containing the feeder layer cells. The lower cell concentration will statistically give 0.5 cells per well, which, with a cloning efficiency of 50% will give approximately 25% of the wells showing cell growth. This is within the statistical probability that the cells in wells showing growth are derived from a single cell (i.e., a clone). The cells plated at the higher density (five cells per well) should show growth in every well, and while obviously not at clonal density, these cultures serve as "back-ups" in case the desired hybridoma cell is not picked up at clonal density. We usually prepare ten plates of each cell density. Every 3 to 5 days, the wells are fed with an additional 50 μℓ of HAT medium, and after the wells are almost filled with medium, 50-μℓ portions are removed before feeding. Seven to ten days after plating, colonies of several hundred cells should be visible with an inverted phase contrast microscope, and by 12 days to 2 weeks, the wells should have enough cells to test for antibody. By 7 to 10 days, the feeder layer cells will have died, but they will have already served their purpose. We always reclone the positive wells at least one additional time. This both insures that we are dealing with cloned cells and eliminates the possibility that the cells from the original cloning might be overrun by the formation of nonproducing revertants. It is an empirical observation that antibody-producing hybridomas can be unstable for about 2 months after their formation, and during that time, they will give off such nonproducing revertants. Performing several cloning steps during this critical period will eliminate the possibility that such revertants might become the predominant members of the population.

G. Expanding the Cell Population and Freezing

At some point, the cells in the antibody-positive wells of the 96-well plates will require transfer into larger culture dishes. They also must be transferred from HAT medium into regular medium. This should be done for the original, uncloned hybridoma cells that test positive for antibody, as well as for the subsequently cloned cells. The reason for expanding the original, uncloned hybridoma cells is that they can be frozen and thus serve as a "back-up" source of cells in case the original cloning fails. The reason for expanding the cloned cell population is to have enough cells to produce useful quantities of antibody and to freeze down for future use.

The important point to remember when expanding a population of cells is to keep the cell density high enough so that cell growth is not inhibited or that the cells do not die. To do this, we transfer the cells to progressively larger culture dishes. For example, we allow a clone in a well of a 96-well plate to grow until the cells cover about one half the surface of the bottom. At that point, we suspend the cells by gentle pipetting and transfer about 90% of the cells to a well in a fresh 96-well plate, keeping the other 10% of the cells in the original plate as a "back-up". Then (depending on the rate of cell growth) every few days we split the new well into 2 or 3 additional wells of the 96-well plate and then continue this process until we have about a dozen wells showing a high cell density. At that point, we pool the cells from about 3 wells and transfer them to a single well of a 24-well plate containing about 1 mℓ of medium. This well is also progressively split every 2 to 3 days to generate more wells, and each of these, when it reaches a high cell density, is transferred to a 35-mm culture dish. The 35-mm dishes are split and subsequently transferred to 60-mm dishes, the 60-mm dishes to 100-mm dishes, and these to 150-mm dishes. Cells in the 100- and 150-mm dishes are in sufficient numbers to obtain antibody from the medium, to inject into mice to obtain ascites fluid antibody, or to freeze. Cells are frozen using the same method previously described for the myeloma parents.

Table 3
MONOCLONAL ANTIBODY TITERS

Monoclonal antibody	Antibody source	Titers ($-\log_2$) Chick I	Chick II
Chick — Type I	Culture medium	6–7	0
	Ascites fluid	15–>17	0
Chick — Type II	Culture medium	0	10.6 ± 0.69
	Ascites fluid	0	21.2 ± 1.95

At some point in the cell expansion procedure, the cells must be removed from the HAT medium and worked into complete medium. We usually do this at the time they are being transferred from 96-well plates to the 24-well plates, but it can be accomplished at an earlier stage. The cells cannot, however, be transferred directly from HAT medium into complete medium. If this is attempted, residual aminopterin within the cells will kill them. Therefore they must first be transferred into HT medium (HAT medium minus the aminopterin) for several generations to ensure the absence of residual, intracellular aminopterin. Then they can be transferred to complete medium for further growth.

H. Antibody Collection

Antibody can simply be collected by precipitation of spent cultured medium with 45% ammonium sulfate or by further purification of DEAE-cellulose columns. A much higher titer antibody can be obtained by injecting the hybridoma cells into the peritoneal cavity of a host mouse and subsequently harvesting the ascites fluid. This also circumvents the need for growing large volumes of cells in vitro. If the myeloma parent is from one strain of mouse and the lymphocyte parent is from another strain, to prevent cell rejection, the host mouse should be an F_1 hybrid of the two parental strains. Alternatively, an athymic nude mouse (nu/nu) can be used. Then 2 to 5 $\times$ 10^6 hybridoma cells are injected into the peritoneal cavity, and ascites fluid is harvested about 2 weeks later. If the tumor has grown properly, 3 to 6 mℓ of ascites can be harvested containing about 5 to 10 mg of antibody per milliliter. Sometimes, it is thought that "priming" the mice with an i.p. injection of 0.5 mℓ of Pristane (2,6,10,14-tetramethylpentadecane) 3 to 5 days before injecting the myeloma cells will improve the yield of ascites, but we have found this to be unnecessary for the nude mouse.

IV. PROPERTIES OF THE ANTICOLLAGEN MONOCLONAL ANTIBODIES

The hybridomas which have already been selected for the production of monoclonal antibodies against collagen Types I and II have been characterized with respect to specificity and titers obtained, immunoglobulin class, and ability to function as immunocytochemical reagents.[15,16]

Antibody titers were determined by passive hemagglutination (see Table 3). For such determinations, 100-μℓ portions of antibody-containing solution were serially diluted twofold with BSA-saline. Then 100-μℓ aliquots of collagen-derivatized RBCs were added to each well and the agglutination end point was determined after an overnight incubation at 4°C. Since the serial dilutions were twofold, the titers are expressed as ($-\log_2$).

Spent culture medium from the hybridoma that had been produced from lymphocytes of a mouse immunized with Type I collagen had a titer of 6 to 7 when tested with RBCs coated with Type I collagen and had no detectable titer when tested against Type II coated cells. Conversely, culture medium from the hybridoma antibody produced from the Type II immunized mouse had a titer of 10 to 11 when tested against Type II coated RBCs and had no detectable titer against the Type I cells. Since these media showed no cross-reactivity between the two collagen types tested, they do not require affinity chromatographic purification.

For both antibodies, a much higher titer could be obtained by injecting the hybridoma cells into the peritoneal cavity of athymic, nude mice (nu/nu) and harvesting the ascites fluid approximately 2 weeks later. The ascites fluid from the Type II hybridoma-containing nude mice had an average titer of 21 ± 1.95 which indicates a detectable average antibody response at an absolute dilution of greater than two million. The ascites fluid from the Type I-containing mice had titers of 15 to >17 which gives an absolute titer of greater than one half million. Still, neither ascites fluid had any detectable activity when tested against the heterologous collagen type.

Limited pepsin digestion of either type of collagen did not affect the titers. Thus, for both antibodies, the antigenic site probably resides within the helical portion of the molecule. To test whether the antigenic site in addition is conformation-dependent and present only when the collagen is in the native state, we performed inhibition of hemagglutination using native and denatured collagens. The results for the Type II antibody are shown in Figure 3, and similar results were obtainable for the Type I antibody. When native collagen was tested using tenfold serial dilutions, a concentration dependence was observed. Collagen solutions of 10^{-1} mg/mℓ lowered the titer from a control value of 7 to an inhibited value of 2, whereas a concentration of 10^{-4} mg/mℓ caused negligible inhibition. Thermally denatured collagen caused no inhibition. Upon renaturation, inhibitory activity was regained. In addition, similar inhibitory activity was found for artificially renatured molecules produced from $\alpha 1$(II)-chains which had been further purified by carboxymethylcellulose chromatography. That the antibody does not bind to denatured collagen was further demonstrated by using thermally denatured collagen on the RBCs surface. Such cells were not agglutinated by antibody (see Figure 3, last two bars). These results are all consistent with ones found for conventional anticollagen antibodies produced in mice.[21]

Information on the immunoglobulin class of both antibodies was determined by double immunodiffusion against antisera specific for mouse IgG, IgM, or IgA, or by immunoelectrophoresis. Both are IgG. An example of the results obtained by immunoelectrophoresis is shown in Figure 4. In this analysis either whole mouse serum (MS) or the immunoglobulin produced by the parental MPC-11 myeloma cells (MPC) or the immunoglobulin produced by the Type I producing hybridoma (I_1B6) were electrophoresed and then allowed to react with antibody against whole mouse serum (anti-MS) or against mouse IgG (anti-IgG). With anti-MS, both the MPC-11 parent and the hybridoma gave a single sharp arc, whereas the whole mouse serum, as expected, gave numerous reaction products. Against the anti-IgG, both parental cell and hybridoma gave single arcs, as did the mouse serum. However, the arc produced by the serum which would contain all the subclasses of IgG was much longer than the arcs produced against the MPC-11 or the hybridoma, both of which are monoclonal.

Both the Type I and Type II antibodies have been shown to react in immunocytochemical procedures. Figure 5 shows fluorescence micrographs of 14-day-old embryonic chick tibia, including both cartilageous and noncartilageous regions

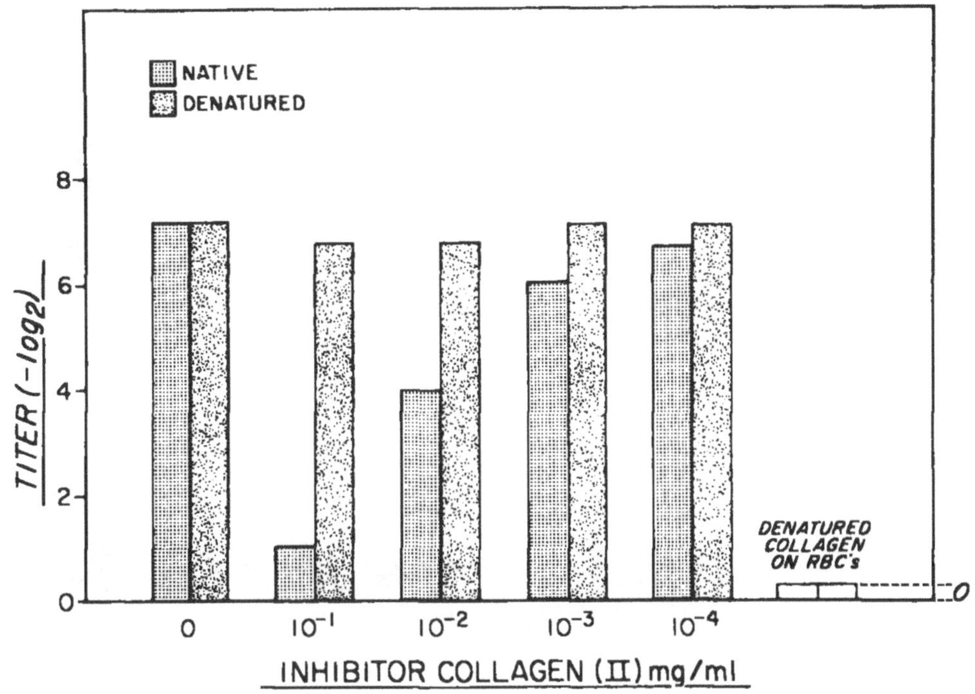

FIGURE 3. Native and thermally denatured (50°C for 30 min) Type II collagens were examined for their ability to inhibit the hemagglutination of Type II-coated RBCs by the Type II monoclonal antibody. Inhibitor collagens at the concentrations shown in the figure were mixed with serial dilutions of antibody and allowed to incubate at 4°C for 1 hr. The RBCs coated with native Type II were added and allowed to react overnight at 4°C. An additional experiment was to thermally denature the Type II collagen on the RBCs *in situ*. The coated RBCs were heated to 50°C for 30 min and after cooling were tested for agglutinability with antibody (see last two bars). (From Linsenmayer, T. and Hendrix, M. J., *Biochem. Biophys. Res. Commun.*, 92, 440, 1980. With permission.)

stained with either Type I or Type II monoclonal antibody using the double-antibody sandwich technique. The second antibody was rhodamine-conjugated rabbit IgG directed against mouse IgG. Sections treated with anti-Type I collagen ascites fluid (see Figure 5A) showed the brightest fluorescence in the perichondrial region, with loose connective tissue also showing bright staining. The matrix in the cartilagenous regions, on the other hand, was negative or at most had a slight background fluorescence. Sections treated with anti-Type II collagen ascites fluid (see Figure 5B) showed bright fluorescence over the cartilage matrix, with little, if any, on the perichondrium and surrounding connective tissue. Uniform staining of the cartilagenous matrix required pretreatment of the sections with hyaluronidase to remove the antibody masking action of proteoglycan.[30] Both ascites-derived antibodies could be diluted at least several 1000-fold with no detectable decrease in staining intensity. The controls, which in all experiments showed only a diffuse background fluorescence, were sections in which the first antibody was replaced with ascites fluid from mice in which the MPC-11 parental myeloma cells had been grown. This provides a source of nonspecific immunoglobulin which is virtually identical to the specific antibody, except for antigen binding activity.

We have used these antibodies to stain sections of developing corneas from selected stages of chick embryos. The chick cornea is a highly organized matrix composed of orthogonal layers of collagen molecules. It develops in two stages (for

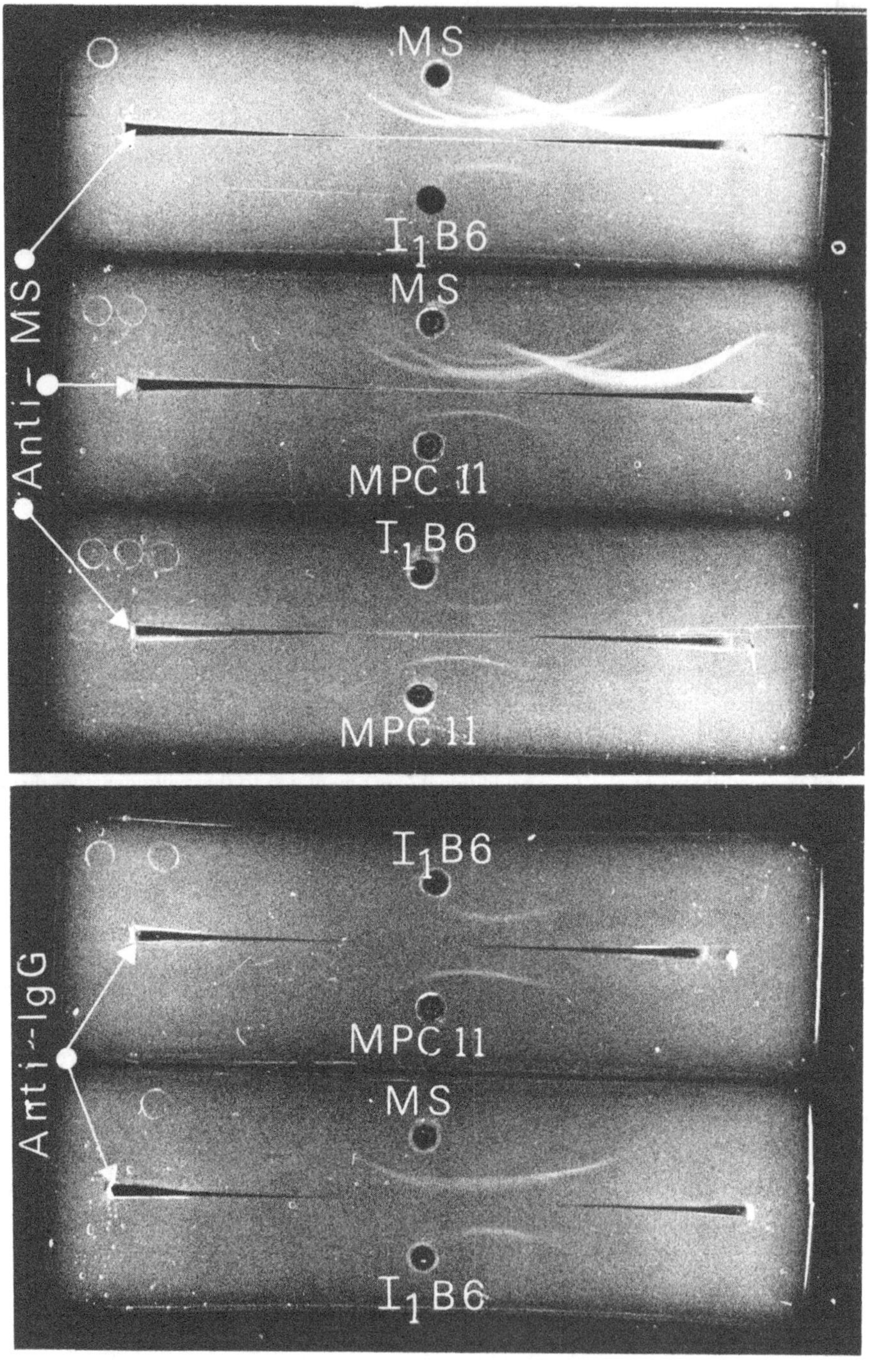

FIGURE 4. Immunoelectrophoretic profiles of whole mouse serum (MS) and concentrated medium from cultures of Type I hybridoma (I_1B6) or MPC 11 parental (MPC) cells. The top three slides were allowed to react with rabbit antiserum made against whole mouse serum (anti-MS); the bottom two were allowed to react with the IgG fraction of goat antiserum made against mouse IgG (anti-IgG).

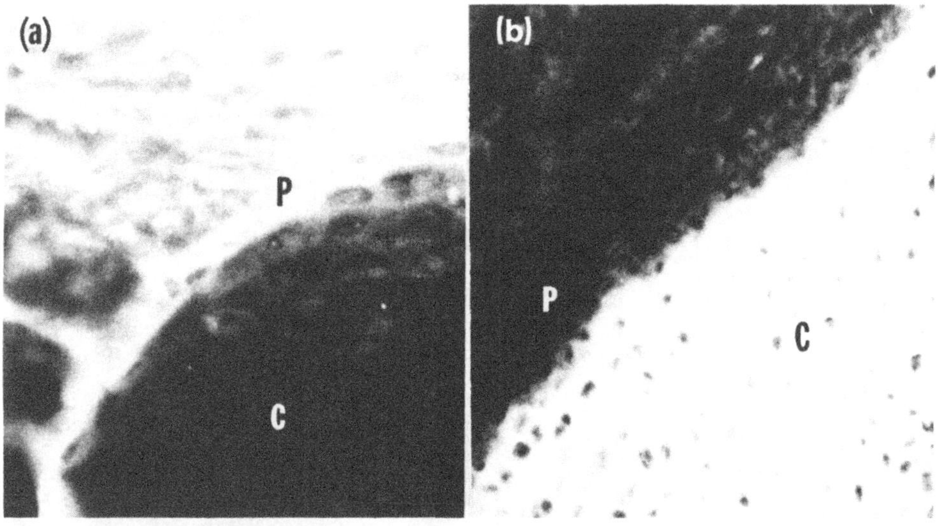

FIGURE 5. Fluorescence micrographs of 14-day-old embryonic chick tibias stained with monoclonal antibody ascites fluid to (A) Type I collagen (magnification × 360) and (B) Type II collagen (magnification × 540) using a double-antibody sandwich technique. The second antibody is rhodamine conjugated rabbit IgG directed against moust IgG; C = cartilage; P = perichondrium.

review, see Hay et al.).[31] The primary stroma is synthesized by the corneal epithelium and is completely acellular during days 4 to 6 of development. From biochemical analyses,[32] it is composed chiefly of collagen Types I and II in a ratio of about 60 to 70% Type I to 30 to 40% Type II. During the 6th day, this primary stroma becomes invaded by mesenchymal cells, first in its posterior portion and subsequently more anteriorly. These cells synthesize the secondary, or mature, stroma which, from biochemical analysis is known to be predominantly Type I collagen. As can be seen in Figure 6 (A and B), the uninvaded, primary stroma stains uniformly for collagen Types I and II. The mesenchymal cells at the periphery are unstained at this stage. In a 10-day-old embryonic cornea (see Figure 6, C and D) which has been invaded by mesenchymal cells, the staining pattern obtained is now distinctly different with the two antibodies. The Type I antibody (see Figure 6C) gives bright fluorescence throughout the stroma with the most intense reaction localized under the corneal epithelium. This staining pattern probably reflects the continued presence of Type I in the uninvaded portion of the primary stroma adjacent to the epithelium, plus new Type I synthesized by the corneal fibroblasts that have invaded. The Type II antibody also gives the brightest staining under the epithelium, with a decreasing gradient of fluorescence across about one fourth the anterior portion of the stroma. The remaining stroma shows little if any staining for Type II, except for a narrow region at the extreme posterior edge just under the corneal endothelium (not shown). This pattern probably reflects the continued presence of Type II in the uninvaded primary stroma under the anterior epithelium and under the endothelium. The decreasing gradient in the anterior portion probably also reflects the presence of the Type II remaining from the primary stroma. The anterior portion is the region most recently invaded by mesenchymal cells, so that Type II staining is still prominant. The more posterior regions, however, have been invaded by mesenchymal cells for progressively longer periods of time, so the large amounts of Type I synthesized by these cells has masked the Type II. Alternatively, the Type II may be progressively

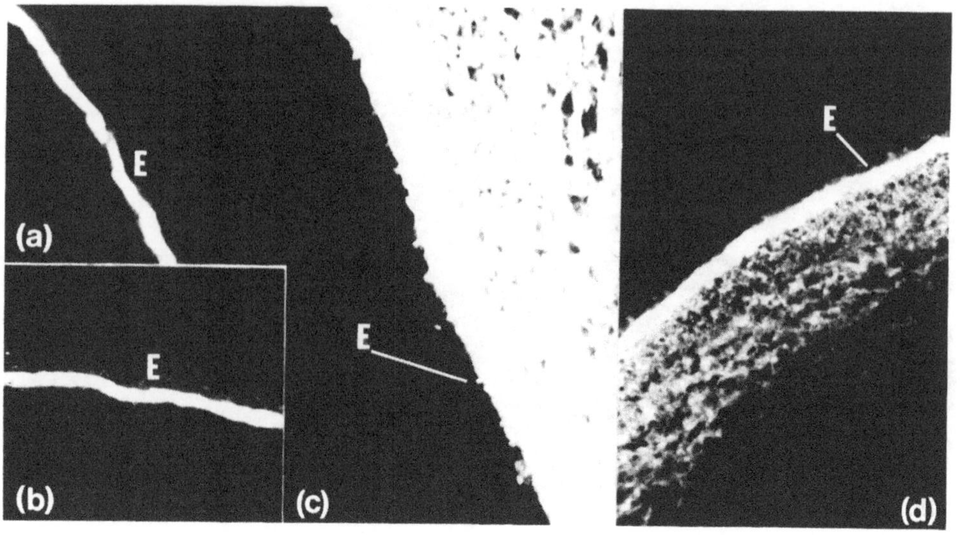

FIGURE 6. Fluorescence micrographs of developing chick corneas from 5-day-old embryos (A,B) (magnification × 180) and 10-day embryos (C,D) (magnification × 170) stained with monoclonal antibody ascites fluid against Type I collagen (A,C) and Type II collagen (B,D) using a double-layer sandwich technique. The second antibody is rhodamine conjugated rabbit IgG directed against mouse IgG; E = epithelium.

degraded by the invading mesenchymal cells. These results are similar to those obtained with conventional antibodies[33] against collagen Types I and II.

Clearly, the monoclonal antibodies can be used to identify collagen types in sectioned material such as in the developing cornea. Thus far, the only negative feature we have found about these monoclonal antibodies is that they seem to precipitate only a fractional portion of the labelled collagen[15] in a radioimmune assay using staphylococcal protein A or a second antibody as a precipitating agent. As yet we do not know why this is the case. It could be that due to their monoclonality they recognize a subpopulation of the collagen molecules. Possibly a single antibody binding site is not strong enough to allow effective precipitation of a large rod-shape molecule like collagen, or the problem may be one of affinity. The hybridomas may have been formed from lymphocytes that themselves were producing low-affinity antibodies. Possibly all anticollagen antibodies have relatively low affinities, but with conventional antibodies, due to the heterogenity of binding sites, the low affinity is not so obvious. Alternatively, as stated earlier, it has been hypothesized that the presence of inactive parental immunoglobulin arms in the antibodies interferes with antibody binding. To discriminate among some of these possibilities, we are currently trying to produce more anticollagen-hybridomas — this time with nonsecretor and nonproducer myeloma cell lines.

V. SUMMARY

The technique of producing monoclonal antibodies by lymphocyte-myeloma cell hybridization has been applied to collagens. In this report, we give detailed descriptions of the methodology we have used for producing these antibodies. In addition, we give some of the characteristics of the antibodies and how they can be used for immunocytochemical staining of tissues such as embryonic chick cartilage and developing corneas.

ACKNOWLEDGMENTS

This is publication number 825 of the Robert W. Lovett Memorial Group for the Study of Diseases Causing Deformities. Original unpublished research and previously published work by the authors was supported by National Institutes of Health grants EY 02261, AM 03564, and HD 00143. Dr. Linsenmayer is recipient of a Research Career Development Award AM00031, and Dr. Hendrix is recipient of Postdoctoral Fellowship HL05682. We would like to thank Eileen Gibney for her excellent technical assistance and Dr. Jerome Gross, Dr. Elizabeth Hay, and Dr. Jack Wands for invaluable discussions.

REFERENCES

1. **Timpl, R.,** Immunological studies on collagen, in *Biochemistry of Collagen*, Ramachandran, G. N. and Reddi, A. H., Eds., Plenum, New York, 1976, 319.
2. **Furthmayr, H. and Timpl, R.,** Immunochemistry of collagens and procollagens, *Int. Rev. Connect. Tiss. Res.*, 7, 1976, 61.
3. **Linsenmayer, T. F.,** Immunology of purified collagens and their use in localization of collagen types in tissues, in *Collagen in Health and Disease*, Jayson, M. I. V. and Weiss, J. B., Eds., Churchill Livingston, Edinburgh, England, in press.
4. **Timpl, R., Wick, G., and Gay, S.,** Antibodies to distinct types of collagens and procollagens and their application in immunohistology, *J. Immunol. Methods*, 18, 165, 1977.
5. **Köhler, G. and Milstein, C.,** Continuous cultures of fused cells secreting antibody of predefined specificity, *Nature*, 256, 495, 1975.
6. **Margulies, D. H., Kuehl, W. M., and Scharff, M. D.,** Somatic cell hybridization of mouse myeloma cells, *Cell*, 8, 405, 1976.
7. **Cotton, R. G. H. and Milstein, C.,** Fusion of two immunoglobulin producing myeloma cells, *Nature*, 244, 42, 1973.
8. **Schwaber, J. and Cohen, E. P.,** Pattern of immunoglobulin synthesis and assembly in a human-mouse somatic cell hybrid clone, *Proc. Natl. Acad. Sci. U.S.A.*, 71, 2203, 1974.
9. **Kung, P. C., Goldstein, G., Reinberg, E. L., and Schlossman, S. F.,** Monoclonal antibodies defining distinctive human T cell surface antigens, *Science*, 206, 347, 1979.
10. **Gasser, D. L., Winters, B. A., Hass, J. B., McKearn, T. J., and Kennett, R. H.,** Monoclonal antibody directed to a B-cell antigen present in rats, mice, and humans, *Proc. Natl. Acad. Sci. U.S.A.*, 76, 4636, 1979.
11. **Koprowski, H., Gerhard, W., and Croce, C. M.,** Production of antibodies against influenza virus by somatic cell hybrids between mouse myeloma and primed spleen cells, *Proc. Natl. Acad. Sci. U.S.A.*, 74, 2985, 1977.
12. **Wiktor, T. J. and Koprowski, H.,** Monoclonal antibodies against rabies virus produced by somatic cell hybridization: detection of antigenic variants, *Proc. Natl. Acad. Sci. U.S.A.*, 75, 3938, 1978.
13. **Melchers, F., Potter, M., and Warner, N., Eds.,** *Lymphocyte Hybridomas*, Springer, New York, 1978.
14. **Moshly-Rosen, D., Fuchs, S., and Eshhar, Z.,** Monoclonal antibodies against defined determinants of acetylcholine receptor, *FEBS Lett.*, 106, 389, 1979.
15. **Linsenmayer, T. F., Hendrix, M. J. C., and Little, C. D.,** Production and Characterization of a monoclonal antibody to chicken type I collagen, *Proc. Natl. Acad. Sci. U.S.A.*, 76, 3703, 1979.
16. **Linsenmayer, T. F. and Hendrix, M. J. C.,** Monoclonal antibodies to connective tissue macromolecules: type II collagen, *Biochem. Biophys. Res. Commun.*, 92, 440, 1980.

17. **Köhler, G., Howe, S. C., and Milstein, C.,** Fusion between immunoglobulin-secreting and nonsecreting myeloma cell lines, *Eur. J. Immunol.*, 6, 292, 1976.
18. **Shulman, M., Wilde, C. D., and Köhler, G.,** A better cell line for making hybridomas secreting specific antibodies, *Nature*, 276, 269, 1978.
19. **Kearney, J. F., Radbruch, A., Liesegang, B., and Rajewsky, K.,** A new mouse myeloma cell line that has lost immunoglobulin expression but permits the construction of antibody-secreting hybrid cell lines, *J. Immunol.*, 123, 1548, 1979.
20. **Littlefield, J. W.,** Selection of hybrids from matings of fibroblasts in vitro and their presumed recombinants, *Science*, 145, 709, 1964.
21. **Nowack, H., Hahn, E., and Timpl, R. E.,** Specificity of the antibody response in inbred mice to bovine type I and type II collagen, *Immunology*, 29, 621, 1975.
22. **Hahn, E., Nowack, D., Götze, D., and Timpl, R.,** H-2-linked genetic control of antibody response to soluble calf skin collagen in mice, *Eur. J. Immunol.*, 5, 288, 1975.
23. **Rohde, H., Nowack, H., and Timpl, R.,** Localization of antigenic activity and immunogenic capacity in different conformational domains of procollagen peptide, *Eur. J. Immunol.*, 8, 141, 1978.
24. **Nowack, H., Hahn, E., and Timpl, R.,** Requirement for T cells in the antibody response of mice to calf skin collagen, *Immunology*, 30, 29, 1976.
25. **Gefter, M. L., Marqulis, D. H., and Scharff, M. D.,** A simple method for polyethylene glycol-promoted hybridization of mouse myeloma cells, *Somatic Cell Genet.*, 3, 231, 1977.
26. **Oi, V. T. and Hertzenberg, L. A.,** Immunoglobulin-producing hybrid cell lines, in *Selected Methods in Cellular Immunology*, Mishell, B. B. and Shugi, S. M., Eds., W. H. Freeman, San Francisco, 1980, chap. 3, p. 51.
27. **Galfre, G., Howe, S. C., Milstein, C., Butcher, G. W., and Howard, J. C.,** Antibodies to major histocompatability antigens produced by hybrid cell lines, *Nature*, 266, 550, 1977.
28. **Schneiderman, S., Farber, J. L., and Baserga, R.,** A simple method for decreasing the toxicity of polyethylene glycol in mammalian cell hybridization, *Somatic Cell Genet.*, 5, 263, 1979.
29. **Beil, W., Furthmayr, H., and Timpl, R.,** Chicken antibodies to soluble rat collagen I characterization of the immune response by precipitation and agglutination methods, *Immunochemistry*, 9, 779, 1972.
30. **von der Mark, H., von der Mark, K., and Gay, S.,** Study of differential collagen synthesis during development of the chick embryo by immunofluorescence: preparation of collagen type I and type II specific antibodies and their application to early stages of the chick embryo, *Dev. Biol.*, 48, 237, 1976.
31. **Hay, E. D., Linsenmayer, T. F., Trelstad, R. L., and von der Mark, K.,** Origin and distribution of collagens in the developing avian cornea, in *Current Topics in Eye Research*, Vol. 1., Zadunaisky, J. A. and Davson, E., Eds. Academic Press, New York, 1979, 1.
32. **Linsenmayer, T. F., Smith, G. N., and Hay, E. D.,** Synthesis of two collagen types of embryonic chick corneal epithelium in vitro, *Proc. Natl. Acad. Sci. U.S.A.*, 74, 39, 1977.
33. **von der Mark, K., von der Mark, H., Timpl, R., and Trelstad, R. L.,** Immunofluorescent localization of collagen types I, II, and III in the embryonic chick eye, *Dev. Biol.*, 59, 75, 1977.
34. **Kessler, S. W.,** Rapid isolation of antigens from cells with a staphylococcal protein A-antibody adsorbent: parameters of the interaction of antibody-antigen complexes with protein A, *J. Immunol.*, 115, 1617, 1975.
35. **Ruoslahti, E.,** Personal communication, 1981.

Chapter 13

RADIOIMMUNOASSAYS IN STUDIES OF CONNECTIVE TISSUE PROTEINS

Rupert Timpl and Leila Risteli

TABLE OF CONTENTS

I. INTRODUCTION

Since the introduction of radioimmunoassays more than 20 years ago,[1,2] this technique became an indispensible tool for immunochemical studies and for the precise measurement of low concentrations of proteins, peptide hormones, and other substances. The basic principles of the test are simple. The antigen is labeled with a radioactive tag and reacted against an appropriate antibody. The amount of antigen-bound to antibody is then determined by separating free antigen from the antigen bound to the immune complexes. For antibodies having high affinity constants, appreciable dissociation of antibody-bound antigen rarely occurs during the separation process, in spite of the low concentrations used in the assays. In the more versatile inhibition assay, labeled and nonlabeled antigens compete for the antibody, and the result is measured as displacement of radioactivity from immune complexes. This assay is particularly useful for analyzing the antigenic structure of complex antigens and for quantitating antigens in unknown samples. Depending on the radioactive label used, the assays show a high sensitivity and work in the range of pico- to nanograms of material. Further advantages are a high precision of the assay not achieved with other serological methods and the universal application to any antigenic substance. Numerous reviews exist on experimental and theoretical aspects of radioimmunoassays, and for a more detailed description the reader is referred to some recent articles.[3-12]

In connective tissue research, radioimmunoassays have been first applied to analyze the antigenic structure of Type I collagen.[13] Reflecting the increasing use of this method, various technical parameters of its application to collagen,[14-17] procollagen peptides,[18,19] noncollagenous proteins,[20,21] and enzymes[22,23] involved in connective tissue metabolism have been studied. Firstly, the present article describes some established procedures for carrying out radioimmunoassays with connective tissue proteins. Secondly, several practical and theoretical aspects are considered for evaluating data obtained with this assay.

II. METHODS AND EXPERIMENTAL PROTOCOLS

A. Labeling of Antigens

Labeling of proteins can be performed with any suitable radioactive isotope, but a high specific activity is most conveniently achieved by the incorporation of iodine. The use of the isotope ^{125}I is superior over that of ^{131}I because of the longer half-life (60 days against 8 days) and higher counting efficiency of ^{125}I.[11] The Chloramine T procedure[24] has been widely used for introducing iodine label into a variety of proteins. Here, presumably, the mildly oxidizing agent produces cationic iodine which mainly substitutes tyrosine residues either as a mono- or di-iodine derivative. Occasionally tryptophan or histidine residues may also participate in the reaction.[25] The reaction is terminated by adding a reducing agent. Radioactive iodine can also be introduced by an iodinated tyrosyl ester capable of reacting with free amino groups in proteins as first described by Bolton and Hunter.[26] This method seems particularly useful for proteins with a low tyrosine content or when oxidizing conditions are harmful for antigenicity.

Less commonly used labeling procedures include iodination with lactoperoxidase.[7] This reaction requires H_2O_2 which is either directly added or generated enzymatically. Due to steric hindrance, lactoperoxidase iodinates only tyrosine residues which are easily accessible. This can be exploited for particular purposes such as selective labeling of plasma membrane surfaces. ^{14}C acetic acid anhydride has also

Table 1
VARIATIONS IN THE EFFICIENCY OF LABELING OF CONNECTIVE TISSUE PROTEINS BY THE CHLORAMINE T METHOD[a]

Antigen (number of experiments)	Tyrosyl residues per 10,000 dalton molecular weight	Total radioactivity incorporated[b] (cpm/ng)	Protein-bound radioactivity[c] (% of total)	Nonspecifically absorbed radioactivity[d] (% of protein bound)
Aminopropeptide Type I (5)	2	8,000–15,400	88–93	0.4–1.0
Aminopropeptide Type III (7)	3	5,900–13,000	64–77	0.5–1.6
Collagen Type II (2)	0.5	3,600–4,100	69–91	3.2–3.8
Collagen Type IV (7)	1	3,800–8,300	77–90	1.8–8.6
Laminin (6)	3	3,800–12,700	85–93	4.3–12.7

[a] About 25 μg of protein labeled with 0.5 mCi of ^{125}I according to the protocol described in the text. Ranges are based on data collected in our laboratory using several batches of antigen in the number of experiments indicated

[b] After removal of free ioding by dialysis

[c] Determined by precipitation with 10% trichloroacetic acid

[d] Bound to 0.5% normal rabbit serum in the double antibody assay

been used for labeling collagen.[15] As expected[11] specific radioactivity (5 × 10^6 dpm/mg) was less by a factor of 1000 when compared to iodination (see Table 1). A low specific radioactivity is also found for proteins metabolically labeled with ^{3}H or ^{14}C amino acids, even though such material may be useful for certain purposes (see Section III.D). Another method which has been applied to prolyl hydroxylase[22] involves reductive alkylation with formaldehyde and tritiated sodium borohydride. Alternatively, oxidation with periodate[27] may be used. In this instance, label is introduced mainly via carbohydrate groups.

1. Labeling with ^{125}I by the Chloramine T Procedure[21,28]

About 25 μg of protein dissolved in 25 to 50 μℓ of 0.1 *M* sodium phosphate buffer, pH 7.2, is mixed with 0.5 to 1 mCi of Na ^{125}I (0.2 to 0.4 nmol) in 60 μℓ of 0.1 *M* phosphate, pH 7.2, and 100 μℓ of Chloramine T (Na salt of the *N*-monochloro-derivative of *p*-toluene sulfonamide) solution (3 mg/5 mℓ of 0.1 *M* phosphate, pH 7.2) and incubated for 5 to 10 min at room temperature. The reaction is stopped by adding 120 μℓ of sodium metabisulfite ($Na_2S_2O_5$; 3 mg/5 mℓ H_2O) and 500 μℓ of 0.1 *M* phosphate, pH 7.2. The mixture is then diluted to 2.5 mℓ by adding 0.04% Tween 20 (polyoxyethylene sorbitan monolaurate) in phosphate buffered saline, pH 7.2 (PBS), and dialyzed at room temperature against 2 to 3 × 2ℓ Tween 20/PBS (2 to 4 days) in order to remove free iodine. Alternatively, free iodine may be removed on a BioGel® P2 column (see below). The solutions of Chloramine T and sodium metabisulfite should be prepared freshly each day and stored at 4°C. In the case of labeling interstitial collagens, sufficient NaCl (0.2 to 0.4 *M*) should be added to the phosphate buffer in order to keep the protein in solution. Labeled collagens may also be dialyzed against 0.1 *M* acetic acid for storage.

2. Labeling with the Bolton-Hunter Reagent[29]

The reagent is usually available in benzene solvent which is evaporated prior to use under a stream of N_2. About lmCi (0.67 nmol) of the dry ester (3-[^{125}I] iodo-1*p*-hydroxyphenyl propionic acid *N*-hydroxysuccinimide ester) is then dissolved in 50

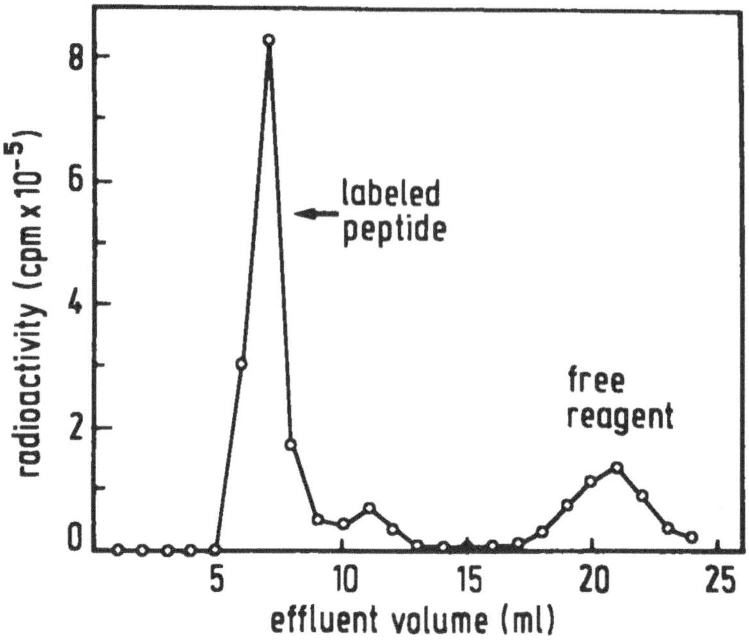

FIGURE 1. Separation of the chain fragment P1 of collagen α1(IV) chain labeled with Bolton-Hunter reagent[30] from nonreacted reagent on BioGel® P-2. The column (1 × 15 cm) was equilibrated in Tween/PBS, and the effluent was collected in 1-mℓ fractions. The reaction product was obtained from 25 μg of protein and 1 mCi of Bolton-Hunter reagent. Radioactivity was determined in 5-μℓ aliquots of the column effluent.

μℓ of 0.1 *M* sodium borate buffer, pH 8.5, containing 25 μg of the protein and incubated for 45 min in an ice bath. The reaction is stopped by adding 0.2 mℓ of 0.2 *M* glycine, 0.1 *M* borate, pH 8.5, and further operations are then carried out at room temperature. Unreacted reagent is separated from labeled protein on a column of BioGel® P2 (1 × 15 to 20 cm) equilibrated and eluted with 0.04% Tween®/PBS. An example is shown in Figure 1. A somewhat different procedure specifically adapted for labeling collagen has been published.[16]

Labeled products are stored at 4°C as solutions containing about 10 μg/mℓ of protein. Addition of serum albumin (0.5%) may help to reduce absorption of the labeled material to the container.

Radioactivity incorporated by these procedures was in the range 4,000 to 15,000 cpm/ng protein (see Table 1) which corresponds to the introduction of 0.1 to 1 mol of iodine per mole of protein. Since a reasonable working range requires about 2,000 to 10,000 cpm per test tube, as little as 0.2 to 1 ng of antigen may be used per single assay. The sensitivity of the assay can be increased by preparing antigens with tenfold or higher specific radioactivity by increasing the amount of radioactive isotope or lowering the amount of protein, but following otherwise the protocols as described. This, however, may result in poor recovery of protein (e.g., loss during handling) or in a large decrease of the activity and stability of the labeled product (lack of trace labeling).

B. Quality and Stability of Labeled Proteins

Freshly labeled products should be routinely checked for the amount of protein-bound radioactivity and for the nonspecific association of radioactivity under the working conditions of a binding assay. Protein-bound radioactivity can be determined by precipitation with 10% trichloroacetic acid in the presence of carrier protein and was found for several proteins (see Table 1) to vary in the range 63 to 93% of the total radioactivity measured after purifying the labeled material (see Section II.A).

1. Rapid Determination of Protein-Bound Radioactivity

The labeled protein is diluted to 0.5 to 1 ng in 0.2 mℓ of 10% normal rabbit serum and mixed with 0.2 mℓ of 20 % trichloroacetic acid dissolved in water. After incubation at 4°C for 1 hr, the precipitated protein is collected by centrifugation and washed twice with 0.4 mℓ of 10% trichloroacetic acid. Measuring the radioactivity prior to precipitation and in the precipitate allows then to calculate

$$\text{protein-bound radioactivity (\% of total)} = \frac{\text{cpm in precipitate}}{\text{total cpm}} \times 100 \qquad (1)$$

Some proteins may not be sufficiently insoluble in 10% trichloroacetic acid to allow a correct estimate by this procedure. The choice of another precipitating reagent (e.g., acetone) or of more elaborate separation methods (e.g., electrophoresis, molecular sieve chromatography) may circumvent this problem. It is also possible to determine protein-bound radioactivity by precipitation with an excessive amount of antibody. Further increase of antibody should not change the fraction of radioactivity precipitated.

The common observation that protein-bound radioactivity rarely equals total radioactivity in a sample of labeled protein can be due to insufficient separation from nonreacted isotope. Very often it reflects dissociation of the label (particularly when associated with histidine) or destruction of the antigen during storage. Usually the decrease during 4 to 12 weeks of storage does not exceed 10 to 20% of the original value determined after labeling and is smaller than the decrease in radioactivity due to the half-life of iodine isotopes. Depending on these parameters and the particular species of protein, individual batches of labeled material may be efficiently used for a period of 8 to 12 weeks.

2. Determination of Nonspecific Binding of Labeled Protein

The protocol follows exactly that described for the binding assay (Section II.D) except that antiserum but not the nonimmune serum carrier is omitted from the reaction. Counting the radioactivity in the precipitate allows to calculate

$$\text{non-specific binding (\% of total)} = \frac{\text{cpm precipitated in the presence of non-immune serum}}{\text{cpm protein-bound radioactivity}} \times 100 \qquad (2)$$

A preferable range is 1 to 2%, but nonspecific binding up to 10% (see Table 1) may not seriously influence the quality of data. Values above that range are likely to interfere with the precision of the assay. Nonspecific binding is mainly due to absorption of labeled protein to the test tubes, but may also vary with the volume of precipitate, depending on the amount of carrier added. Including nonionic detergents such as Tween® 20[31] in the reaction mixtures may help to keep the amounts low. Further improvement can be achieved by using other reagents, e.g., 0.1% Triton® X-100, urea up to a concentration of 1 *M*, 0.5 to 1% bovine serum albumin, or by using tubes with low protein absorption or after prewashing them with solutions of appropriate proteins.

The labeling of proteins may also cause destruction of antigenicity. Damage could be due to substituting amino acid residues involved in particular antigenic determinants (avoided by trace labeling), to chemical oxidation, to irradiation effects, or to noxious substances present in the batch of radioisotope used.[11] The antigenic quality of a labeled protein can be evaluated by either using a different serological method or by comparing mixtures of labeled and nonlabeled antigens in radioimmunoassay binding. For appropriate control of losses during handling, the nonlabeled preparation should be carried through the labeling procedure except that all modifying reagents are omitted. The comparison of labeled and nonlabeled aminopropeptides of procollagen by passive hemagglutination inhibition (see Figure 2) showed no difference in antigenicity within the accuracy of that assay ($\pm 50\%$, equal to one dilution step). This could be corroborated by comparing labeled antigens and mixtures of labeled and nonlabeled antigens in binding assays (see Figure 3). Lack of change in antigenicity after labeling is expected for aminopropeptides, since most of the tyrosine is found in the C terminal region which does not significantly contribute to the immunological properties.[32]

It will be frequently important to show by chemical methods that radioactivity is bound to a particular protein and not to impurities. This seems important for antigens with a low content of amino acid residues required to attach the label (e.g., tyrosine in collagen) or when the homogeneity of the antigen cannot be assessed properly. Electrophoresis using autofluorography for visualizing labeled peptide bands (see Figure 4) is a convenient way to analyze the distribution of radioactive label among various proteins. This approach has also been used to monitor destruction of the antigen in the case of ^{125}I-labeled procollagen peptides.[18,19] The distribution of label within a given protein such as collagen could also be analyzed by chromatographic separation of α-chains[14] characterization of CNBr peptides,[16] or by proteolysis for telopeptide regions.[14,15]

C. Separation of Bound and Free Antigen

The most versatile and universally employed method is the use of a second antibody directed against the immunoglobulin of the animal species from which the antiserum (first antibody) under study was obtained. This double antibody assay requires an excess of the second antibody in order to ascertain that all immunoglobulin (first antibody and nonimmune immunoglobulin) is removed from the reaction (see Figure 5). Such second antisera usually contain mainly antibodies directed against the constant domains of the immunoglobulin molecule and do not interfere with the labeled or nonlabeled antigen bound to the Fab domain. The procedure allows the rather complete separation of bound from free antigen. In the standard method, the second antibody is used to precipitate the immunoglobulin from solutions containing the first antibody and the antigen. This reaction requires some time for completion (usually overnight incubation at 4°C) and the addition of a fixed amount of nonimmune immunoglobulin in order to keep the amount of precipitate constant over a wide range of antiserum dilutions. Alternative methods include coating the test tubes with second antibody (accomplished by absorption) or the use of the second antibody in particular form such as after coupling to agarose.[35] Second antibodies against immunoglobulins from a variety of laboratory animals are commercially available, but rather expensive. For extensive radioimmunoassay studies, it is recommended to prepare these antibodies preferentially in goat or sheep.

1. Immunization of Goats with Immunoglobulin G[36]

A goat is injected i.c. or s.c. with 25 μg rabbit immunoglobulin dissolved in 0.2 mℓ of 0.1 *M* phosphate buffer, pH 7.0, and mixed with an equal volume of complete

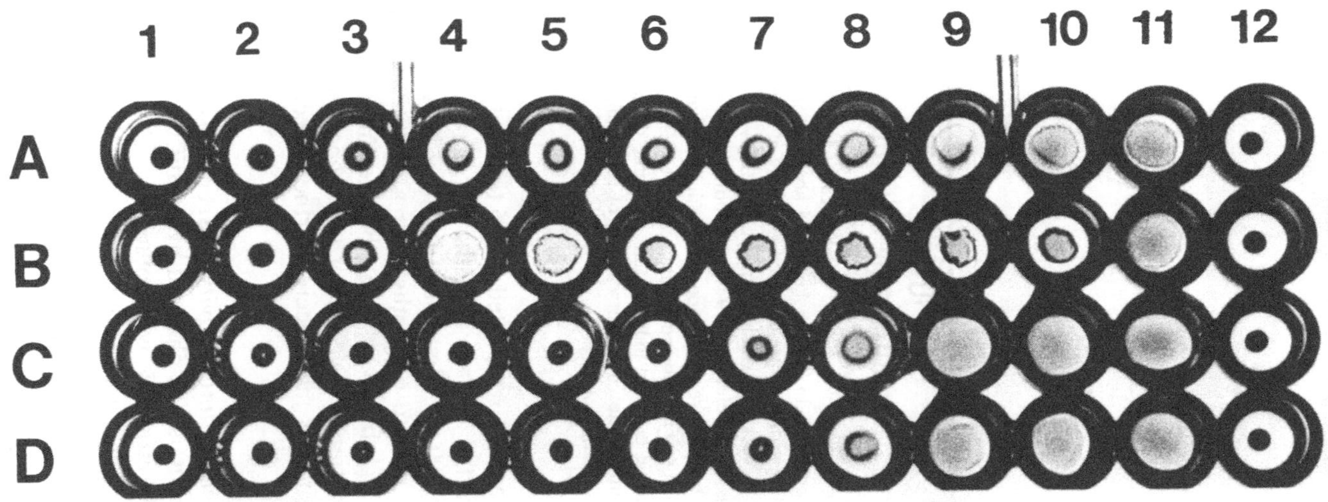

FIGURE 2. Comparison of ^{125}I-labeled (A, C) and nonlabeled (B, D) aminopropeptides of Types I (A, B) and III (C, D) procollagens by passive hemagglutination inhibition (see Chapter 3). Serial dilutions of the antigens from Well number 1 to 10 were incubated with a constant amount of antibody, and residual antibody activity was then detected by adding red cells coated with the respective procollagens. Well 11 shows agglutination by antibody in the absence of inhibitors, and Well 12 shows the normal settling pattern in the absence of antibody. Labeled and nonlabeled aminopropeptides show comparable inhibiting activity up to Well 3 (A, B) or Well 7 (C, D), indicating lack of damage by iodine labeling. The starting concentrations of aminopropeptides in Well 1 were 10 μg/mℓ.

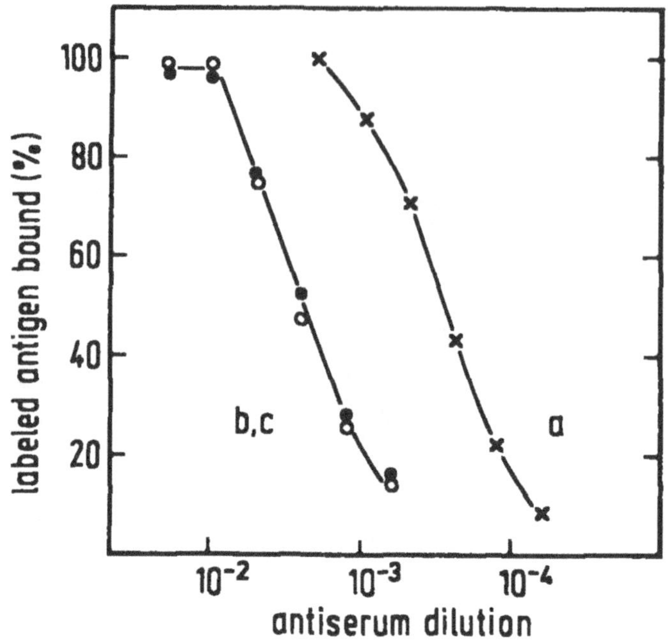

FIGURE 3. Examination of antigenic damage caused by ^{125}I labeling using binding to antibody in the presence and absence of nonlabeled antigen. The binding profiles were obtained with 1 ng of labeled antigen (a, ×——×), with 10 ng of labeled antigen (b, ○——○), or with a mixture of 1 ng of labeled and 9 ng of nonlabeled antigen (c, ●——●). Curves b and c are superimposable, indicating antigenic identity between labeled and nonlabeled product. A shift of Curve c towards the left side of Curve b would indicate that antigenic damage has occurred.[11] The example shown is the reaction of antibodies to Type I procollagen with its aminopropeptide.

Freund's adjuvant. This injection is repeated 4 to 6 times at 1- to 2-week intervals. The first serum can be collected 20 days after the last booster. If the antibody titer is sufficiently high, 400- to 500-mℓ blood samples may be withdrawn upon single bleedings done in intervals of 4 to 6 months over a period of several years. The antibody level can be kept high by giving single booster injections 3 weeks prior to a new bleeding. In our experience, two out of three goats show a good antibody response. Low-responder animals may receive further injections by increasing the antigen doses to 1 mg.

2. Determination of the Precipitating Activity of Second Antibody

The quality of the antiserum can be checked by gel diffusion in 1% agarose using microequipment (Miles Laboratories®; well volume, 44 μℓ; distance between wells, 4 mm). The center well is filled with rabbit immunoglobulin (1 mg/mℓ) and the peripheral wells are filled with serial dilutions (1:2 to 1:64) of the goat antiserum. A positive precipitation reaction at dilution 1:32 after 2 days incubation at room temperature indicated that 10 μℓ of goat antiserum is capable to precipitate completely the immunoglobulin of 0.5 μℓ of rabbit serum. More precise but laborious ways to evaluate the precipitating potency are based on quantitative precipitation assays[7] using either labeled or nonlabeled antigen. It is often more practical to analyze binding profiles of a standard first antiserum with labeled antigen by varying

FIGURE 4. Electrophoretic analysis of aminopropeptides of Type III (1,2) and Type I (3,4) procollagen in nonlabeled form (1,3) and after labeling with ^{125}I (2,4). The proteins were run in slab gels (3% stacking gel; 12.5% separating gels; top of the latter denoted by arrow) using a standard procedure.[33] Nonlabeled proteins (about 10 μg) were stained with coomassie blue. Labeled proteins (about 125,000 cpm) were visualized by exposure to Kodak® preflashed[34] X-Omate® R film for 48 hr at −70°C. Note the complete correspondance of both patterns, indicating the lack of substantial labeling of impurities. The upper band in Lanes 3 and 4 represent an aggregated form of the aminopropeptide.

either the amount of carrier immunoglobulin or the amount of second antiserum as shown in Figure 5. The minimal amount of carrier or second antibody found to produce an optimal binding profile can in most cases then be used for systems containing other first antisera and antigens.

Numerous other procedures have been described for the separation of bound and free antigen.[7,35] Similar to second antibody is the use of staphylococcal protein A which shows high affinity for the Fc portion of certain immunoglobulin isotypes and is preferably employed in form of a conjugate with an insoluble carrier.[37] Various precipitation methods have also been applied, but require a better solubility of the

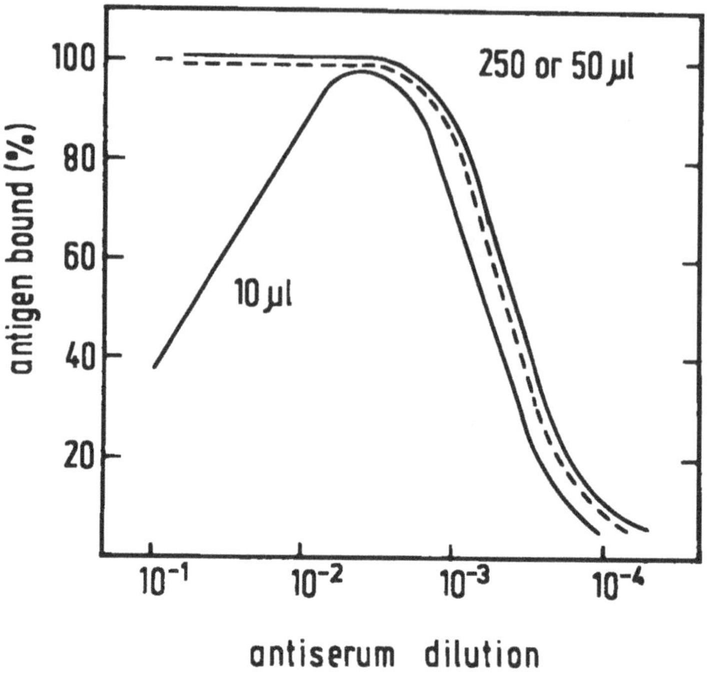

FIGURE 5. Effect of the amount of second antibody on the binding profiles in the double antibody assay. The same antiserum was serially diluted from 1:10 to 1:10,240 using 0.5% normal serum as diluent and incubated with a constant amount of labeled antigen. Separation of free from bound antigen was then carried out with second antibody (goat antiserum to rabbit IgG, titer in gel diffusion, 1:32) adding either 250-, 50-, or 10-µℓ aliquots. Note in the experiment with 10 µℓ a drop in the fraction of precipitated antigen in the dilution of the antiserum from 1:100 to 1:10 (carrier dilution 1:200!). This is apparently due to insufficient precipitation of the total immunoglobulin from the mixture rather than due to insufficient antigen binding by antiserum, since increasing the amount of second antibody produced binding profiles where 100% of the antigen was bound by high concentrations of the antiserum.

antigen as compared to immunoglobulin. This will rarely be the case in the work with connective tissue proteins. An often-used method (Farr test)[12] exploited precipitation of immunoglobulin with ammonium sulfate at 40 to 50% saturation. Such methods are inexpensive, but may be influenced by a variety of additional parameters, such as total protein concentration in the test tube and volume of the precipitate. It is also occasionally observed that antiserum binding is lower in the Farr assay than with the double antibody method.[19] For small antigens such as hormones, the removal of free instead of bound antigen (e.g., by absorption onto charcoal) may provide a better alternative. Other methods used include more laborious chromatographic or electrophoretic separations of bound from free antigen.

D. Binding Assay

Radioimmunoassays detect primary antigen-antibody complexes, and it is sufficient that the labeled antigen combines with a single antibody binding site. Binding is usually determined by reacting various dilutions of antiserum with a constant amount of labeled antigen. Although in most cases little time is required to reach equilibrium, incubation of the reaction is routinely carried out overnight at 4°C. The most crucial step is the separation of bound from free antigen. Precipitation and washing of the immune complexes have to be done under conditions which do not favor dissociation of the complexes and establishment of a new equilibrium.

1. Procedure for Determining Binding of Labeled Antigen by Antibody

All dilutions are carried out in phosphate buffered saline, pH 7.2 (PBS), which is prepared by dissolving 5.45 g of KH_2PO_4, 20.8 g of $Na_2HPO_4 \cdot 2H_2O$, 70.1 g of NaCl, and 2.0 g of sodium azide as preservative in 10 ℓ of distilled water. Finally, 0.4 mℓ of Tween 20 is added to 1 ℓ to reduce nonspecific binding during the assay. Serial dilutions of rabbit antiserum (first antibody) are prepared in 0.5% normal rabbit serum which serves as a carrier.

Aliquots (0.1-mℓ) of each dilution are then pipetted into tubes (70 × 10 mm) set up in duplicate or triplicate. Control tubes for determining nonspecific binding receive 0.1 mℓ of 0.5% normal serum without antibody. The content of each of these test tubes is then mixed with 0.1 mℓ of a working dilution of labeled antigen (usually containing 1 ng to provide 2,000 to 10,000 cpm protein-bound radioactivity). After addition of 0.2 mℓ of Tween®/PBS, the samples are incubated for about 16 hr at 4°C. Then a pretested amount of second antiserum (e.g., 50 μℓ, see Section II.C, Figure 5) is rapidly mixed with the content of each tube, and incubation is continued for another 16 to 24 hr at 4°C. Precipitates are collected in a refrigerated (4°C) centrifuge (30 min, 1500 × g) and the supernatant fluid is carefully sucked off by aspiration on a water pump. The precipitates are then washed two times by adding 1.0 mℓ of ice-cold Tween®/PBS, mixing the contents briefly on a vortex mixer, and by centrifugation as above. During all steps, the tubes should be kept in an ice bath in order to minimize solubilization of the precipitates. Radioactivity of the precipitates is determined by counting in a well-type scintillation crystal (e.g., Beckman® gamma 4000 counting system) at a setting of 0 to 80 keV for ^{125}I-labeled antigens. The results are expressed as percent of total antigen added bound by antibody:

$$\text{antigen bound (\%)} = \frac{a - b}{c - b} \times 100 \qquad (3)$$

a = cpm in precipitates containing first antibody
b = cpm in precipitates containing only non-immune carrier (non-specific binding, equation 2).
c = total protein-bound radioactivity added (see equation 1).

The time required to separate bound from free antigen is crucial for maintaining the initial equilibrium (see Section III.A). The use of separating agents (second antibody, protein A) in insoluble form may considerably shorten the incubation time, but at the expense of assay sensitivity.[7] Washing of precipitates may be avoided by adding a second radioactive isotope (e.g., ^{22}Na) as a volume marker to the reaction.[38] It is also possible to add the labeled antigen to preformed complexes of first and second antibody,[39] but this may change antigen binding.

As discussed later (see Section III.A), the profile of a binding curve is not only dependent on the concentration and affinity of antibodies, but also in many cases on the amount of labeled antigen added. An example is shown in Figure 6 for the titration of antibodies against Type III procollagen using labeled aminopropeptide. Lowering the antigen increases the sensitivity of the assay for detecting low concentrations of antibody, but only when they are of high affinity. The limits and precision of the assay at low antigen concentration are, however, determined by the

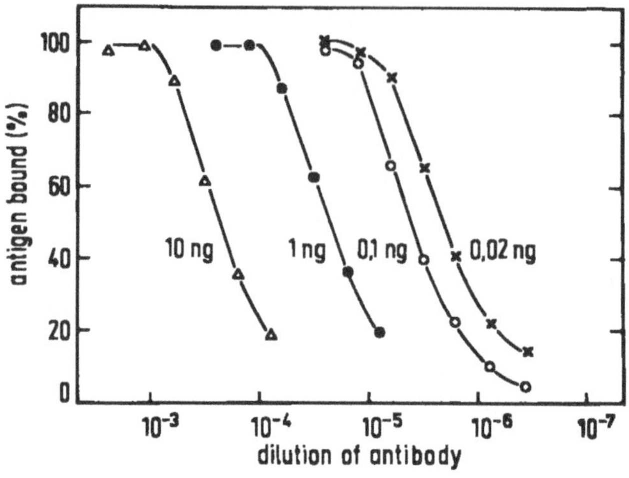

FIGURE 6. Shift in binding profile of antibodies to Type III procollagen with ^{125}I-labeled aminopropeptide Type III as a function of the amount of labeled antigen added. The starting concentration of purified antibody added was 120 μg/0.1 mℓ; the total reaction volume was 0.4 mℓ.

extent of label which can be incorporated into the antigen without changing the antigenic structure.

Several forms of binding curves may be observed depending on the quality of the antiserum and the labeled antigen (see Figure 7). Antisera with high affinity usually show about 100% antigen binding in the excess antibody zone and a rather sharp fall of binding in the range 10 to 95% binding when plotted against the logarithm of antiserum dilution. Deviations from this profile may indicate low affinity of the antibodies.[7,11] Particular antisera may show less than 100% binding of the antigen, but nevertheless a plateau in the antibody excess region. This observation indicates for various reasons (see Figure 7) that the antigen is of poor quality.

E. Inhibition Assay

The inhibition assay measures the competition of labeled and nonlabeled antigens for the same antibody combining sites and is conveniently carried out in the presence of constant amounts of antiserum and of labeled antigen. The result of these reactions is then evaluated by determining the amount of residual radioactivity remaining in the precipitate. Three principal ways for analyzing such interactions have been devised.[5,6]

1. Equilibrium assay — Constant amounts of labeled antigen and varying amounts of nonlabeled antigen are mixed prior to the addition of antibody. After reaching equilibrium, free and bound antigen are separated.
2. Sequential saturation assay — The antibody is first incubated with nonlabeled antigen under equilibrium conditions. The addition of a constant amount of labeled antigen for a shorter period insufficient to allow reequilibrium precedes the separation step.
3. Dissociation assay — Constant amounts of antibody and labeled antigen are incubated under equilibrium conditions. The replacement of label from immune complexes is then measured by adding nonlabeled antigen, varying its amount and/or the time of incubation.

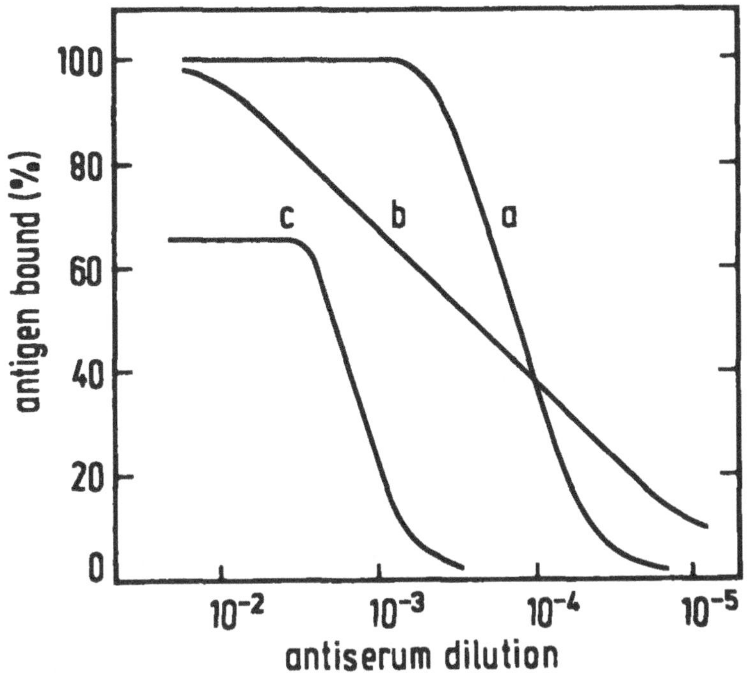

FIGURE 7. Schematic representation of various types of binding curves observed in double antibody radioimmunoassay with ^{125}I-labeled antigens. Curve a shows a plateau of about 100% binding in the region of excess antiserum and a sharp decline from 80 to 90% to 10 to 20% binding upon further dilution by a factor 5 to 10. This indicates an antiserum of high affinity (avidity); examples of this type are shown in Figures 3 and 8 for antisera to procollagens. Curve b also starts with 100% binding, but has a distinctly less-steep slope compared to Curve a. This profile indicates low affinity of antibodies and/or a broad population of antibodies with different affinities.[7,11] Experimental examples can be found in immunological studies with Type IV collagen.[40] Curve c resembles Curves a or b except that the plateau of maximal antigen binding is distinctly below 100%. It indicates that part of protein-bound label (35% in the example shown) does not bind to antibody. It may reflect damage of the antigen by labeling or the presence of labeled contaminants not participating in the reaction with the particular antiserum.

1. Protocol of an Inhibition Assay Depending on Sequential Saturation

It follows basically the steps described in the binding assay (see Section II.D) and is carried out in duplicate or triplicate test tubes. Each test tube receives 0.1 mℓ of antiserum dilution (first antibody) sufficient to bind about 50% of the labeled antigen added later. The test tubes (blank) for determining nonspecific binding of labeled antigen receive only carrier immunoglobulin. The antiserum-containing test tubes are then mixed with varying amounts of nonlabeled antigen (inhibitor) dissolved in 0.2 mℓ of Tween®/PBS, while the blanks and two or three tubes with antibody (antigen-binding in the absence of inhibitor) receive 0.2 mℓ of Tween®/PBS without inhibitor. These mixtures are incubated for 16 hr at 4°C and after addition of 0.1-mℓ of labeled antigen (as determined in the binding assay) are further incubated for 6 to 8 hr at 4°C. The addition of second antibody, the collection, washes and determination of radioactivity of the precipitates follow the protocol described for the

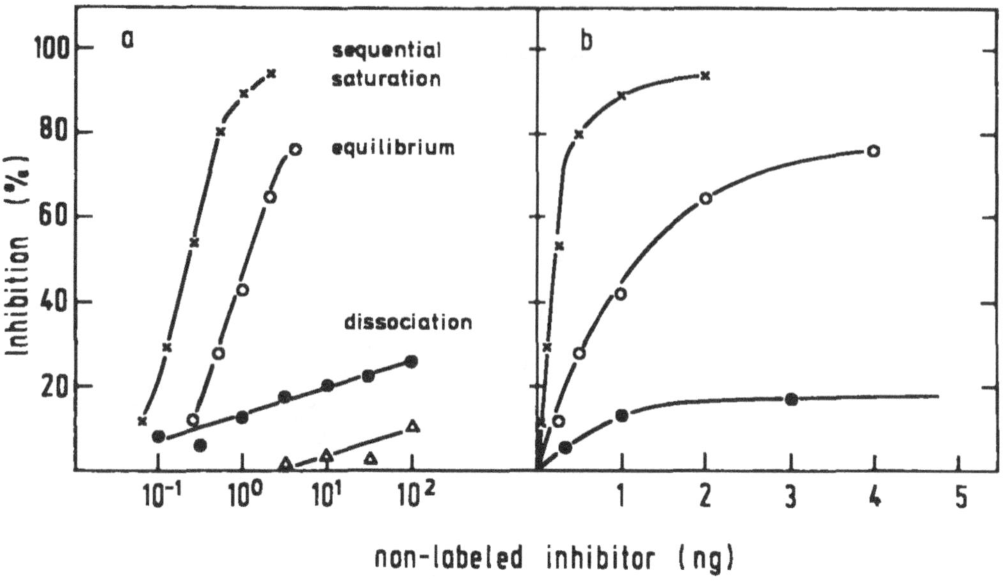

FIGURE 8. Inhibition assay with antibodies to Type III procollagen and ^{125}I-labeled aminopropeptide Type III. The assays were carried out in three different ways:

1. Sequential saturation (×——×) — Antibody and nonlabeled aminopropeptide were incubated together (16 hr, 4°C) prior to addition and further incubation (8 hr, 4°C) with labeled aminopropeptide.
2. Equilibrium (○——○) — Antibody was added to mixtures of labeled and nonlabeled aminopropeptide and incubated for 16 hr at 4°C.
3. Dissociation — Antibody and labeled aminopropeptide were incubated together (16 hr, 4°C) prior to the addition of nonlabeled aminopropeptide and further incubation at 4°C for 8 (△——△) or 56 hr (●——●).

Results were expressed as percent inhibition (see Equation 4) and plotted against amount of nonlabeled aminopropeptide in a semilogarithmic (a) or linear (b) way. The amounts of antibody (3 ng) and labeled aminopropeptide (1 ng) were kept constant. Reaction volume: 0.4 mℓ.

binding assay. The results may be calculated in the following ways:

$$\text{inhibition of antigen binding (\%)} = \left(1 - \frac{c - b}{a - b}\right) \times 100 \qquad (4)$$

or

$$\text{fraction of residual labeled antigen bound to antibody} = \frac{c - b}{a - b} \qquad (5)$$

a = cpm in precipitate containing first antibody and only labeled antigen (antigen binding in the absence of inhibitor)

b = cpm in precipitates containing only non-immune carrier (nonspecific binding, blank)

c = cpm in precipitates containing first antibody and labeled and nonlabeled (inhibitor) antigen.

Similar protocols can be followed for analyzing inhibition by nonlabeled antigens in the form of an equilibrium or a dissociation assay just by changing the order of additions of the various reagents and the incubation times (see Figure 8). A linear and semilogarithmic plot of inhibition and dissociation data obtained with antibodies against the aminopropeptide of Type III procollagen is shown in Figure 8. The inhibition assay following sequential saturation showed the highest sensitivity and

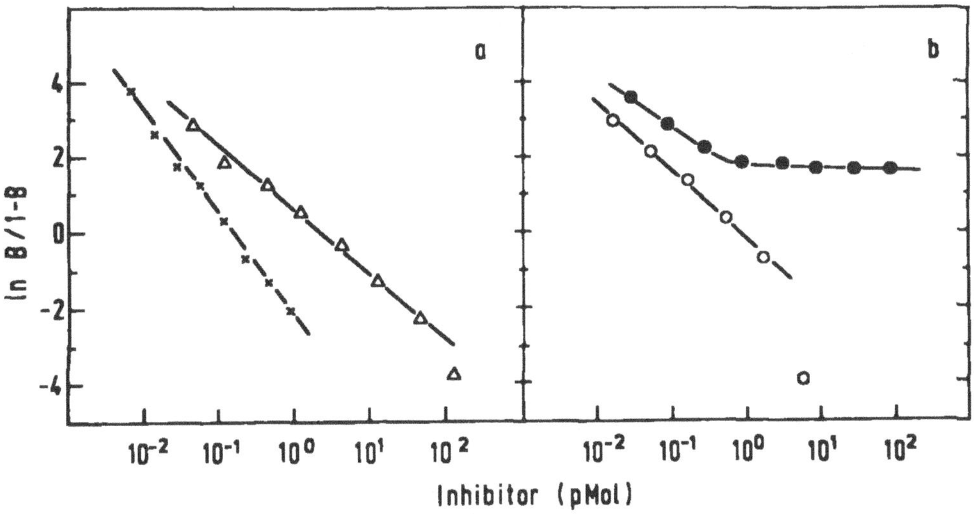

FIGURE 9. Logit plots of inhibition assays (equilibrium type) with antibodies to Type III procollagen (a) and to laminin (b). The ^{125}I-labeled test antigens (1 ng) were the aminopropeptide Type III in (a) and laminin in (b). Inhibitors of the reaction were the aminopropeptide (×——×) and its globular domain as released by collagenase[41] (△——△), laminin (○——○), and laminin fragment P1 (●——●).[21] The apparently different cross-reaction patterns shown in (a) resemble those of Figure 14c, and patterns in (b) resemble those of Figure 14d, both obtained by the sequential saturation technique. Note further that the logit plot is essentially linear over a large range of inhibitor concentrations except sometimes for extreme values. Deviations from linearity can in part reflect the accuracy of measuring the bound fraction (B) in the absence of inhibitor and the nonspecific binding of label (see Equation 2), but may also be due to other reasons.[5,8]

was linear over a large range of the dose response curve. This is a common observation for high-affinity antibodies. The equilibrium inhibition assay showed a typical saturation curve in the linear plot. After semilogarithmic transformation, a sigmoidal shape of the curve is obtained with a nearly linear section in the range 20 to 70% inhibition. A better linearization is obtained by logit-log transformation.[5,10]

$$\text{logit}\,\frac{B}{B_o} = \ln\left(\frac{B/B_o}{1 - B/B_o}\right) \qquad (6)$$

where B is the bound labeled antigen, B_o is the bound labeled antigen in the absence of nonlabeled antigen, and B/B_o is the fraction bound (calculated by Equation 5). An example of this plot is shown in Figure 9. For many other ways to plot inhibition data, including a Scatchard type of analysis, the reader is refered to previous reviews.[9,10]

Dissociation tests are less commonly used because of the low sensitivity due to the long time required to reach equilibrium. This is clearly illustrated in Figure 8, also indicating (see curve for 8-hr dissociation) that under the condition of the sequential saturation assay, no appreciable amount of nonlabeled antigen has dissociated from antibody after addition of the labeled antigen (see Section III.B).

III. APPLICATIONS AND INTERPRETATION OF DATA

Radioimmunoassays have been applied for studying a variety of connective tissue proteins (see Table 2). They were used for measuring antigen concentrations in

Table 2
RADIOIMMUNOASSAYS USED IN STUDIES OF CONNECTIVE TISSUE PROTEINS

Antigen	Labeling procedure	Ref.
Type I collagen	^{125}I iodination with Chloramine T	13, 14, 42
	^{125}I iodination with Bolton-Hunter reagent	16
	^{125}I iodination with lactoperoxidase	17
	^{14}C-labeling with acetic acid anhydride	15
Type II collagen	^{125}I iodination with Chloramine T	42
	^{125}I iodination with Bolton-Hunter reagent	16
Type III collagen	^{125}I iodination with Chloramine T	43
	^{125}I iodination with Bolton-Hunter reagent	16
Type IV collagen	^{125}I iodination with Chloramine T	30, 40, 44
Type V collagen	^{125}I iodination with Bolton-Hunter reagent	16
7-S collagen	^{125}I iodination with Chloramine T	45
Type I procollagen[a]	^{125}I iodination with Chloramine T	18, 19
Type III procollagen [a]	^{125}I iodination with Chloramine T	41, 46
Fibronectin	^{125}I iodination with Chloramine T	20
Laminin	^{125}I iodination with Chloramine T	21, 47
Collagenase	^{125}I iodination with Chloramine T	23, 48, 49
Proplyl 4-hydroxylase	Reductive alkylation with formaldehyde and sodium ^{3}H-borohydride	22, 50
Proteoglycan, link protein	^{125}I iodination with Chloramine T or lactoperoxidase	51, 52

[a] including procollagen peptides

Table 3
PARAMETERS OF ANTIBODIES AND ANTIGENIC STRUCTURE AS A REASON FOR VARIABLILTY OF DATA IN IMMUNOCHEMICAL ASSAYS

1. Polyclonal vs. monoclonal antibody response
   variations in antibody isotype
2. Variations in affinity and size of antigenic determinants
3. Differences between antisera obtained from different animal species
4. Variations between different individual antisera
5. Variations introduced by technical parameters (e.g., antigen injection, early and hyperimmune antisera, etc.)

biological specimens, for the quantitation of biosynthetically labeled products, or for a detailed immunochemical analysis of antigenic structures. Data obtained in radioimmunoassays will to a certain extent be influenced by the quality of the antisera used. The various parameters which could have caused variability of the results in previous immunochemical studies with collagens are compiled in Table 3. Antisera obtained by conventional immunization usually contain a population of antibodies (polyclonal response) against a variety of antigenic determinants with a large range of affinity constants of the antibody binding sites. Monoclonal antibodies with homogenous affinity can be obtained by cell hybridization and have been recently introduced in studies on collagen.[53,54] Such antibodies may overcome the problems encountered due to the heterogeneity and variability of polyclonal antibodies. On the other hand, monoclonal antibodies may be too restricted in their fine specificity (e.g., distinguish antigens otherwise classified as a homogenous protein) and there-

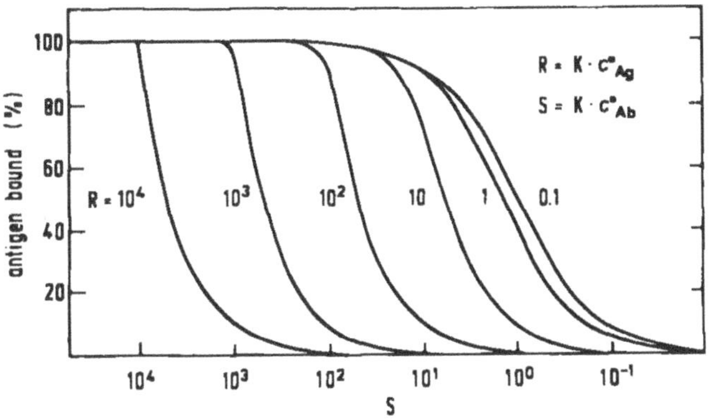

FIGURE 10. Theoretical binding profiles calculated according to Equation 9 as a function of concentration of binding sites of antibody (c^o_{Ab}) and labeled antigen (c^o_{Ag}) K = affinity constant. (From Engel, J. and Schalch, W., *Mol. Immunol.*, 17, 675, 1980. With permission.)

fore be of limited use in experiments aimed at the quantitation of antigens. Since other factors may also determine the quality of an antiserum (see Table 3), the appropriate choice for a particular problem is rather empirical.

A. Information from Binding Assay

Most interpretations of antigen (Ag) — antibody (Ab) interaction in radioimmunoassay assume that it follows the mass law and can be written as

$$Ag + Ab \underset{k_2}{\overset{k_1}{\rightleftarrows}} AgAb \tag{7}$$

and

$$K = \frac{k_1}{k_2} = \frac{[AgAb]}{[Ag][Ab]} \tag{8}$$

with k_1, k_2 as rate constants and K as equilibrium (affinity) constant. By assuming that the equilibrium (Equation 7) is established in solution and not disturbed by the precipitation reaction, a recent theoretical analysis[55] showed that in a binding assay with labeled antigen:

$$\text{antigen bound (\%)} = \left(\frac{1 + R + S - \sqrt{(1 + R + S)^2 - 4RS}}{2R}\right) \times 100 \tag{9}$$

with $R = K\,c^o_{Ag}$ and $S = K\,c^o_{Ab}$. The amounts of antigen and antibody are here expressed as total concentration of antigenic sites and antibody binding sites, respectively. A plot of the fraction of bound antigen as a function of S for various values of R (see Figure 10) showed identical binding curves for $R \ll 1$, e.g., when $c^o_{Ag} \ll K^{-1}$. Under these conditions, K may be determined at the point of 50% antigen bound by

$$K = \frac{1}{c^o_{Ab}} \tag{10}$$

For $R \gg 1$ ($c^o_{Ag} \gg K^{-1}$), the binding curves are shifted to the left, and the shift equals the change in c^{o}_{Ag} used to determine binding profiles at two different concentrations of antigen. The affinity constants cannot be determined under these conditions, but concentration of antibody or antigen binding sites[55] because of $c^{o}_{Ab} = 0.5\ c^{o}_{Ag}$ at 50% antigen binding for $R \gg 1$. The binding profiles shown for antibodies against Type III procollagen (see Figure 6) closely resemble the theoretical curves. Assuming that the profile obtained with 0.02 ng of antigen is already approaching $R \ll 1$, then the antibodies have an apparent affinity constant of $1.2 \times 10^{11}\ l \cdot mol^{-1}$.

The evaluation of binding constants by Equation 9 or by Scatchard plots is only possible if the precipitation with the second antibody does not shift equilibrium (see Equation 7). This is the case when the time needed for analysis τ_A (separation of free from bound antigen, about 16 to 20 hr in the double antibody assay) is smaller than the time τ_R needed to shift equilibrium (see Equation 7). Under most conditions, $\tau_R \approx k_2^{-1}$. This means that the dissociation rate constant k_2 must be smaller than $10^{-6}\ sec^{-1}$ in order to prevent reequilibrium which is only expected for extremely stable immune complexes. Rate constants $k_2 < 10^{-6}\ sec^{-1}$ are usually found for immune complexes in which both binding sites of the first antibody have reacted with the antigen. The equilibrium (see Equation 7) may not be disturbed even for $\tau_A > \tau_R$ when the solubilities of AgAb and Ab in the precipitate obtained with the second antibody are equal. In general this will not be the case. It was shown[55] that Equation 9 still describes the binding curves, but that instead of the true binding constant K, an apparent binding constant K_{app} which may be smaller or larger than K is obtained with Equation 10.

Antigen binding curves determined by radioimmunoassays are frequently used to compare the antigen binding capacity of various antisera. A typical way[12] to calculate this value at the point of 33% bound antigen (ABC-33) is shown in Figure 11. From the theoretical considerations discussed above, such comparisons are only meaningful when determined with large doses of labeled antigen ($c^{o}_{Ag} \gg K^{-1}$). It can be difficult also to compare and interpret binding data for two different labeled antigens with a single antiserum. The shift of the binding curves of one of the antigens to the left of the other may indicate that this antigen reacts only with some of the antibodies present in the serum, but it could also mean that all antibodies react with the second antigen but with lower affinity. Furthermore, superimposable binding profiles do not necessarily indicate complete antigenic identity. Such identical binding profiles were, for example, observed for laminin and its large pepsin fragment P1, while inhibition assays showed that P1 reacted only with a portion of the antibodies[21] (compare Figures 9b and 14d). The failure to detect such differences in binding assays is in part due to the precision of the binding assay and the low accuracy of determining the amount of labeled antigen after separation from nonbound label.

B. Inhibition Assays in the Study of Antigenic Structure

For the comparison of different antigens, inhibition assays are more versatile and provide more reliable results than binding studies. For many purposes, it is sufficient to study the reaction with a single labeled antigen using several nonlabeled inhibitors. Comparisons are made against a reference inhibitor which is normally the same antigen used for labeling. If the assay is sensitive enough, the entire working range may include four to six orders of magnitude of molar concentration of the inhibitors. The application can be extended by reacting the antiserum not only against the labeled component resembling the antigen used for immunization, but also against labeled cross-reacting antigens including antigenic fragments.

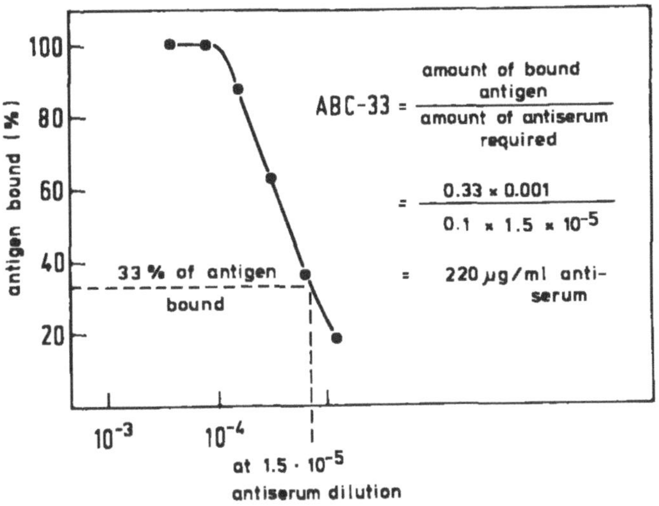

FIGURE 11. Calculation of antigen binding capacity (ABC-33) of an antiserum from radioimmunoassay binding.[12] The assay was carried out with 0.1-mℓ antiserum dilutions and 0.001 µg of labeled antigen in a total volume of 0.4 mℓ. The experimental curve is that shown in Figure 3.

The reaction products formed from labeled antigen (Ag) nonlabeled inhibitor (I) and antibody (Ab) can be formally written as

$$\begin{array}{c} I \\ + \\ Ag + Ab \underset{}{\overset{K_{Ag}}{\rightleftharpoons}} AgAb \\ \updownarrow K_I \\ AbI \end{array}$$

The mathematical treatments of such theoretical models which are similar to that used for the direct binding assay have defined a range of optimal conditions for yielding inhibition assays of high precision and sensitivity.[3,5,8,10] The models are based on the assumptions that the reactions are governed by the mass law, involve homogenous binding sites with no cooperativity, and that the reaction is in equilibrium which is not changed during the separation process. The inhibition reaction can be described by

$$\frac{F}{B} = (c^{o}_{Ab} - B) = K^{-1} \qquad (12)$$

or in the form suitable for a Scatchard plot[9] as

$$\frac{B}{F} = (c^{o}_{Ab} - B)\,K \qquad (13)$$

where B and F are the concentrations of bound and free antigenic sites (labeled antigen plus inhibitor), c^{o}_{Ab} is the concentration of antibody binding sites, and K is the affinity (equilibrium) constant assumed to be identical for both labeled antigen and inhibitor. Since the bound labeled antigen is measured as a function of the

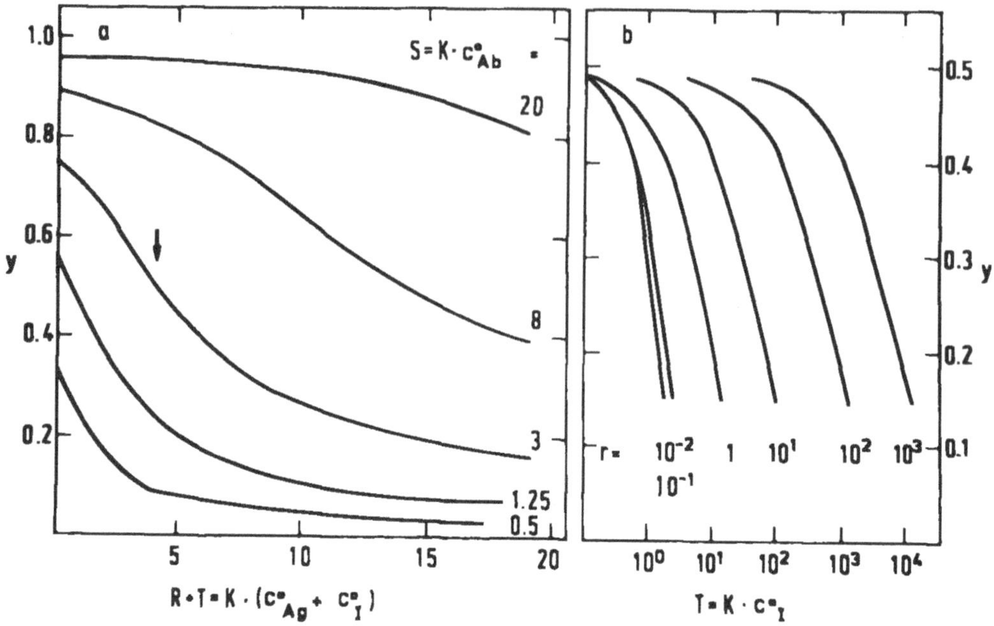

FIGURE 12. Theoretical inhibition profiles based on the equilibrium technique. (a) Dependence of the fraction of labeled antigen bound (y) on concentration of antibody (Ab), labeled antigen (Ag), and inhibitor (I). K denotes the affinity constant. (b) Shift in inhibition profile as a function of inhibitor concentration and the ratio r of affinity constants of labeled antigen and inhibitor (K_{Ag}/K_I). The arrow in (a) denotes 50% antigen binding achieved in the absence of inhibitor in an optimized assay with $c^o_{Ag} = 4K^{-1}$ and $c^o_{Ab} = 3K^{-1}$. If at this point y is set at 1.0 and c^o_I at 0, then experimental data can be expressed according to Equation 5. The plots in (b) are based on $c^o_{Ag} = 4K^{-1}$ and $c^o_{Ab} = 3K^{-1}$. Note the almost parallel shift of curves for r > 1, while they become nearly superimposable for r > 10^{-2}. (From Walter, W. H. C. and Keane, P. M., in *Handbook of Radioimmunoassay*, Abraham, G. E., Ed., Marcel Dekker, New York, 1977, chap. 3. With permission.)

nonlabeled inhibitor, a useful transformation is given by

$$c^o_I = \frac{c^o_{Ab}}{y} - \frac{K^{-1}}{1 - y} - c^o_{Ag} \qquad (14)$$

or

$$R + T = \frac{S}{y} - \frac{1}{1 - y} \qquad (15)$$

where c^o_I and c^o_{Ag} are the total concentrations of binding sites of the inhibitor and labeled antigen, respectively, and y is the ratio between bound labeled antigen and total labeled antigen added (c_{Ag}/c^o_{Ag}), and with $R = Kc^o_{Ag}$, $S = Kc^o_{Ab}$ and $T = Kc^o_I$. These equations give a series of dose-response curves where the concentrations of reactants are expressed in multiples of the affinity constant (see Figure 12a). If the origin of the ordinate in Figure 12a is displaced to the right by subtracting the fixed amount of c^o_{Ag}, then a standard inhibition curve is obtained which relates y to c^o_I. At the point of $c^o_I = 0$, the value of y may be set to 1.0 for obtaining inhibition curves according to Equations 5 or 4 (percent inhibition). From these data it is apparent that sensitivity of an inhibition assay (see Section III.C) is directly proportional to the affinity constant and can be influenced by changing the fixed amounts of labeled antigen and antibody.

The inhibition pattern also changes when two antigens with different affinity compete for the same antibody binding sites. This may be the case for the labeled antigen (Ag) in comparison to the reference inhibitor (I). A mathematical treatment[10] shows (see Figure 12b) that the inhibition profile changes differently with the ratio of the affinity constants expressed as $r = K_{Ag}/K_I$ (see Equation 11). For $r > 1$, the inhibition curves are displaced to the right to higher inhibitor concentrations. The curves show an almost parallel shift, and the distance is proportional to the decrease in affinity of I. The opposite displacement is observed for $r < 1$, but the curves become superimposable at $r < 10^{-2}$. This data indicates that the interaction between labeled antigen and antibody is crucial for determining the sensitivity of the assay and an increase in affinity of the inhibitor by a factor of more than 10 ($r < 10^{-1}$) does essentially not improve this. The parallel shift of the curves for $r > 1$ is also of importance for interpreting cross-reactions of two different inhibitors when analyzed separately in tests using the same labeled antigen and antibody (see below). This is, however, only valid in the case of two inhibitors for which the assumptions discussed above (e.g., identical species of homogenous binding site) can be applied.

In inhibition assays carried out under equilibrium conditions, maximum sensitivity and precision is achieved with concentrations of labeled antigen and antibody set at $c^o_{Ag} = 4K^{-1}$ and $c^o_{Ab} = 3K^{-1}$.[5,10,56] In the absence of any inhibitor, 50% of the labeled antigen is bound by antibody (see Figure 12a). Other variants[5] with $c^o_{Ag} \ll K^{-1}$ and $c^o_{Ab} = 0.5K^{-1}$ (33% binding in the absence of inhibitor) are also recommended. Both conditions allow the detection of low levels of inhibitor in the order of magnitude of K^{-1}, require high specific activity of the labeled antigen, but may show frequently sigmoidal deviations of the logit plot $\ell n \frac{B/B_0}{1 - B/B_0}$ vs. logarithm of inhibitor concentration. With labeled antigens of low specific radioactivity, inhibition assays may be used with c^o_{Ab} and c^o_{Ag} being multiples of $3K^{-1}$ and $4K^{-1}$, respectively. Here the system is highly saturated, requires high concentrations of inhibitor, and shows a straight line in the logit plot even at extremely low or high concentrations of inhibitor.[5]

At least with high-affinity antibodies, inhibition assays applying sequential saturation are more sensitive than equilibrium assays as defined by a mathematical model[6,57] and are used quite frequently (see Figure 8). As the first step, inhibitor and antibody are incubated together until equilibrium is reached. As the second step, residual antibody is reacted with a fixed amount of labeled antigen at conditions which essentially eliminates competition with the bound inhibitor. This requires that significant dissociation of the inhibitor-antibody complex does not occur during the second step. This can be checked by dissociation assays (see Figure 8). The largest increase in sensitivity is usually achieved with high concentrations of reactants (e.g., $c^o_{Ab} = 50K^{-1}$), but lower amounts can also be used. This is illustrated in the example of Figure 8, where the concentration expressed as binding sites of labeled antigen was about $15K^{-1}$ and of antibody about $9K^{-1}$. A more detailed discussion of the advantages and disadvantages of the sequential saturation vs. the equilibrium technique can be found in a review.[6]

Optimization of inhibition assays is important for the quantitative measurement of antigens (see Section III.C), and although not essential for many immunochemical studies, it may extend the range of useful information. It is often done without knowing the affinity constant K of the system by the trial and error method. The

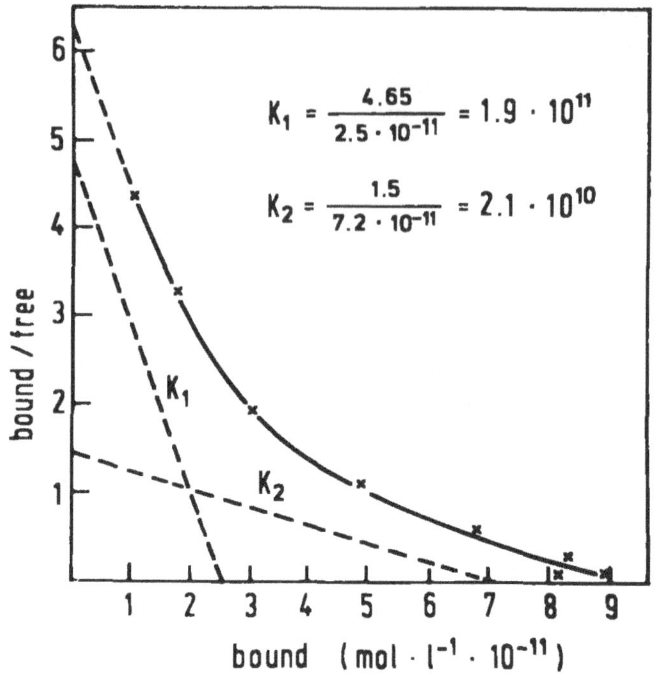

FIGURE 13. Scatchard plot of binding of ^{125}I-labeled aminopropeptide to a constant amount of antibodies against procollagen Type I. Various concentrations of labeled antigen (1.3 to 167 $\times$ 10^{-11} mol $\times$ l^{-1}) were incubated with antibodies for 16 hr at 4°C prior to separation of bound from free antigen by precipitation with second antibody. Because of the curved plot, the data were analyzed by assuming two different population of antibodies with the affinity constants K_1 and K_2.[58] The actual heterogeneity is presumably larger, but the approximation used provides a reasonable range for the apparent average affinity constant.

determination of an average affinity constant is feasible, e.g., by binding studies as shown in Figures 6 and 10 if a reasonable estimate can be made for the amount of antibody in the antiserum. It is also possible to determine K by a Scatchard type of analysis using binding of various amounts of labeled antigen to a constant amount of antiserum. The working range may be extended by addition of increments of nonlabeled antigen as in inhibition assays. A plot of B/F vs. B (see Equation 13) yields a straight line with the slope $\Delta y/\Delta x = -K$ for homogenous binding sites. This has been found particularly with small antigens.[10] The more common observation is a curved Scatchard plot without maximum as shown for antibodies to Type I procollagen (see Figure 13). The lack of linearity reflects heterogeneity of binding sites, but does still estimate an average affinity constant.[5,58,59] Nonlinearity of Scatchard plots can also be due to other reasons, such as nonequilibrium conditions during the assay or cooperative binding of both antibody binding sites to the same antigen molecule.

Immunochemical studies on large proteins are faced with the complexity of binding sites (antigenic determinants) on the antigen and as a consequence with an enormous heterogeneity in affinity constants of the antibodies. The theoretical approaches discussed before were based on less complex models, but were of help to support the more empirical interpretation of inhibition data observed in radioimmunoassays

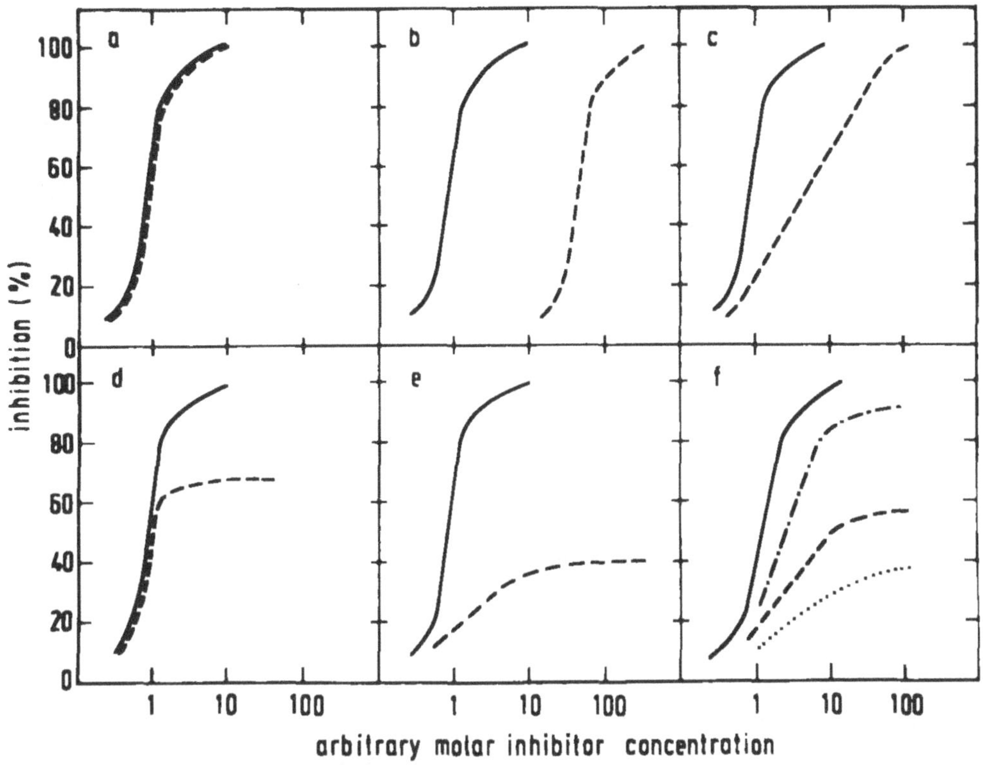

FIGURE 14. Common types of inhibition curves observed in immunochemical comparison of antigens and interpretations of these patterns. These schemes include the curve produced by the reference inhibitor (full line) which usually resembles the labeled test antigen and (but not necessarily) the antigen used for immunization and the inhibition profiles of the second antigens (interrupted lines) used for comparison. (a) Pattern of complete antigenic identity. Both inhibition curves show the same slope and approach 100% inhibition at identical molar concentrations. (b) Both inhibition curves show the same shape, but are not superimposable when plotted against molar concentration. In many cases, this pattern is due to contamination of the second antigen by material resembling the reference inhibitor, and the distance in the parallel shift can be used to quantitate this impurity (see Section III.C). The other possibility would be a true but weak cross-reaction of the second antigen which is difficult to evaluate (see text and Figure 12b). (c) The inhibition curve of the second antigen approaches 100%, but the slope is less steep than that shown by the reference inhibitor. Apparently the antigen is able to react with all antibodies, but with a lower affinity. (d) Pattern of partial antigenic identity. Both inhibition curves are superimposable over a certain range, but the profile for the second antigen levels off into a plateau region below 100% inhibition. The antigen apparently reacts with only a portion of the antibodies, but with an affinity resembling that of the reference inhibitor. (e) Pattern of partial antigenic identity as shown in (d), but the affinity for the cross-reacting portion of antibodies is lower when compared with the reference inhibitor (different slope of the dose-response curve between 10 to 40% inhibition). (f) Additivity of partial antigenic activities. The pattern includes two antigens (– – –, . . .) which exhibit partial antigenic identity as shown in (e). An equimolar mixture of both antigens (·–·–·–) is a distinctly better inhibitor, indicating that each single antigen reacts with a different population of antibodies.

with proteins. When comparing two antigens in the sequential saturation technique, six different examples of inhibition patterns are commonly observed. They are schematically illustrated in Figure 14a to f.

1. Pattern A: Reaction of Complete Antigenic Identity

Both inhibition curves are superimposable within the precision of the assay, indicating identical antigenic determinants as defined by the particular antiserum used

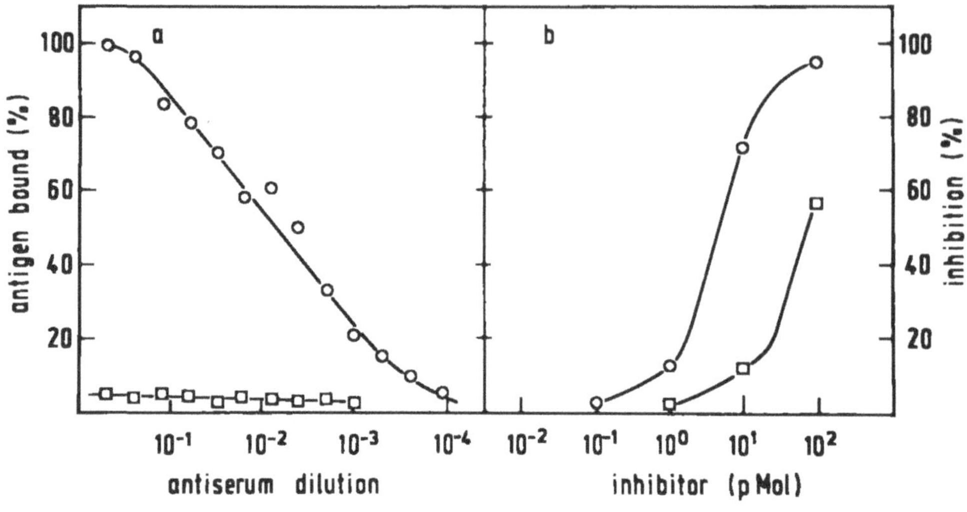

FIGURE 15. Comparison of the reactions of mouse antiserum to Type I collagen with Types I and II collagens by binding (a) and inhibition (b) assays. The ^{125}I-labeled antigens used in (a) were Type I (O——O) and Type II (□——□) collagens. The same, nonlabeled antigens were used as inhibitors (b) of the reaction between antibody and ^{125}I-labeled Type I collagen. (From Nowack, H., Hahn, E., and Timpl, R., *Immunology*, 29, 621, 1975. With permission.)

in the assay. Examples of this type have been reported when aminopropeptides and procollagen[31] or the basement membrane component 7-S collagen obtained from different tissues[45] were compared.

2. Pattern B: Parallel Shift of the Inhibition Curve

This is the most difficult pattern to interpret, since it may be due to lower affinity of the second inhibitor where the distance of the shift equals the difference in affinity (see Figure 12b), or it may indicate that the second inhibitor is in fact inactive and the reaction observed is caused by contamination with the first inhibitor. Actually, inhibition profiles with antigens of different affinities are not completely parallel.[10] This may, however, be obscured by poor precision of the assay, or it may escape detection due to an insufficient number of inhibitor concentrations examined. The alternative explanation, namely shift of the curve due to contamination, will be quite frequently encountered in comparing antigens obtained from the same or similar sources or when studying antigenic fragments. In highly sensitive inhibition assays, contamination of the antigen far below 1% may even be detected, and it will be practically impossible to verify or disprove this interpretation by independent methods.

The combination of binding and inhibition studies can provide the means to possibly decide between the two alternatives as has been shown previously in a study of mouse antibodies against Type I collagen.[42] The comparison of Types I and II collagen in the inhibition assay showed a parallel shift of the curve for Type II collagen by a factor of 20 (see Figure 15b). In contrast, the binding assay (using both collagens after labeling with ^{125}I) indicated a strong reaction with Type I collagen, but only marginal reaction not exceeding 5% with Type II collagen over a wide range (1:2 to 1:1000) of antiserum dilutions (see Figure 15a). Efficient labeling of Type II collagen was demonstrated by showing strong binding to antibodies against Type II collagen. These data are compatible with the interpretation that the Type II collagen used in the assay contained about 5% Type I collagen. This interpretation is more likely than making the assumption of a distinct cross-reaction between the two collagens with an affinity ratio of $r = 20$.

Other systems exhibit weak although distinct reactions in both binding and inhibition assays. This has been observed in the analysis of two forms of 7-S collagen.[45] The possibility that this pattern reflects weak but complete cross-reaction cannot be easily excluded and eventually has to be confirmed by showing the same differences in affinity in both assays.

3. Pattern C: Complete Cross-Reaction with Lower Affinity

Both inhibition curves approach 100% inhibition, but differ distinctly in their slope. Even though an appropriate theoretical explanation for this type does not exist, it very likely reflects different affinities of inhibitors presumably based on heterogenous binding sites. Various examples of this pattern can be found in the literature.[7,11] Typical for this pattern is the behavior of the aminopropeptide of Type III procollagen when compared to the globular domain of this peptide.[41] By using this system, similar differences in slope could also be detected by the equilibrium type of inhibition assay (see Figure 9a).

4. Pattern D: Partial Antigenic Identity with Similar Affinity

One of the inhibitors reacts only with a fraction of the antibodies since a plateau region is reached below 100% inhibition. This pattern is indicative for lack of some of the antigenic determinants present on the other inhibitor. The shared antigenic determinants show a comparable affinity for the antibodies. This pattern is often found when an antigen is compared with fragments derived from it, such as laminin and the pepsin fragment P1.[21] A similar type of inhibition pattern could be demonstrated for this system by the equilibrium inhibition technique showing a sharp bend in the logit plot for fragment P1 (see Figure 9b).

The correct interpretation of the data can be confirmed in two ways. If the antigenically deficient fragment can be labeled, inhibition assays with antibodies to the whole antigen should show comparable or superimposable inhibition profiles for both the antigen and the antigen fragment.[21] Alternatively, the antibodies specific for the fragment will have to be separated from the rest of the antibodies by affinity chromatography. The purified antibodies should then be inhibitable by nonlabeled antigen and antigen fragment in a similar fashion when reacted against labeled intact antigen.

5. Pattern E: Partial Antigenic Identity with Lower Affinity

This type resembles a combination of Patterns c and d, showing a plateau of maximum inhibition below 100% and a different slope of the curve for one of the inhibitors. This pattern is frequently observed in studies on interspecies cross-reactions between chemically related antigens such as fragments of mouse and human laminin[60] or carboxypropetides obtained from various sources.[18,61]

6. Pattern F: Evaluation of Differences in Partial Antigenic Activities

By comparing various inhibitors, a variety of inhibition curves resembling either Pattern D or E may be obtained and are frequently found in studies with antigen fragments. Such fragments often are derived from different structural domains of the original antigen. They are likely to react with different populations of antibodies in the antiserum, and their inhibiting activities should be additive, resulting in a distinct increase in the plateau region (see Figure 14f). An example of this pattern has been reported for the collagenase-resistant and pepsin-resistant domains of Type

IV collagen.[40] Even if the fragments are related to each other, they may show different inhibition profiles as reported for two aminopropeptides of chicken Type I procollagen.[62,63] Such differences may be due to additional antigenic determinants (increase in the plateau region) or to a higher affinity (shift in slope). A test for additivity would be to determine the effects of equimolar mixtures of the inhibitors. No substantial change in the inhibition profile is expected when comparison is made with the better inhibitor alone.

Occasionally, even more complex and inconsistent inhibition patterns are observed. These indicate either considerable complexity of the antigenic structures, based on a combination of the simple types shown in Figure 14, or low affinity of the antibodies which results in nonequilibrium conditions during the assay. Variation of technical parameters, such as incubation time and temperature, concentration of labeled antigen or antibody, and the separation system may help to overcome these problems. It is also important, particularly for quantitative studies (see Section III.C), to determine the amount of the inhibitor added with reasonable accuracy (±20%). This can be accomplished by hydrolyzing aliquotes (25 to 100 μg) of inhibitor stock solutions with 6 *M* HCl (16 hr, 110°C) and quantitative amino acid analysis on an amino acid analyzer.

C. Quantitation of Antigens

Radioimmunoinhibition assays have been originally developed for the purpose of measuring low concentrations of antigen in complex biological samples. The main factors influencing this type of analysis are the slope of the inhibition curve obtained with the unknown inhibitor and the sensitivity and precision of the assay achieved with the reference inhibitor. The latter parameters are a function of the sum of manipulative errors (e.g., pipetting) and counting errors in measuring radioactivity, but depend also on the slope of the inhibition curve.[8,10] Sensitivity is usually defined as the lowest amount of inhibitor which can be detected in the assay and equals the precision obtained in the absence of any inhibitor. It is also dependent on an optimal concentration of labeled antigen (see Section III.B).

The extensive use of quantitative inhibition assays on a routine basis (e.g., as a clinical test) should include a regular quality control both within and between assay.[9,64–66] For established systems, the intra-assay coefficient of variation is commonly 4 to 15%, and the interassay variability of precision is 8 to 20%.[10] For a comprehensive consideration of basic principles of designing a quantitative assay, the reader should consult appropriate reviews[8–10] as well as more specialized articles on statistical analysis[3] and standardization[67] of such assays.

Radioimmunoassays which are quite sufficient for analyzing the antigenic structure of a given protein usually have to be adjusted to the particular requirements of a sensitive and quantitative assay. Insufficient sensitivity and nonparallel inhibition curves are the two major problems encountered. Since sensitivity is a function of antibody affinity and inhibitor concentrations distinctly below K^{-1} cannot be measured, the selection of an appropriate antiserum is important. This may require screening of several individual antisera as well as different bleedings from an individual animal (see Figure 16). Sensitivity may also vary with the immunogen (e.g., procollagen or procollagen peptide) used for immunization. Other parameters for optimizing the assay have been discussed already in Section III.B.

Assays of sufficiently high sensitivity have been developed for several connective tissue proteins as shown by the examples in Figure 17. They measured these proteins at concentrations as low as 0.2 to 10 ng/mℓ and showed in many cases parallel inhibition profiles produced by both the reference inhibitor and the unknown inhib-

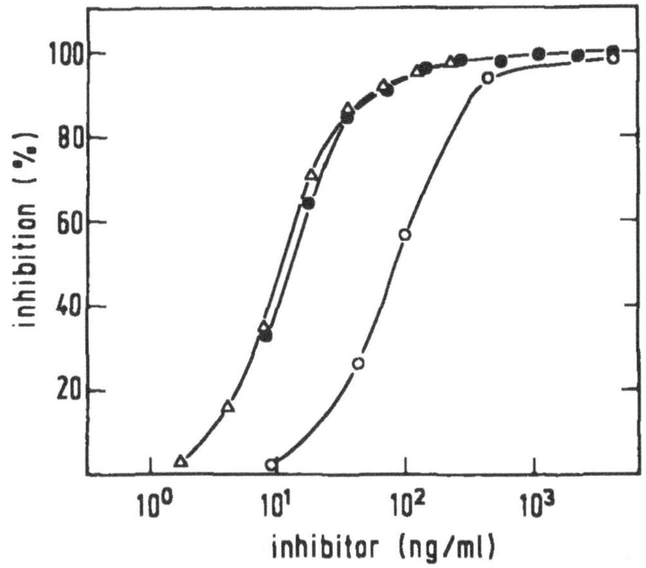

FIGURE 16. Variation in sensitivity of an inhibition assay (sequential saturation) depending on the batch of antiserum used. The antiserum was against fragment P1 of human laminin[60] (1-mg injections on day 0 and 30), and serum was collected on day 40 (○——○), day 72 (●——●), and day 97 (△——△). Note the about sevenfold increase in sensitivity from bleeding on day 40 to day 72.

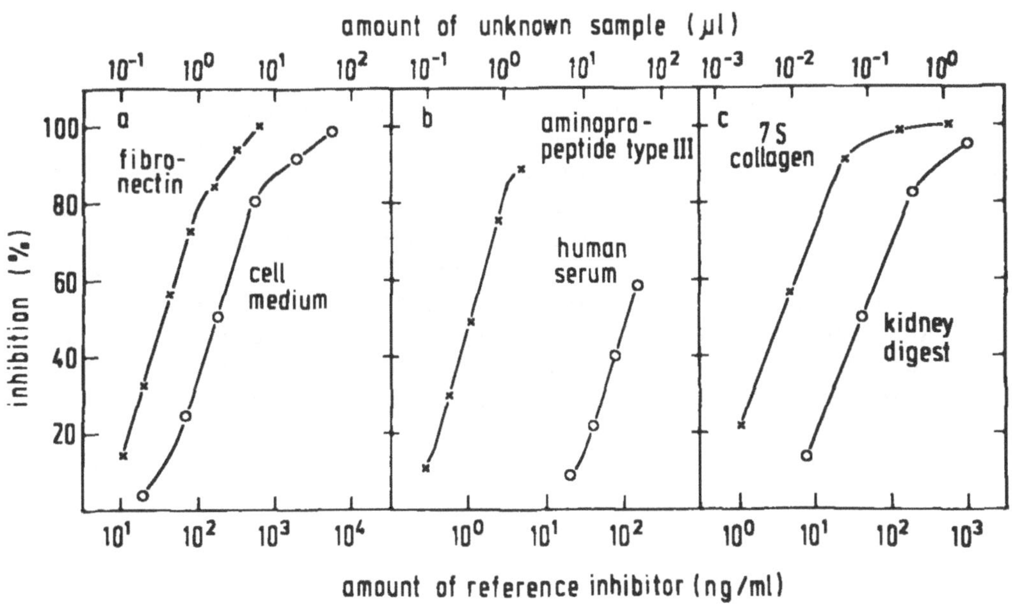

FIGURE 17. Use of radioimmuneinhibition assay (sequential saturation) for measuring connective tissue proteins in cell culture, serum and tissue. (a) Fibronectin in skin fibroblast medium using antibodies to plasma fibronectin.[68] (b) Aminopropeptide of Type III procollagen in normal human serum.[41] (c) The basement membrane component 7-S collagen in a complete protease digest of rat kideny.[69] The concentrations determined in these samples were 4 µg/mℓ fibronectin, 6.7 ng/mℓ aminopropeptide, and 8.1 µg/mℓ 7-S collagen by comparison with the amount of reference inhibitor required to give 50% inhibition. Note that the dose-response curves for the reference inhibitors and unknown samples are parallel over a wide range of concentrations.

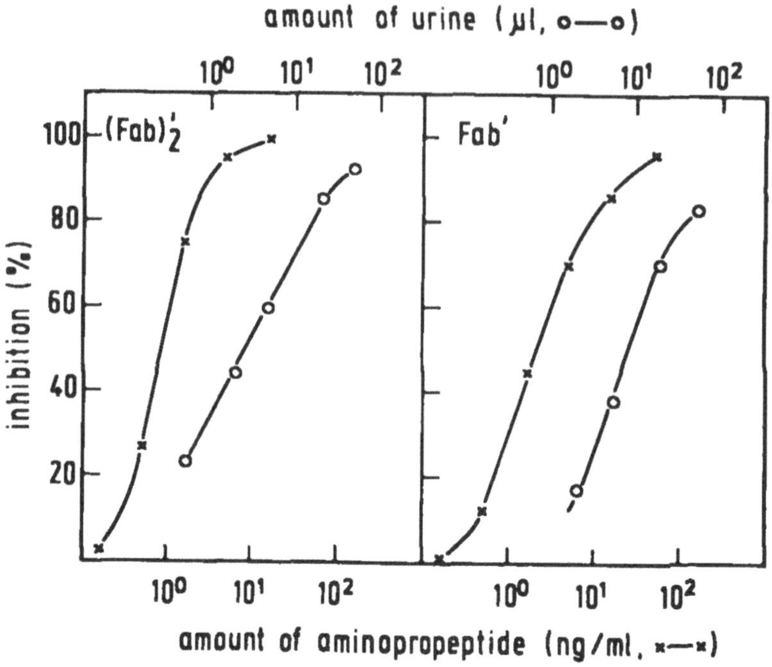

FIGURE 18. Inhibition assay with bivalent $(Fab)_2'$ and monovalent Fab' antibody fragments against Type III procollagen and comparison with cross-reacting material present in human urine. The aminopropeptide was used as labeled antigen and as the reference inhibitor. The different slopes of the inhibition curves of procollagen peptide and the urinary antigen with $(Fab)_2'$ are changed to parallel slopes with Fab' fragments.[72]

itor. Lower sensitivity (factor 10 to 50) is commonly obtained in assays for collagens.[14,16,42] Similarity of the inhibition profiles is of course not necessarily proof for identity of the components to be compared. In many cases, it will not be possible to isolate the unknown inhibitor in quantities sufficient for chemical characterization. Chromatographic or electrophoretic separation of biological samples and tracing of the inhibitor by radioimmunoassay were used to identify aminopropeptide antigens of different sizes in various human body fluids.[41] Antigenic properties of a component may not necessarily be related to other properties such as enzymic activity. An example is prolyl hydroxylase which is enzymatically active only as a tetramer, while both monomeric and tetrameric forms are antigenic and react in inhibition assays.[22,70,71]

Nonparallel inhibition profiles are occasionally observed and do not allow unambiguous identification of the unknown inhibitor. One major reason for this finding is cross-reactivity between two antigens with different affinity (see Figure 14c). A typical example was seen with the aminopropetides of Type III procollagen from human serum and urine.[41] The two peptides differ in size and affinity and gave a logit plot similar to that shown in Figure 9a. Provided the cross-reacting antigen can be isolated, the problem may be overcome by using the cross-reactive material as the labeled antigen and/or reference inhibitor in the assay. As shown by a recent finding[72] which has not yet been theoretically explored, the change from a bivalent antibody to a monovalent Fab fragment (see Figure 18) may also help to circumvent this problem.

Many other factors not fully understood may contribute to nonparallel inhibition patterns.[7,11] They include nonfavorable conditions for antigen-antibody reactions (high salt, extreme pH), proteases, and "incubation damage" which is a common term to refer to interference by high protein concentration such as found in undiluted serum or plasma. Dialysis of samples, high dilutions of samples (requires high sensitivity of the assay), addition of protease inhibitors, a change in the inhibition technique (equilibrium vs. sequential saturation), or a change in the batch of first or second antibody are some of several empirical procedures suggested to cope with these problems. It is also recommended to add the undiluted biological sample as a blank or damage control to the reference inhibitor which requires that samples essentially devoid of inhibitor substance are available. Nonparallel inhibition curves may also be used for evaluations in studies where only a relative change in the amount of unknown inhibitor is important. Here comparisons should be made at a fixed point (usually 50% inhibition) which requires examination of quite a number of different dilutions of the unknown sample.

Quantitative inhibition assays have been applied to study procollagen peptides[18,19,41,61,68] (see Figure 17b), fibronectin[20,68,73–76] (see Figure 17a), basement membrane components,[69,77] prolyl hydroxylase,[22,70,71,78–81] and collagenase[23,48,49,82] in serum and cell cultures. Such assays may be of potential use in studies on the composition of extracellular matrices. This would require complete solubilization of tissues by techniques preserving a particular antigenic structure and the development of special radioimmunoassays for such resistant fragments. This has been attempted for 7-S collagen which resists digestion by pepsin and bacterial collagenase[45] and which could be measured in protease digests of rat kidneys[69] (see Figure 17c).

There is also a demand for noninvasive methods to monitor connective tissue diseases, e.g., by quantitating antigens in the circulation. Such analyses have been reported for carboxypropeptides in Paget's disease,[83] aminopropeptides[41] and prolyl hydroxylase in liver fibrosis,[84,85] rheumatic diseases,[81] and skin diseases.[79,80] Some of the connective tissue antigens apparently are released into the circulation as larger fragments, but others may be degraded. The design of assays can be similar to those discussed above for matrix analysis. Increased serum levels of 7-S collagen have been demonstrated recently in streptozotocin diabetes of rats.[86] Radioimmunoassays for connective tissue proteins have also been applied in screening for autoantibodies in a variety of human autoimmune diseases.[15,87–89]

D. Analysis of Metabolically Labeled Proteins

A variety of connective tissue proteins can be obtained in labeled form from cell and organ cultures commonly carried out in the presence of ^{3}H- or ^{14}C-proline or glycine (for labeling collagenous protein) or of ^{35}S-methionine (for labeling noncollagenous protein). Such labeled antigens have been used in immunochemical studies of Type I collagen,[13] Type I procollagen,[62,90] and Type IV collagen.[91,92] In accordance with the low specific radioactivity[62] of these antigens, inhibition assays were less sensitive at least by one order of magnitude than those with ^{125}I-labeled antigens. It is often not possible to determine exactly the amount of metabolically labeled antigens which makes it difficult to evaluate binding studies with various forms of antigens, e.g., hydroxylated and nonhydroxylated collagens.[92] In spite of these disadvantages, such studies may still be valuable for antigens that cannot be obtained in sufficient chemical quantities or when external labeling changes antigenicity.[13]

Binding of labeled material to increasing amounts of antibodies is a useful method to quantitate a particular antigen in terms of its radioactivity in a mixture of biosynthetic products (see Figure 19). A similar technique was used in several other studies.[94–96] In a perfect system (e.g., laminin), the binding profile should show a

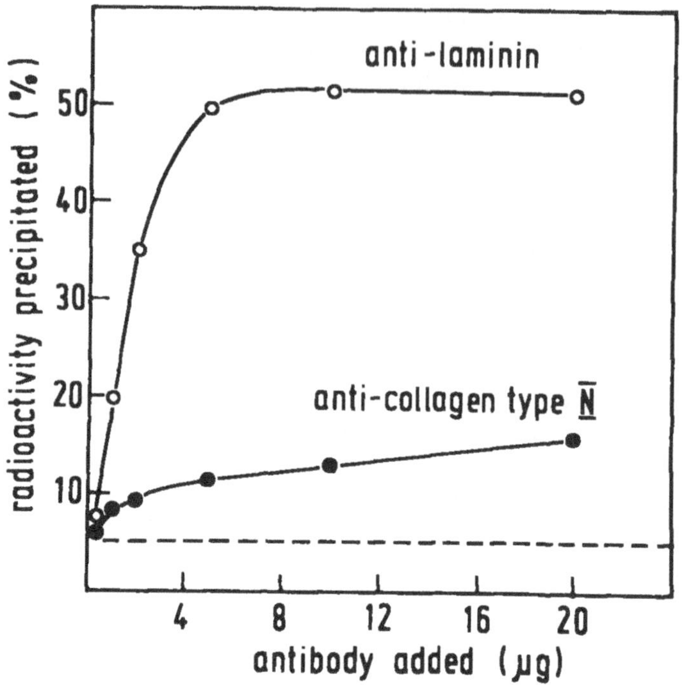

FIGURE 19. Quantitative precipitation by antibodies to Type IV collagen or laminin of proteins labeled with ^{3}H-proline. Aliquots 0.1 mℓ containing 48,000 cpm of a culture medium obtained from a teratocarcinoma cell line[94] were incubated for 1 hr at room temperature with increasing amounts of purified antibody (0.1 mℓ) in PBS/Tween containing 0.5% bovine serum albumin. The total amount of IgG was adjusted to 20 µg by adding normal IgG. The radioactivity bound to IgG was then removed by binding to protein A-Sepharose® (4 mg suspended in 0.2 mℓ) for 2 hr at room temperature, the adsorbent was washed three times with 1 mℓ of PBS/Tween, and proteins were then dissolved in 2 mℓ of 8 *M* urea, pH 7, 1% sodium dodecylsulfate, and 0.5% mercaptoethanol for 30 min at 56°. The radioactivity solubilized was determined in a β scintillation counter. The dashed line indicates background binding by 20 µg of normal IgG.

straight plateau region in the antibody excess zone. Deviations from this pattern as demonstrated for Type IV collagen (see Figure 19) usually indicate that the antigen is not homogenous or that the antiserum is not sufficiently specific for a single antigen. The main problem encountered in these studies is the high nonspecific absorption of radioactive material which can be in the range 3 to 15% of the total. Accordingly, the determination of components in this range becomes rather unprecise. There are several ways (see Section II.B and Reference 97) to reduce background binding of labeled antigens.

Electrophoretic analysis of labeled polypeptides bound to antibody is a common approach to judge the identification of a particular antigen[74,94,95] and is useful as a quality control for any type of radioimmunoassay. Under optimal conditions, complete separation of various antigens present in a culture medium can be achieved (see Figure 20). The individual steps in this analysis are solubilization of the antigens bound to antibody under denaturing and reducing conditions, separation of peptides usually carried out by slab gel electrophoresis in the presence of sodium dodecyl-

FIGURE 20. Identification of proteins synthesized by rhabdomyosarcoma cells by precipitation of medium with specific antibody and electrophoretic analysis of the bound antigen. Medium proteins (M) labeled with ^{14}C-proline showed five major polypeptide bands. The two major bands could be precipitated with antibodies to Type III procollagen (III) and resemble the intact and a truncated form of the pro-α1(III)-chain. Two weaker bands were precipitated with antibodies to Type IV collagen (IV) and correspond in mobility to the α1(IV)- and α2(IV)-chain. Antifibronectin antibodies (F) precipitated only one band. Electrophoretic analysis[33] was carried out after reduction with mercaptoethanol, and the bands were visualized by fluorography.[34] (From Krieg, T., Timpl, R., Alitalo, K., Kurkinen, M., and Vaheri, A., *FEBS Lett.*, 104, 405, 1979. With permission.)

sulphate, and visualization of labeled peptide bands by fluorography using preexposed X-ray films.[34] The density of labeled bands is a function of exposure time and specific radioactivity. Densitometric scanning of such patterns, as shown for example in a study on intra- and extracellular fibronectin[98] or on elastin biosynthesis,[99] can be useful for the quantitation of radioactive antigens and corrects nonspecific binding to antibody.

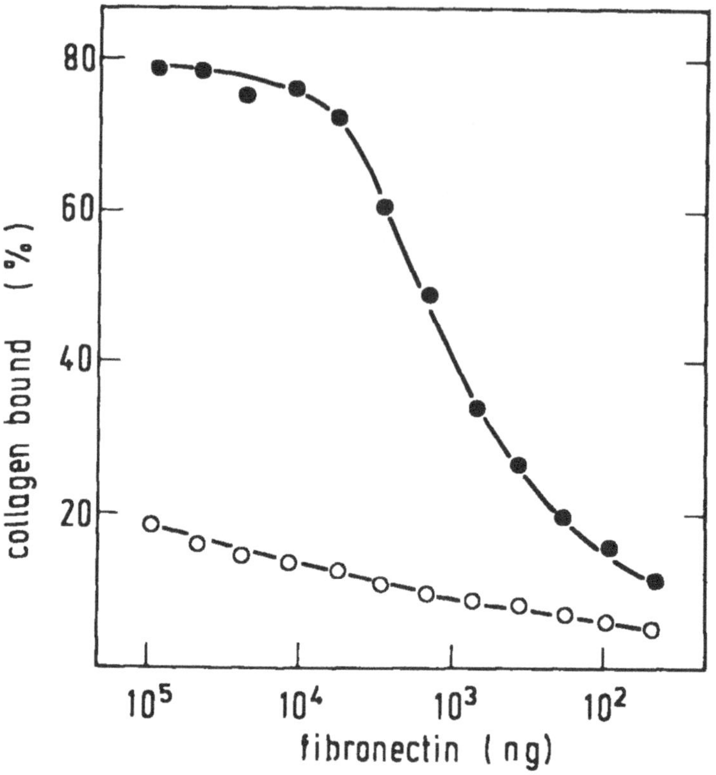

FIGURE 21. Binding of collagen to fibronectin determined by an assay similar to radioimmunoassay. The reaction included 100 ng of native (○——○) or denatured (●——●), ^{125}I-labeled collagen Type III and varying amounts of nonlabeled plasma fibronectin in a volume of 0.4 mℓ. The complexes formed between collagen and fibronectin after 20 hr of incubation at 4°C were then separated from free collagen by precipitation with an excess of antiserum to fibronectin. (From Jilek, F. and Hörmann, H., *Hoppe-Seylers Z. Physiol. Chem.*, 359, 247, 1978. With permission.)

E. Other Applications

The interaction between two different proteins can generally be studied by assays resembling the radioimmunoassay. They are based on similar methods for the separation of the free and bound fraction of the labeled reactant. This has been demonstrated for the binding of fibronectin to several types of native and denatured collagen[100,101] (see Figure 21). In these studies, a constant amount of collagen labeled with ^{125}I was incubated with varying concentrations of fibronectin. The bound collagen was then separated from free collagen by precipitation with excess antibodies to fibronectin. The binding profiles were similar to those found for antigen-antibody reactions, suggesting that these interactions could be subjected to similar mathematical treatments as discussed in Sections III.A and B. It was also possible to inhibit this reaction[101,102] as well as to examine the interactions with glycosaminoglycans in binding assays.[103]

In applying this method, one should be aware of two basic technical differences to antigen-antibody reactions. The first is the lack of an appropriate equivalent to nonimmune immunoglobulin. This will introduce inaccuracy due to the lack of a precise estimate of nonspecific absorption of labeled material, and it may also cause poor yields when nonlabeled reactant at low concentrations is precipitated. The latter is usually overcome in the double antibody assay by adding an inert carrier. The second is due to the possibility that the antibody interferes with the primary interaction by competing for similar binding sites. This possibility was indicated in reactions between collagen, fibronectin, and antifibronectin.[76]

The disadvantages of radioimmunoassays are the low stability of labeled products and general problems arising from the work with radioactive substances. A variant of increasing popularity uses enzymes bound to antigen or antibody (see Chapter 14) and has been named enzyme-immunoassay or enzyme-linked immunosorbent assay (ELISA). In the standard procedure, antigen is absorbed under slightly alkaline conditions to the wall of plastic tubes and serves as target for the first and second antibody, the latter usually carrying the enzyme. A recent study[104] showed that enzyme-immunoassays for connective tissue proteins may approach the sensitivity of radioimmunoassays. It is also possible to study other interactions, such as those between fibronectin and collagen.[105] Enzyme-immunoassays may suffer from the requirement of highly purified antigens and/or antibodies, since impurities may become preferentially attached during the absorption step and may dominate the reaction if antibodies against the contaminants are present in the first antiserum. Such pitfalls can be more easily avoided when the primary reaction takes place between soluble reactants (see Sections II.B and III.D). Fluid phase variants for the assay are possible,[106] but may be influenced by steric hindrance due to the bulky enzyme label. Other assays are based on fluorescent markers.[107] They have been developed recently and their full range of potential applications not yet explored.

ACKNOWLEDGMENTS

We wish to thank Dr. J. Engel, Basel, for his helpful discussion and comments. As far as the author's own work is concerned, it was supported by grants of the Deutsche Forschungsgemeinschaft.

REFERENCES

1. **Yalow, R. S. and Berson, S. A.,** Assay of plasma insulin in human subjects by immunological methods, *Nature*, 184, 1648, 1959.
2. **Farr, R. S.,** A quantitative immunochemical measure of the primary interaction between I^+ BSA and antibody, *J. Infect. Dis.*, 103, 239, 1958.
3. **Rodbard, D. and Frazier, G. R.,** Statistical analysis of radioligand assay data, *Methods Enzymol.*, 37, 3, 1979.
4. **Skelley, D. S., Brown, L. P., and Besch, P. K.,** Radioimmunoassay, *Clin. Chem.*, 19, 146, 1973.
5. **Zettner, A.,** Principles of competitive binding assays (saturation analyses). I. Equilibrium techniques, *Clin. Chem.*, 19, 699, 1973.
6. **Zettner, A. and Duly, P. E.,** Principles of competitive binding assays (saturation analyses). II. Sequential saturation, *Clin. Chem.*, 20, 5, 1974.
7. **Felber, J. P.,** Radioimmunoassay of polypeptide hormones and enzymes, *Methods Biochem. Anal.*, 22, 1, 1974.
8. **Ekins, R. P.,** Radioimmunoassay and saturation analysis. Basic principles and theory, *Br. Med. Bull.*, 30, 3, 1974.
9. **Walker, W. H. C.,** An approach to immunoassay, *Clin. Chem.*, 23, 384, 1977.
10. **Walker, W. H. C. and Keane, P. M.,** Theoretical aspects of radioimmunoassay, in *Handbook of Radioimmunoassay*, Abraham, G. E., Ed., Marcel Dekker, New York, 1977, chap. 3.

11. **Hunter, W. M.,** Radioimmunoassay, in *Handbook of Experimental Immunology*, 3rd ed., Weir, D. M., Ed., Blackwell Scientific, Oxford, 1978, chap. 14.
12. **Minden, P. and Farr, R. S.,** Ammonium sulphate method to measure antigen-binding capacity, in *Handbook of Experimental Immunology*, 3rd ed., Weir, D. M., Ed., Blackwell Scientific, Oxford, 1978, chap. 13.
13. **Lindsley, H., Mannik, M., and Bornstein, P.,** The distribution of antigenic determinants in rat skin collagen, *J. Exp. Med.*, 133, 1309, 1971.
14. **Adelmann, B. C., Gentner, G. J., and Hopper, K.,** A sensitive radioimmunoassay for collagen, *J. Immunol. Methods*, 3, 319, 1973.
15. **Menzel, J.,** Radioimmunoassay for anticollagen-antibodies using ^{14}C-labelled collagen, *J. Immunol. Methods*, 15, 77, 1977.
16. **Roll, F. J., Madri, J. A., and Furthmayr, H.,** A new method of iodinating collagens for use in radioimmunoassay, *Anal. Biochem.*, 96, 489, 1979.
17. **Clague, R. B., Brown, R. A., Weiss, J. A., and Holt, P. J. L.,** Solid-phase radioimmunoassay for the detection of antibodies to collagen, *J. Immunol. Methods*, 27, 31, 1979.
18. **Taubmann, M. B., Goldberg, B., and Sherr, C. J.,** Radioimmunoassay for human procollagen, *Science*, 186, 1115, 1974.
19. **Rohde, H., Nowack, H., Becker, U., and Timpl, R.,** Radioimmunoassay for the amino-terminal peptide of procollagen pα1 (I) chain, *J. Immunol. Methods*, 11, 135, 1976.
20. **Kuusela, P., Ruoslahti, E., Engvall, E., and Vaheri, A.,** Immunological interspecies cross-reactions of fibroblast surface antigen (fibronectin), *Immunochemistry*, 13, 639, 1976.
21. **Rohde, H., Wick, G., and Timpl, R.,** Immunochemical characterization of the basement membrane glycoprotein laminin, *Eur. J. Biochem.*, 102, 195, 1979.
22. **Tuderman, L., Kuutti, E. R., and Kivirikko, K. I.,** Radioimmunoassay for human and chick prolyl hydroxylases, *Eur. J. Biochem.*, 60, 399, 1975.
23. **Bauer, E. A., Eisen, A. Z., and Jeffrey, J. J.,** Radioimmunoassay of human collagenase. Specificity of the assay and quantitative determination of in vivo and in vitro human skin collagenase, *J. Biol. Chem.*, 247, 6679, 1972.
24. **Greenwood, F. C., Hunter, W. M., and Glover, J. S.,** The preparation of ^{131}I-labelled human growth hormone of high specific radioactivity, *Biochem. J.*, 89, 114, 1963.
25. **Roholt, O. A. and Pressman, D.,** Iodination — isolation of peptides from the active site, *Methods Enzymol.*, 25, 438, 1972.
26. **Bolton, A. E. and Hunter, W. M.,** The labelling of proteins to high specific radioactivities by conjugation to a ^{125}I-containing acylating agent, *Biochem. J.*, 133, 529, 1973.
27. **Remy-Heintz, N., Magous, R., and Bali, J. P.,** Comparative labelling of α1-acid glycoprotein by (^{3}H-)-borohydride or (^{125}I)-iodide for use in a radioimmunoassay, *Biochem. Biophys. Res. Commun.*, 93, 1217, 1980.
28. **McConahey, P. J. and Dixon, F. J.,** Method of trace iodination of protein for immunological studies, *Int. Arch. Allergy Appl. Immunol.*, 29, 185, 1966.
29. **Tanswell, P., Reiter, H., and Timpl, R.,** Structure of antigenic determinants in the amino-terminal region of bovine fibrinogen Aα chain, *Eur. J. Biochem.*, 88, 565, 1978.
30. **Risteli, J., Schuppan, D., Glanville, R., and Timpl, R.,** Immunochemical distinction between two different chains of type IV collagen, *Biochem. J.*, 191, 517, 1980.
31. **Etzrodt, H., Schleyer, M., Schröder, K. E., and Pfeiffer, E. F.,** The use of Tween 20 in the double antibody radioimmunoassay, *Hoppe-Seylers Z. Physiol. Chem.*, 357, 1205, 1976.
32. **Rohde, H., Becker, U., Nowack, H., and Timpl, R.,** Antigenic structure of the aminoterminal region in type I procollagen. Characterization of sequential and conformational determinants, *Immunochemistry*, 13, 967, 1976.
33. **Laemmli, U. K.,** Cleavage of structural proteins during the assembly of the head of bacteriophage T4, *Nature*, 227, 680, 1970.
34. **Bonner, W. M. and Laskey, R. A.,** A film detection method for tritium-labelled proteins and nucleic acids in polyacrylamide gels, *Eur. J. Biochem.*, 46, 83, 1974.
35. **Ratcliffe, J. G.,** RIA and saturation analysis. Separation techniques in saturation analysis, *Br. Med. Bull.*, 30, 32, 1974.
36. **Lotter, H. and Timpl, R.,** Antigenic structure of the cyanogen bromide peptide F-CB3 from fibrinogen α chain, *Eur. J. Biochem.*, 60, 221, 1975.
37. **Gupta, R. K. and Morton, D. L.,** Double-antibody method and the protein-A-bearing *Staphylococcus aureus* cells compared for separating bound and free antigen in radioimmunoassay, *Clin. Chem.*, 25, 752, 1979.
38. **Gotschlich, E. C.,** Simplification of the radioactive antichain-binding test by a double label technique, *J. Immunol.*, 107, 910, 1971.

39. **Cerasi, E., Della Casa, L., Luft, R., and Roovete, A.**, Human growth hormone (HGH) in plasma by a double antibody radioimmunoassay, *Acta Endocrinol.*, 53, 101, 1966.
40. **Timpl, R., Glanville, R. W., Wick, G., and Martin, G. R.**, Immunochemical study on basement membrane (type IV) collagens, *Immunology*, 38, 109, 1979.
41. **Rohde, H., Vargas, L., Hahn, E., Kalbfleisch, H., Bruguera, M., and Timpl, R.**, Radioimmunoassay for type III procollagen peptide and its application to human liver disease, *Eur. J. Clin. Invest.*, 9, 451, 1979.
42. **Nowack, H., Hahn, E., and Timpl, R.**, Specificity of the antibody response in inbred mice to bovine type I and type II collagen, *Immunology*, 29, 621, 1975.
43. **Becker, U., Nowack, H., Gay, S., and Timpl, R.**, Production and specificity of antibodies against the aminoterminal region in type III collagen, *Immunology*, 31, 57, 1976.
44. **Gunson, D. and Kefalides, N. A.**, The use of the radioimmunoassay in the characterization of antibodies to basement membrane collagen, *Immunology*, 31, 563, 1976.
45. **Risteli, J., Bächinger, H. P., Engel, J., Furthmayr, H., and Timpl, R.**, 7S collagen: characterization of an unusual basement membrane structure, *Eur. J. Biochem.*, 108, 239, 1980.
46. **Nowack, H., Olsen, B. R., and Timpl, R.**, Characterization of the aminoterminal segment in type III procollagen, *Eur. J. Biochem.*, 70, 205, 1976.
47. **Timpl, R., Rohde, H., Gehron Robey, P., Rennard, S. I., Foidart, J. M., and Martin, G. R.**, Laminin — a glycoprotein from basement membranes, *J. Biol. Chem.*, 254, 9933, 1979.
48. **Bauer, E. A., Eisen, A. Z., and Jeffrey, J. J.**, Radioimmunoassay of collagenase, in *Handbook of Radioimmunoassay*, Abraham, G. E., Ed., Marcel Dekker, New York, 1977, chap. 17.
49. **Eisen, A. Z., Nepute, J., Stricklin, G. P., Bauer, E. A., and Jeffrey, J. J.**, Solid-phase radioimmunoassay of rat skin collagenase. Comparative studies and in vivo measurement of the enzyme, *Biochim. Biophys. Acta*, 350, 442, 1974.
50. **Risteli, J., Tuderman, L., and Kivirikko, K. I.**, Intracellular enzymes of collagen biosynthesis in rat liver as a function of age and in hepatic injury induced by dimethylnitrosamine. Purification of rat prolyl hydroxylase and comparison of changes in prolyl hydroxylase activity with changes in immunoreactive prolyl hydroxylase, *Biochem. J.*, 158, 369, 1976.
51. **Caterson, B., Baker, J. R., Levitt, D., and Paslay, J. W.**, Radioimmunoassay of the link proteins associated with bovine nasal cartilage proteoglycan, *J. Biol. Chem.*, 254, 9363, 1979.
52. **Wieslander, J. and Heinegård, D.**, Immunochemical analysis of cartilage proteoglycans. Radioimmunoassay of the molecules and the substructures, *Biochem. J.*, 187, 687, 1980.
53. **Linsenmayer, T. F., Hendrix, M. J. C., and Little, C. D.**, Production and characterization of a monoclonal antibody to chicken type I collagen, *Proc. Natl. Acad. Sci. U.S.A.*, 76, 3703, 1979.
54. **Linsenmayer, T. F. and Hendrix, M. J. C.**, Monoclonal antibodies to connective tissue macromolecules: type II collagen, *Biochem. Biophys. Res. Commun.*, 92, 440, 1980.
55. **Engel, J. and Schalch, W.**, Determination of antibody binding constants from Farr test and other radioimmunoassays. A theoretical and experimental analysis, *Mol. Immunol.*, 17, 675, 1980.
56. **Ekins, R. P. and Newman, G. B.**, Theoretical aspects of saturation analysis, *Acta Endocrinol. (Copenhagen)*, 64 (Suppl. 147), 11, 1970.
57. **Rodbard, D. and Lewald, J. E.**, Computer analysis of radioligand assay and radioimmunoassay data, *Acta Endocrinol. (Copenhagen)*, 64 (Suppl. 147), 79, 1970.
58. **Nørby, J. G., Ottolenghi, P., and Jensen, J.**, Scatchard plot: common misinterpretation of binding experiments, *Anal. Biochem.*, 102, 318, 1980.
59. **Pinckard, R. N.**, Equilibrium dialysis and preparation of hapten conjugates, in *Handbook of Experimental Immunology*, 3rd ed., Weir, D. M., Ed., Blackwell Scientific, Oxford, 1978, chap. 17.
60. **Risteli, L. and Timpl, R.**, Isolation and characterization of pepsin fragments of laminin from human placental and renal basement membranes, *Biochem. J.*, 193, 749, 1981.
61. **Taubmann, M. B. and Goldberg, B.**, Antigenic relatedness of primate procollagens as determined by a competitive radioimmunoassay, *Comp. Biochem. Physiol.*, 59b, 47, 1978.
62. **von der Mark, K., Click, E. M., and Bornstein, P.**, The immunology of procollagen. I. Development of antibodies to determinants unique to the proα1 chain, *Arch. Biochem. Biophys.*, 156, 356, 1973.
63. **Pesciotta, D. M., Silkowitz, M. H., Fietzek, P. P., Graves, P. N., Berg, R. A., and Olsen, B. R.**, Purification and characterization of the amino-terminal propeptide of proα1 (I) chains from embryonic chick tendon procollagen, *Biochemistry*, 19, 2447, 1980.
64. **Rodbard, D., Bridson, W., and Rayford, P. L.**, Rapid calculation of radioimmunoassay results, *J. Lab. Clin. Med.*, 74, 770, 1969.
65. **Healy, M. J. R.**, Statistical analysis of radioimmunoassay data, *Biochem. J.*, 130, 207, 1972.
66. **Rodbard, D.**, Statistical quality control and routine data processing for radioimmunoassays and immunoradiometric assays, *Clin. Chem.*, 20, 1255, 1974.

67. **Bangham, D. R. and Cotes, P. M.**, Radioimmunoassay and saturation analysis. Standardization and standards, *Br. Med. Bull.*, 30, 12, 1974.
68. **Fleischmajer, R., Perlish, J. S., Krieg, T., and Timpl, R.**, Variability in collagen and fibronectin synthesis by scleroderma fibroblasts in primary culture, *J. Invest. Dermatol.*, 76, 400, 1981.
69. **Risteli, J., Rohde, H., and Timpl, R.**, Sensitive radioimmunoassays for 7-S collagen and laminin. Application to serum and tissue studies of basement membranes, *Anal. Biochem.*, 113, 372, 1981.
70. **Tuderman, L.**, Developmental changes in prolyl hydroxylase activity and protein in chick embryo, *Eur. J. Biochem.*, 66, 615, 1976.
71. **Risteli, J., Tuderman, L., Tryggvason, K., and Kivirikko, K. I.**, Effect of hepatic injury on prolyl 3-hydroxylase and prolyl 4-hydroxylase activities in rat liver and on immunoreactive prolyl 4-hydroxylase concentrations in the liver and serum, *Biochem. J.*, 170, 129, 1978.
72. **Rohde, H. and Timpl, R.**, Unpublished.
73. **Mosher, D. F. and Vaheri, A.**, Thrombin stimulates the production and release of a major surface-associated glycoprotein (fibronectin) in cultures of human fibroblasts, *Exp. Cell Res.*, 112, 323, 1978.
74. **Krieg, T., Timpl, R., Alitalo, K., Kurkinen, M., and Vaheri, A.**, Type III procollagen is the major collagenous component produced by a continuous rhabdomyosarcoma cell line, *FEBS Lett.*, 104, 405, 1979.
75. **Mosher, D. F., Saksela, O., Keski-Oja, J., and Vaheri, A.**, Distribution of a major surface-associated glycoprotein, fibronectin, in cultures of adherent cells, *J. Supramol. Struct.*, 6, 551, 1977.
76. **Ruoslahti, E., Vuento, M., and Engvall, R.**, Interaction of fibronectin with antibodies and collagen in radioimmunoassay, *Biochim. Biophys. Acta*, 534, 210, 1978.
77. **Risteli, J., Risteli, L., Rohde, H., and Timpl, R.**, Radioimmunoassays for circulating basement membrane proteins: laminin and 7S collagen, *Fresenius Z. Anal. Chem.*, 301, 122, 1980.
78. **Tuderman, L. and Kivirikko, K. I.**, Immunoreactive prolyl hydroxylase in human skin, serum and synovial fluid: changes in the content and components with age, *Eur. J. Clin. Invest.*, 7, 295, 1977.
79. **Kuutti-Savolainen, E. R. and Kero, M.**, Enzymes of collagen biosynthesis in skin and serum in dermatological diseases. I. Enzymes of the skin, *Clin. Chim. Acta*, 96, 43, 1979.
80. **Kuutti-Savolainen, E. R.**, Enzymes of collagen biosynthesis in skin and serum in dermatological diseases. II. Serum enzymes, *Clin. Chim. Acta*, 96, 53, 1979.
81. **Kuutti-Savolainen, E. R., Kivirikko, K. I., and Laitinen, O.**, Serum immunoreactive prolyl hydroxylase in inflammatory rheumatic diseases, *Ann. Rheum. Dis.*, 39, 217, 1980.
82. **Uitto, J., Bauer, E. A., and Eisen, A. Z.**, Scleroderma. Increased biosynthesis of triple helical type I and type III procollagens associated with unaltered expression of collagenase by skin fibroblasts in culture, *J. Clin. Invest.*, 64, 921, 1979.
83. **Taubmann, M. B., Kammerman, S., and Goldberg, B.**, Radioimmunoassay of procollagen in serum of patients with Paget's disease of bone, *Proc. Soc. Exp. Biol. Med.*, 152, 284, 1976.
84. **Tuderman, L., Risteli, J., Miettinen, T. A., and Kivirikko, K. I.**, Serum immunoreactive prolyl hydroxylase in liver disease, *Eur. J. Clin. Invest.*, 7, 537, 1977.
85. **Kuutti-Savolainen, E. R., Risteli, J., Miettinen, T. A., and Kivirikko, K. I.**, Collagen biosynthesis enzymes in serum and hepatic tissue in liver disease. I. Prolyl hydroxylase, *Eur. J. Clin. Invest.*, 9, 89, 1979.
86. **Risteli, J., Draeger, V. E., Regitz, G., and Neubauer, H.**, Increase in circulating basement membrane protein in diabetic rats and effects of insulin treatment, *Diabetol.*, in press, 1982.
87. **Mahieu, P. M., Lambert, P. H., and Maghuin-Rogister, G. R.**, Primary structure of a small glycopeptide isolated from human glomerular basement membrane and carrying a major antigenic site, *Eur. J. Biochem.*, 40, 599, 1973.
88. **Timpl, R., Wick, G., and Granditsch, G.**, Reticulin autoantibodies in childhood coeliac disease not directed against type III collagen, *Clin. Exp. Immunol.*, 28, 546, 1977.
89. **Wick, G. and Timpl, R.**, Study on the nature of the Goodpasture antigen using a basement membrane producing mouse tumour, *Clin. Exp. Immunol.*, 39, 733, 1980.
90. **Nist, C., von der Mark, K., Hay, E. D., Olsen, B. R., Bornstein, P., Ross, R., and Dehm, P.**, Location of procollagen in chick corneal and tendon fibroblasts with ferritin-conjugated antibodies, *J. Cell Biol.*, 65, 75, 1975.
91. **Gunson, D. E., Arbogast, B., and Kefalides, N. A.**, Rat parietal yolk sac basement membrane. An investigation of the antigenic determinants using a radioimmunoassay, *Immunology*, 31, 577, 1976.
92. **Arbogast, B. W., Gunson, D. E., and Kefalides, N. A.**, The role of hydroxylation of proline in the antigenicity of basement membrane collagen, *J. Immunol.*, 117, 2181, 1976.
93. **Risteli, L., Rohde, H., Krieg, T., Müller, P. K., and Timpl, R.**, in preparation.
94. **Crouch, E., Balian, G., Holbrook, K. Duksin, D., and Bornstein, P.**, Amniotic fluid fibronectin, characterization and synthesis by cells in culture, *J. Cell Biol.*, 78, 701, 1978.

95. **Crouch, E., Sage, H., and Bornstein, P.,** Structural basis for apparent heterogeneity of collagen in human basement membranes: type IV procollagen contains two distinct chains, *Proc. Natl. Acad. Sci. U.S.A.*, 77, 745, 1980.
96. **von der Mark, K. and von der Mark, H.,** Immunological and biochemical studies of collagen type transition during in vitro chondrogenesis of chick limb mesodermal cells, *J. Cell Biol.*, 73, 736, 1977.
97. **Hasilik, A. and Neufeld, E. F.,** Biosynthesis of lysosomal enzymes in fibroblasts; synthesis as precursors of higher molecular weight, *J. Biol. Chem.*, 255, 4937, 1980.
98. **Choi, M. G. and Hynes, R. O.,** Biosynthesis and processing of fibronectin in NIL. 8 hamster cells, *J. Biol. Chem.*, 254, 12050, 1979.
99. **Burnett, W., Eichner, R., and Rosenbloom, J.,** Correlation of functional elastin messenger ribonucleic acid levels and rate of elastin synthesis in the developing chick aorta, *Biochemistry*, 19, 1106, 1980.
100. **Dessau, W., Jilek, F., Adelmann, B. C., and Hörmann, H.,** Similarity of antigelatin factor and cold insoluble globulin, *Biochim. Biophys. Acta*, 533, 227, 1978.
101. **Jilek, F. and Hörmann, H.,** Cold-insoluble globulin (fibronectin). IV. Affinity to soluble collagens of various types, *Hoppe-Seylers Z. Physiol. Chem.*, 359, 247, 1978.
102. **Dessau, W., Adelmann, B. C., Timpl, R., and Martin, G. R.,** Identification of the sites in collagen α chains that bind serum anti-gelatin factor (cold-insoluble globulin), *Biochem. J.*, 169, 55, 1978.
103. **Jilek, F. and Hörmann, H.,** Fibronectin (cold-insoluble globulin). VI. Influence of heparin and hyaluronic acid on the binding of native collagen, *Hoppe-Seylers Z. Physiol. Chem.*, 360, 597, 1979.
104. **Rennard, S. I., Berg, R., Martin, G. R., Foidart, J. M., and Gehron Robey, P.,** Enzyme-linked immunoassay (ELISA) for connective tissue components, *Anal. Biochem.*, 104, 205, 1980.
105. **Engvall, E., Ruoslahti, E., and Miller, E. J.,** Affinity of fibronectin to collagens of different genetic types and to fibrinogen, *J. Exp. Med.*, 147, 1584, 1978.
106. **Belanger, L.,** Alternative approaches to enzyme immunoassays, *Scand. J. Immunol.*, 8 (Suppl. 7), 33, 1978.
107. **Soini, E. and Hemmila, J.,** Fluoroimmunoassay: present status and key problems, *Clin. Chem.*, 25, 353, 1979.

Chapter 14

ENZYME LINKED IMMUNOASSAY (ELISA) FOR CONNECTIVE TISSUE PROTEINS: TYPE I COLLAGEN

Stephen I. Rennard, George R. Martin, and Ronald G. Crystal

TABLE OF CONTENTS

I. INTRODUCTION

The classical approaches for the quantitation of collagen take advantage of unique features of collagen such as the presence of hydroxyproline[1] or the sensitivity of the protein to bacterial collagenase.[2] However, since these features are common to the different molecular species of collagen,[3] other methods are needed to selectively measure each collagen type. One approach to this problem is to capitalize on the fact that the different collagen types are immunologically distinct and antibodies with high binding affinity specific for each collagen type can be produced.[4] Such antibodies can be utilized to quantify human type I collagen in an enzyme-linked immunoassay (ELISA)[5,6] which is sensitive, specific and easily performed. This assay offers several practical advantages over the more familiar radioimmunoassays.[7,8]

There are two basic types of ELISA tests, direct and inhibition. Both utilize the sensitivity of an enzyme reaction to measure the antigen antibody interaction. Since very sensitive enzyme-substrate-product combinations are available, the test is very powerful.

In the direct test, this is accomplished by first binding the sample antigen to an insoluble support. An enzyme linked antibody is incubated with the insolubilized antigen and will bind specifically to it. The enzyme is thus bound through the antibody to the insoluble support in proportion to the amount of antigen present. Unbound material is removed by rinsing, and the bound enzyme is then measured by adding a substrate specific for the enzyme and quantifying its reaction product.

In the inhibition test, the antigen is bound to an insoluble support as a means of separating bound and free antibody. In contrast to the direct test, the sample antigen (inhibitor) is in solution and reacts with the specific antibody in solution. This interaction between the sample antigen and the specific antibody causes a proportional reduction in the amount of specific antibody which can bind to the antigen on the insoluble support. By incubating the mixture of sample antigen and antibody with the antigen on the insoluble support, residual free antibody is bound to the insoluble support, and separated from the antigen antibody complexes by rinsing. The amount of antibody bound to the support is then measured with a second antibody directed against the first antibody. The amount of second antibody is determined by virtue of having an enzyme conjugated to it. Enzyme substrate is added and the product of the enzyme reaction is measured. Since the inhibitor reduces the binding of specific antibody, less second antibody enzyme conjugate will bind and less product will result. By comparing the amount of product with standard antigen samples, unknowns can be measured.

Each type of test has advantages and disadvantages, and both will be outlined. However, the inhibition test is more applicable to the measurement of collagen types, and this will be described in detail for human Type I collagen.

II. MATERIALS

A. Antigen

Human Type I collagen can be extracted from spleen with acetic acid or from skin by incubation with pepsin; these extracts are then purified by amino acid analysis and gel electrophoresis.

B. Antisera

Goats are immunized with 100 μg of Type I collagen in complete Freund's adjuvant injected intramuscularly. At three week intervals booster injections of 100 μg of the antigen are given in incomplete Freund's adjuvant and sera collected weekly after the first booster injection.

C. Solid Support

All immunoassays must, at some point, separate the unbound antibody from that bound to antigen. One of the powerful advantages of the ELISA test is the binding of the reactants to an insoluble support. In this context, a single 96 well microtiter plate can be emptied and rinsed in one operation. This is the equivalent of 96 centrifugations in an analogous radioimmunoassay and thus represents a great simplification in the handling of samples. In order to bind the collagen to an insoluble support, we capitalize on the nonspecific adsorption of proteins to the walls of individual wells of polystyrene plastic microtiter plates (MicroELISA Plates, Dynatech Laboratories, Alexandria, VA) by diluting the desired amount of collagen from a stock solution of 1 mg/mℓ in 0.5 *M* acetic acid into Voller's buffer (1.59 gm Na_2CO_3, 2.93 gm $NaHCO_3$, 0.2 gm NaN_3 per 100 mℓ).[11] The diluted collagen solution is then immediately pipetted into the microtiter plate (0.2 mℓ/well) and allowed to adsorb overnight at 4°C. Once coated, plates are stable for as long as one year if kept at 4°C in a humid environment. Nonspecific adsorption in subsequent steps is prevented by including 0.05% Tween-20 (polyoxysorbitan monolaurate) in all subsequent buffers. This detergent does not elute the collagen from the surface of the plastic well, nor does it interfere with specific antigen antibody interactions.[12]

Since the inhibition test utilizes an anti-IgG with bound enzyme as a "second antibody" to bind to the specific antibody, it is necessary to standardize batches of enzyme-anti-IgG-conjugate. This is accomplished by adsorbing goat IgG to plastic wells by diluting goat IgG (Miles Laboratories, Elkart, IN) to 500 ng/mℓ in Voller's buffer and allowing it to adsorb to polyvinyl plastic plates (Dynatech Laboratories, Alexandria, VA).[11] (The further steps necessary to standardize the enzyme-anti-IgG conjugate are described below.)

D. Pipets

Another convenience of the microtiter plate method is that the many repetitious pipetting steps necessary for immunoassays can be simplified. Positive displacement multichannel pipettors (Titertek, Flow Laboratories, McLean, VA), spaced to fit the distance between the rows of the microtiter plates, are used. These pipettors are accurate within 5%, and greatly simplify the execution of the ELISA. Dilutors which depend on capillary action cannot be used because of the detergent included in most buffers.

E. Enzyme

Measurement in all ELISA tests depends on the presence of an active enzyme bound to one of the components of the antigen antibody reaction.[5,6] It is the product of the enzyme reaction which is actually measured; this is analogous to the measurement of radioactivity in the radioimmunoassay to quantitate the antibody bound. In the inhibition test, a "double antibody" method is used. This greatly simplifies the preparation of reagents when several different ELISAs are being designed since the same second antibody can be used for different specific antisera and raised in the same species. For the measurement of Type I collagen, anti-type I antibody is raised in goats and the amount of goat antibody bound to the wall of the plastic microtiter well is determined by adding horseradish peroxidase conjugated rabbit anti-goat IgG (Miles Laboratories, Elkhart, IN). The optimal concentration of each lot of conjugate for the final assay is first determined using the method of Voller.[11]

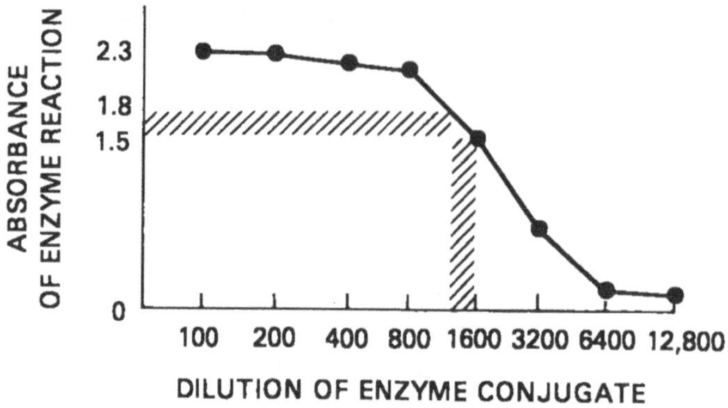

FIGURE 1. Titration of enzyme-linked rabbit anti-goat IgG conjugate. To define the titer and enzyme activity of each lot of peroxidase conjugated rabbit anti-goat IgG, microtiter wells are coated with 0.1 μg of goat IgG. After rinsing, serial dilutions of enzyme-antibody conjugate are added and the amount of colored product produced is quantitated by measuring the absorbance of the product of 480-nm. The absorbance is then plotted on the vertical axis and the dilution of conjugate on the horizontal axis. The concentration producing the desired amount of color is shown by the cross hatched area.

Briefly, polyvinyl microtiter plates are coated with 100 ng/well of goat IgG. After adsorption, each well is rinsed three times with phosphate (0.02 *M*) buffered 0.9% saline pH 7.2 and supplemented with Tween-20 (PBST) by filling each well and allowing it to remain for three min at room temperature. This removes unbound material and coats each microtiter well with Tween to prevent further nonspecific adsorption.

Serial dilutions from 1:100 to 1:12,800 of the enzyme linked-rabbit anti-goat IgG are then prepared in PBST, and 0.2 mℓ are pipetted into the wells of the microtiter plate. After a 90 min incubation at room temperature, unbound enzyme-antibody conjugate is rinsed away with PBST as described above. The substrate for the enzyme (see below) is added and the reaction allowed to proceed for 45 min. The product of the enzyme reaction is measured and is plotted against the dilution of the enzyme linked rabbit anti-goat IgG (Figure 1). From this plot, the appropriate dilution of the enzyme-antibody conjugate can be chosen (Figure 1, shaded region). It is important to utilize the steep range of this curve, where the amount of product most accurately reflects the amount of enzyme and therefore of bound antibody.

F. Substrate Reaction

To measure the amount of enzyme adsorbed, a colorless substrate is added. This is converted into a chromophore by the enzyme reaction and the product is quantitated spectrophotometrically. The colorless substrate, *O*-phenylenediamine[13] (Aldrich, Milwaukee, Wis.) is always prepared just prior to use. It is first dissolved in methanol at a concentration of 10 mg/mℓ and then diluted 1:100 with deionized water. Next, 0.01 mℓ of 30% H_2O_2 is added per 100 mℓ and the substrate is gently mixed. At room temperature, in the absence of enzyme, the substrate solution remains colorless for several hours. To measure the amount of enzyme bound to the surface of a microtiter well, 200 μℓ of this substrate solution is added to each well. This is then incubated at room temperature until, in the darkest well, about one half of the substrate is utilized and converted to a yellow brown product by the enzyme. This usually takes 30 to 90 min, and with experience, the time to terminate the reaction can easily be judged visually. The reaction is stopped by the addition of 50 μℓ of 8N H_2SO_4. The "developed" plates can be stored for several weeks at 4°C without further alteration in the color.

G. Spectrophotometer

Since the oxidation product of *O*-phenylenediamine is yellow brown, in contrast to the clear starting solution, the product can be quantified at 480 nm. For our studies, we utilize a specially designed spectrophotometer that reads the plates *in situ*[14] (NIH Instrumentation Laboratory, Bethesda, MD). Similar machines are commercially available.

H. Data Analysis

Since there is no theoretical difference between an enzymatically generated product due to binding of antibody and bound radioactive label,[15] the data can be analyzed with a program originally designed for radioimmunoassays using the method of Rodbard.[15] A computer program facilitates statistical analysis, but manual graphing allows adequate quantitation.

III. INHIBITION ELISA FOR TYPE I COLLAGEN

The inhibition type of the ELISA test is relatively unaffected by minor contaminants in either antigen or antibody. Since contaminants are a major problem in both collagen preparations and anticollagen antibodies, the inhibition test is best suited to measure Type I collagen. To perform an inhibition ELISA for Type I collagen, collagen is allowed to adsorb to the surfaces of each well of the microtiter plate (Figure 2, step 1a). In separate containers, goat anti-type I collagen antiserum is mixed with various amounts of soluble Type I collagen (Figure 2, step 1b). This allows the soluble Type I collagen to bind to the antibody and, when the mixture is subsequently transferred to the previously coated plastic well (Figure 2, step 2), the prior interaction of antibody with the soluble collagen prevents some of the antibody from binding to the collagen coating the well. The wells are then rinsed, leaving behind only the antibody bound to the collagen coating the well. Then rabbit anti-goat IgG with covalently coupled peroxidase is added (Figure 2, step 3). This antibody will bind in proportion to the goat anti-collagen adherent to the wall of the well of the microtiter plate. After rinsing away the unbound antibody-enzyme conjugate, the substrate for the enzyme is added (Figure 2, step 4). The enzyme converts this substrate into a product which is easily measured. In summary, prior incubation of the anticollagen antibody with soluble collagen (Figure 2, step 1a) blocks binding of the antibody to the collagen on the surface of the microtiter well (Figure 2, step 2) resulting in less antibody-enzyme bound (Figure 2, step 3) and therefore, less enzyme reaction product (Figure 2, step 4). The amount of collagen in various samples can be determined from a graph of enzyme product (absorbance) vs. the amount of collagen in comparison to known standards.

The optimal amount of collagen used for coating the wells and concentration of specific antibody used to incubate with sample are determined empirically. It is necessary to find the best combination since the sensitivity and reproducibility of the test depend on these parameters. The following method is used to determine the best combination. First, serial dilutions of collagen, from 32 μg/mℓ to 0.25 μg/mℓ are made in Voller's buffer.[19] Each microtiter plate contains twelve horizontal rows and eight vertical columns, i.e., each row contains eight wells. Aliquots (0.2 mℓ) of

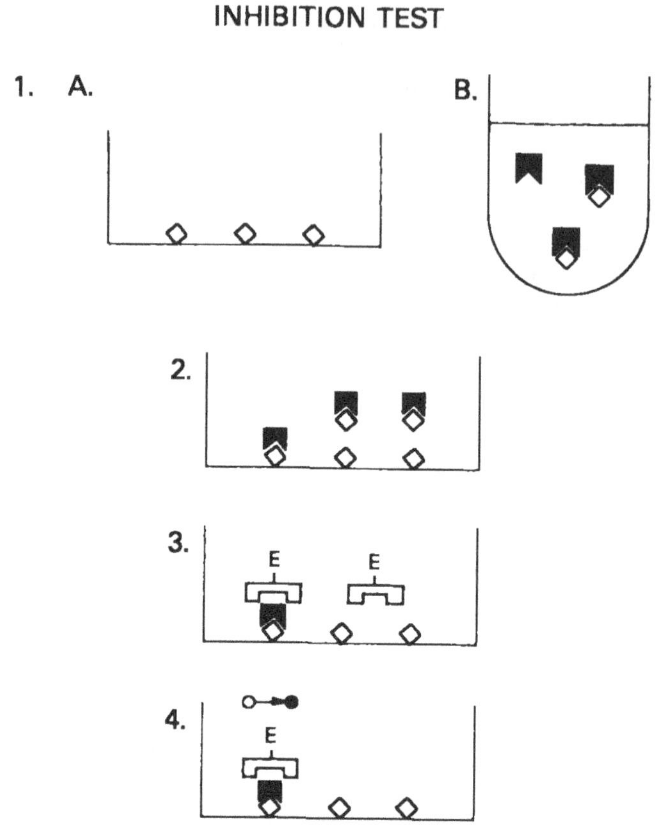

FIGURE 2. Schematic procedure of an ELISA test for human Type I collagen. Between each step reagents not bound to the surface of the plastic well are rinsed away as described in methods. Step 1a: The surface of a well of a plastic microtiter plate is allowed to adsorb collagen (◇) by incubation in the cold in Voller's buffer. Step 1b: In a separate microtiter well, goat anti-type I collagen antibody (■) is incubated with and allowed to bind to Type I collagen in buffer containing Tween to prevent non-specific adsorption. Step 2: The mixture of anti-collagen antibody and collagen is transferred to the collagen coated plastic well and free antibody (■) will bind to the plastic well, but antibody previously bound to collagen in solution (■) is prevented from binding. Step 3: Rabbit anti-goat IgG coupled to peroxidase (E) is added and binds to the goat anticollagen present in the well. Step 4: Substrate (○) is added and is converted to product (●) by the enzyme. The product is then measured.

each dilution are then pipetted into polystyrene microtiter wells with each of the eight wells in a given row receiving the same dilution of collagen. Thus, the first row of wells receive 6.4 μg/well, the second row 3.2 μg/well, etc. The plate is then incubated overnight to allow adsorption. After rinsing away unbound collagen with PBST, 0.2 mℓ aliquots of serial dilutions of antisera prepared in PBST are added followed by incubation for 30 min at room temperature. Unbound antibody is rinsed away with PBST, and the second antibody, rabbit anti-goat coupled to peroxidase, is added and incubated for 90 min. The amount of this second antibody bound to the plate is quantified by adding enzyme substrate (see above). After stopping the reaction with 8 N H_2SO_4, the absorbance in each well is read spectrophotometrically.

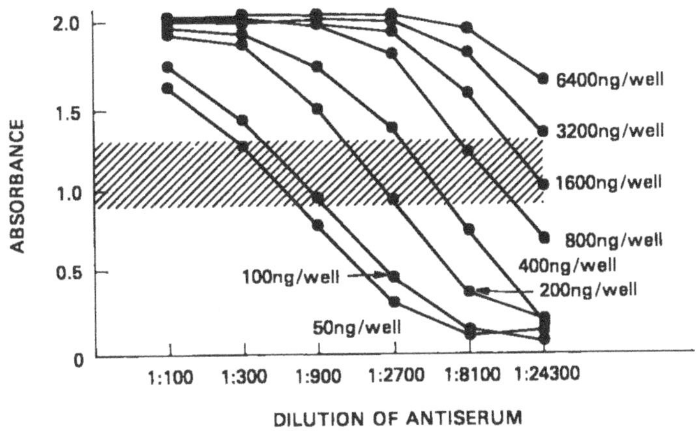

FIGURE 3. Titration of Type I collagen coating and goat anti-type I collagen antibody. Each well in a row of microtiter plates is coated with the same amount of collagen, each row receiving one concentration in a series of dilutions. After adsorption, unbound collagen is rinsed away and serial dilutions of goat anti-collagen antibody are added so that each column receives the same dilution. After binding, the second antibody with its enzyme conjugate is added, followed by substrate. The product of the enzyme reaction is then measured at 480-nm and plotted vs. the dilution of antiserum. This produces a family of curves. The "workable" combinations are indicated by the shaded area.

In general, an increase in both the concentration of specific antibodies and the amount of collagen used to coat the plate results in a greater amount of enzyme bound and therefore of color produced (Figure 3). Any combination of collagen coating and antiserum concentration which produces 0.8 to 1.3 absorbance units will produce a "workable" assay. However, the "best" of these assays should be selected empirically by testing all of the reasonable combinations in an inhibition test. In this way, the sensitivity and reproducibility of each pair of conditions is determined, and the most appropriate test can be chosen.

Empirical testing of a series of conditions is done as follows. Rows of microtiter plates are coated with collagen in Voller's buffer as before. Next, known dilutions of human Type I collagen in PBST are prepared starting with 20 μg/mℓ and making serial dilutions at a ratio of 1 : 3. A convenient method is to add 0.165 mℓ of a solution of 20 μg/mℓ to each well in the first column of a round bottomed microtiter plate and then make the serial dilutions by pipetting 0.055 mℓ into 0.11 mℓ of PBST. A multichannel pipettor can be used to perform this step. Next, 0.11 mℓ of antiserum at a two-fold higher concentration than determined above is added to each row. It is necessary to use a two-fold higher concentration of antiserum since the addition of antiserum to sample further dilutes the antibody. In this way, a "standard curve" will be generated for each antibody concentration. Once the dilutions are completed, the collagen coated microtiter plate and the mixtures of collagen and anti-collagen antiserum (in a separate microtiter plate) are incubated overnight at 4°C. The collagen coated plate is then rinsed with PBST to remove any unbound collagen, and to coat the remaining surface of the plastic well to eliminate further nonspecific adsorption. Next, 0.2 mℓ of the collagen-antiserum mixture is transferred to each well of the coated microtiter plate. Each row of the plate will thus contain a given amount of collagen and a constant amount of added anti-collagen antiserum. The presence of soluble collagen at a series of concentrations in the wells of a row with a constant concentration of antibody will prevent some of the antibody from binding to the collagen coated surface of the well. After 30 min, the plates are rinsed with PBST and the antibody-enzyme conjugate is added and incubated for 90 min to allow for binding of the second antibody. Unbound antibody is then rinsed away, and enzyme substrate is added as described above. A "standard curve" is produced for each row, one for each antigen-antibody combination. Note that the amount of colored product generated depends on the amount of enzyme present. This, in turn, depends on the binding of the second antibody to the anticollagen antibody that was not blocked by the soluble collagen and thus was free to bind to the collagen coating the plastic well. Thus the color produced varies inversely with the amount of collagen in the sample.

Some concentrations of antibody are used more than once to screen for appropriate assay conditions since each antibody concentration can be used for different amounts of collagen coating. Data obtained in such a manner are shown in Figure 4 which illustrates standard curves for three different conditions summarized in Table 1. Also shown (shaded areas) are the computer calculated standard errors for each curve obtained by the method of Rodbard.[15] Note that decreasing the amount of antibody decreases the amount of antigen required to inhibit the assay by 50% unless the limit defined by the binding constant of the antiserum for collagen is reached. However, the reproducibility of the test also decreases with decreasing antiserum concentration and the slope parameter (see Table 1) gets "flatter" with a loss in accuracy.

In general, the best test represents a balance between accuracy, reproducibility and the amount of antibody present; these concepts have been discussed theoretically by Rodbard.[15,16] In the example given for the Type I collagen ELISA, optimal conditions included 400 ng of collagen per well and a 1:800 dilution of antibody; these conditions allowed accurate assays sensitive to 34 ng/mℓ (Table 1).

In summary, a human Type I collagen inhibition ELISA test can be carried out as follows (Figure 2)

Step 1. A. Plates are coated with 0.4 μg Type I collagen in Voller's buffer

B. The sample to be assayed or a sample of known collagen content to be used as standard is mixed with an equal volume of 1:800 dilution of goat anti-collagen antibody to a final volume of 0.22 mℓ and incubated overnight. Parallel samples of known collagen content are used as standards

Step 2. The collagen coated plate is rinsed and 0.2 mℓ of the mixture of sample and anti-collagen antibody is transferred to the coated plate and incubated for 30 min.

Step 3. The plate is rinsed and 0.2 mℓ of 1:1200 dilution of antibody to goat IgG made in rabbits and coupled to peroxidase is added and incubated for 90 min.

Step 4. The plate is rinsed a final time and enzyme substrate is added and incubated for about 45 min. The reaction is stopped with 8N H_2SO_4 and the product of the enzyme reaction is measured.

Table 1

Antigen coating each well (ng)	Antibody dilution	Slope parameter∮	Midrange of test ng/mℓ‡	Minimal detectable amount of type 1 collagen (ng/mℓ)∯
100	1:100	2.07	856	412
	1:200	1.38	953	277
	1:400	1.13	620	188
	1:800	.55	330	79
200	1:100	2.44	881	379
	1:200	1.36	434	65
	1:400	1.16	648	84
	1:800	.93	577	94
400	*1:200	1.28	592	81
	1:400	1.06	335	38
	*1:800	.96	406	34
	1:1600	.68	239	10
	*1:3200	.84	268	22
	1:6400	1.01	198	90
800	1:800	.74	700	6
	1:1600	.70	351	14
	1:3200	.74	273	17
	1:6400	.77	88	298
	1:12800	1.04	99	23
	1:25600	1.63	235	180
1600	1:800	.86	229	11
	1:1600	.64	383	5
	1:3200	.72	313	16
	1:6400	.53	193	16
	1:12800	.43	87	27
3200	1:1600	.60	572	7
	1:3200	.82	781	236
	1:6400	1.26	537	238
	1:12800	.82	563	72
6400	1:3200	.69	381	29
	1:6400	.75	238	76

Note: Test parameters for an entire series of "workable" Type I collagen ELISAs. A standard curve is generated for each workable antigen-antibody combination. These standard curves are then analyzed to determine the slope parameter (steepness), the midrange, and the sensitivity of each test. Data for all tests studied is shown. Those tests indicated by an asterisk (*) are shown graphically in Figure 4.

† = antibody diluted into PBST and mixed with known concentrations of human Type I collagen. The same serum was used for all tests. Different antisera, even different antiserum from the same animal, require a separate standardization procedure.

∮ = the slope parameter is calculated by the method of Rodbard.[15,16] This reflects the steepness of the test and is the calculated value for $-(dy/dx)$ at the steepest portion of the test where y is the enzyme product and x is the logarithm of the collagen concentration. Steeper curves are more accurate near the midrange of the test since smaller differences in collagen concentration result in larger differences in enzyme product.

‡ = the midrange of the test is that collagen concentration required to reduce the antibody binding by 50% when compared to maximal inhibition; this is also the condition under which the slope parameter is greatest.

∯ = the minimal detectable amount of collagen is defined as the lowest concentration of collagen that will produce a statistically significant inhibition in the ELISA.

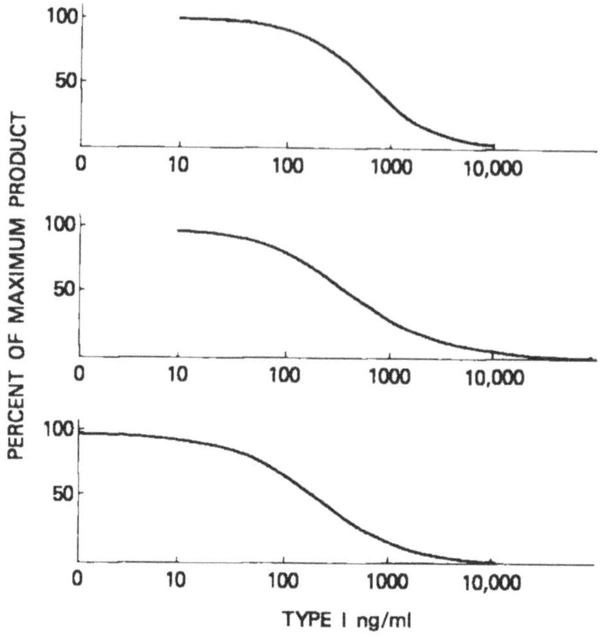

FIGURE 4. Standard curves for selected antigen-antibody combinations. Wells of a microtiter plate, coated with 0.4 μg of Type I collagen, are used to generate standard curves. The enzyme product is plotted on the vertical axis (expressed as % of maximum) and the collagen concentration used for inhibition is plotted on the horizontal axis. The upper panel shows the standard curve generated with a 1:200 dilution of antibody, the middle panel shows the standard curve generated with a 1:800 dilution of antibody, and the lower panel was generated by a 1:3200 dilution of antibody. The shaded area indicates the confidence interval for each curve.

IV. GENERAL CONSIDERATIONS OF TEST DESIGN

As described above, quantitative ELISA tests can be done by direct or inhibition methods.

A. Direct

In the direct test (Figure 5), the wells of a microtiter plate are coated with antibody specific for the antigen to be measured. Samples containing this antigen are then added and any antigen present will be bound to the specific antibody coating the microtiter plate. After rinsing, specific collagen-antibody-conjugate is added and is allowed to bind to the antigen on the microtiter well. Finally, substrate for the enzyme reaction is added and the product of this reaction is measured. The amount of product is dependent on the amount of adsorbed antigen (Figure 6a). Using this type of test, very sensitive assays have been designed.[17,18] Unfortunately, if a contaminant is present in the preparation of antigen used to prepare the antiserum, and if this contaminant is also present in the sample to be tested (Figure 7a, 8a), there will be no way to distinguish between "true" antigen and "contaminant" antigen.

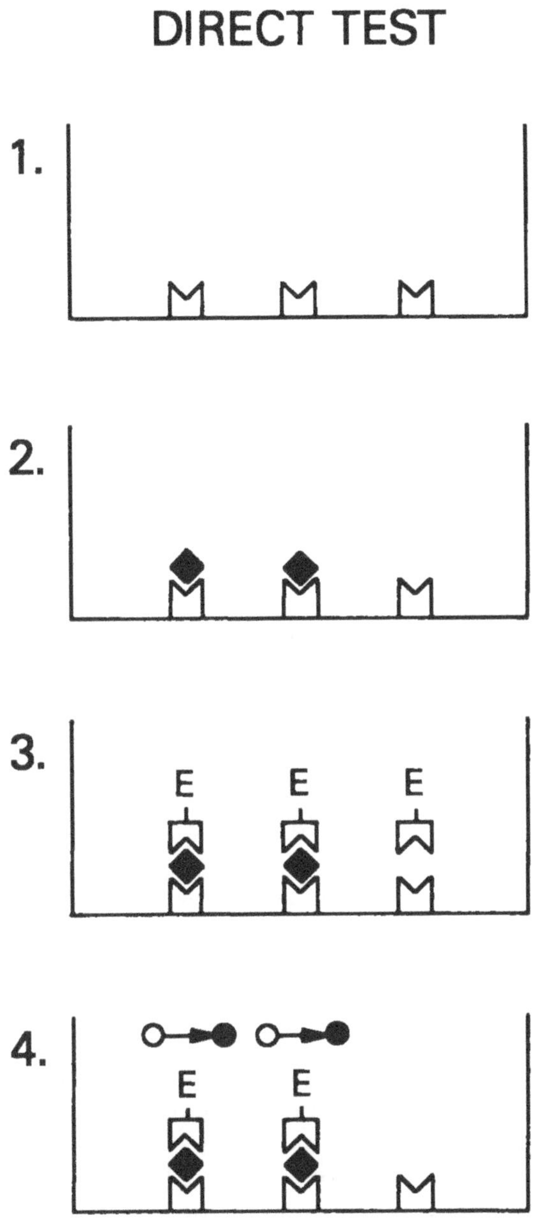

FIGURE 5. Schematic outline for a direct ELISA test. Step 1. Specific antibody (M) is adsorbed to the surface of a microtiter well. Step 2. Soluble antigen (◆) is added and binds to the insolubilized antibody. Step 3. Enzyme conjugated (E) specific antibody is added and binds to the antigen. Step 4. Enzyme substrate (○) is added and converted to product (●) which is measured.

B. Inhibition

Inhibition tests rely on the ability of the test antigen to bind to specific antibodies and therefore block the binding of these antibodies to adsorbed antigen (Figure 2). By combining with some of the antibody, the inhibitor reduces the amount that can interact with the antigen coating the microtiter plate (Figure 2). The amount of bound

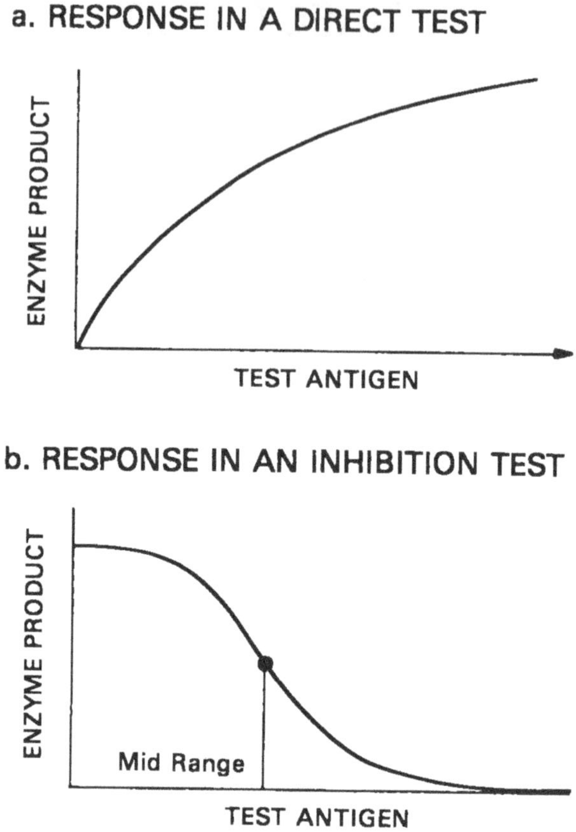

FIGURE 6. Schematic response for ELISA tests. The theoretical response of ELISA tests are shown. a) Direct Test: increasing amounts of antigen result in increasing amounts of enzyme product. b) Inhibition Test: increasing amounts of antigen result in a decrease in the amount of enzyme product.

antibody is then measured with an anti IgG-enzyme conjugate by addition of enzyme substrate (Figure 6b). In these assays the sensitivity is limited by the affinity of the antibody for the antigen[15] and these tests are not as sensitive, theoretically, as direct tests. Nevertheless, inhibition tests offer some advantages in terms of specificity. Contaminants, when present, produce only partial inhibition no matter how much is present (Figure 7b, 8b). This is true because the contaminant only inhibits the binding of contaminant antibody to contaminant antigen, but does not affect the interaction of true antibody and true antigen. Thus, when an unknown sample produces an inhibition curve that is parallel to the standard curve, it implies close antigenic similarity between the sample and the unknown. If the contaminant antigen resembles the true antigen so closely that it can affect their interaction, then the above argument does not apply. However, this situation is very rare, even for the collagen types.

C. Equilibrium Vs. Nonequilibrium

In an inhibition ELISA, there is a definite advantage to utilize nonequilibrium conditions. If antibody and antigen are first mixed in the antigen coated microtiter well, simple competition occurs between the test antigen and the bound antigen for

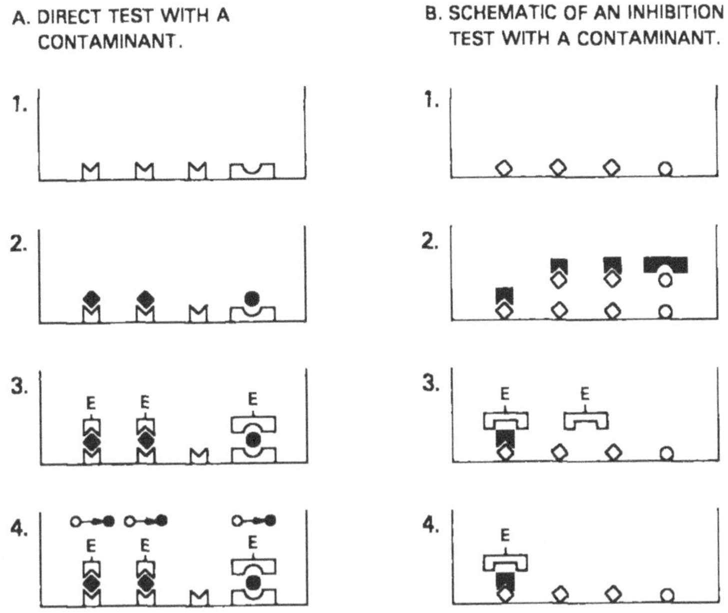

FIGURE 7. Schematic for ELISA tests in the presence of a contaminant. Contaminating antigen (○) and antibody (▯) are contrasted with "true" antigen (◇) and antibody (▯). A) Direct test. Step 1. Both "true" antibody (▯) and "contaminant" antibody (▯) are adsorbed to the solid phase. Step 2. A sample with both "true" antigen (◆) and "contaminant" antigen (●) is added and both bind. Step 4. Substrate is added and converted to product. B) Inhibition test. Step 1. "True" antigen (◇) and "contaminant" antigen (○) are adsorbed to the surface of a microtiter plate. Step 2. "True" (▮) and "contaminant" (▮) antibody is added to soluble antigen. "True" antigen blocks the binding of "true" antibody (▮) and "contaminant" antigen blocks the binding of "contaminant" antibody to the surface of the microtiter well. Step 3. Enzyme conjugate is added. Step 4. Substrate (○) is added and converted to product (●).

antibody. However, if the antibody and the test antigen are mixed and incubated prior to addition to the antigen coated well, binding of the test antigen inhibitor is favored. Although equilibrium will eventually be reached, little competition will take place if the incubation on the coated well is brief. For collagen types, an increase in sensitivity of 5 to 10-fold results when the test antigen and antibody are incubated overnight followed by a 30 min incubation on the coated plate.[19] Shorter incubations result in an undesirable decrease in the reproducibility of the test.[19]

D. Choice of Conjugate

A large number of enzymes and coupling reactions have been used for ELISA tests.[16] For inhibition tests, there is little advantage of one over another since the sensitivity of the test is limited by the affinity of the antibody for the antigen, and not by the enzyme chosen. For the direct test, the activity of the enzyme and the sensitivity of the method used to detect the reaction product both affect the sensitivity of the test. Increasing either will result in an increase in sensitivity of the test (reflected by the initial slope of the response as shown in Figure 2a). Taking advantage of this, extremely sensitive tests have been designed in which radioactive[17] or fluorescent[18] products are generated.

Horse radish peroxidase and alkaline phosphatase-linked second antibodies are

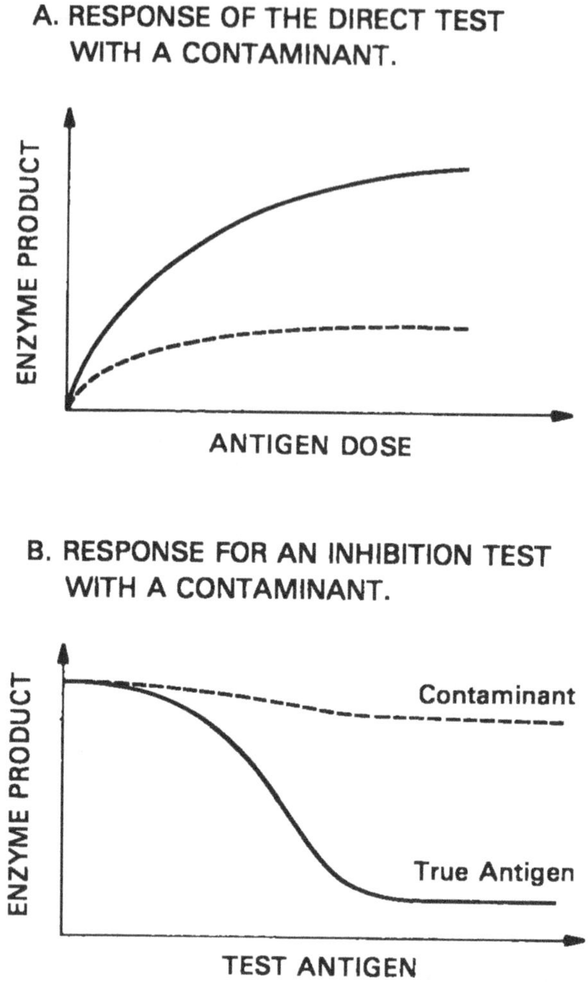

FIGURE 8. Theoretical response for ELISA tests with a contaminant present. A. Direct test. The response to increasing both the true (——) and the contaminant (– – –) is a similar increase in the measured enzyme product. B. Inhibition test. The response to increasing true antigen (——) is a nearly complete inhibition of enzyme product, in contrast to increasing amounts of contaminant (– – –) which can produce only limited inhibition of enzyme product.

available commercially. Both are convenient and are acceptable for use in inhibition tests. Peroxidase has the advantage of producing a darker color allowing ready visual inspection of the reaction. In addition, horse radish peroxidase can be coupled to IgG by a two step reaction;[20] this results in lower molecular weight conjugates which slightly improves the performance of the test.

E. Insoluble Support

While a number of insoluble support systems have been used[6] and are presumably equivalent, the microtiter plate method described by Voller[11] has several advantages. These plates utilize small volumes and can be used with multichannel pipettors and semiautomated spectrophotometers thereby greatly increasing the ease of performing many assays simultaneously. The flat bottomed polystyrene plates are acceptable for many different antigens and allow in situ reading in the ELISA spectrophotometers.

For some antigens, problems with the insoluble support can arise. The binding of antigen to the plastic plates is through hydrophobic adsorption. Although collagen, as well as many other large proteins, present no special problems in this regard, some hydrophilic molecules such as proteoglycan, and to some extent IgG, do not bind as readily to the polystyrene plates. These molecules adsorb somewhat better to a polyvinyl surface. This type of plate is also available commercially. Lastly, although collagen binds well at alkaline pH (as described above), some molecules bind better at neutral or acid pH.

V. CONCLUSION

The preparation of antigen and antibody is a critical part in all ELISA tests. In this context, purification of collagen often presents a problem, and contaminants, albeit to a small extent, are often present. Furthermore, collagen is a relatively poor immunogen. Thus, while it is possible to prepare a collagen sample with less than 2% contamination (a sample that will be judged pure by chemical criteria), a significant proportion of the antibodies obtained by immunizing an animal with such a preparation may be directed against various contaminating proteins. It is possible to purify specific antibodies by affinity chromatography on specific antigen columns, and particularly when the contaminants can be identified, by "cross adsorptions".[4] Monoclonal antibodies directed against a single antigenic determinant would obviate this problem. Nevertheless, by careful test design, the specificity of the ELISA can be optimized and very sensitive tests readily performed. These tests will greatly facilitate the measurement of connective tissue macromolecules in complex biological samples.

REFERENCES

1. **Green, H. and Goldberg, B.,** Collagen and cell protein synthesis by an established mammalian fibroblast line, *Nature*, 204, 347, 1964.
2. **Peterkofsky, B. and Diegelman, R.,** Use of a mixture of proteinase-free collagenases for the specific assay of radioactive collagen in the presence of other proteins, *Biochemistry*, 10, 988, 1971.
3. **Bornstein, P. and Sage, H.,** Structurally distinct kinds of collagen, *Ann. Rev. Biochem.*, 49, 957, 1980.
4. **Furthmayr, H. and Timpl, R.,** Immunochemistry of collagens and procollagens, *Int. Rev. Connect. Tissue Res.*, 7, 61, 1976.
5. **Engvall, E. and Pearlmann, P.,** Enzyme-linked immunoassay (ELISA) quantitative assay of immunoglobulin G, *Immunochemistry*, 8, 871, 1971.
6. **Schuurs, A. H. W. M. and Van Weeman, B. K.,** Enzyme-immunoassay, *Clin. Chem. Acta*, 81, 1, 1977.
7. **Taubman, M. D., Goldberg, B., and Sharr, C. J.,** Radioimmune assay for human procollagen, *Science*, 186, 1115, 1974.
8. **Von der Mark, K., Click, E. M., and Bornstein, P.,** The immunoassay of procollagen, *Arch. Biochem. Biophys.*, 156, 356, 1973.
9. **Epstein, E.,** $(\alpha 1(III))_3$ Human skin collagen released by pepsin digestion and preponderance in fetal life, *J. Biol. Chem.*, 249, 3225, 1974.
10. **Laemmli, U. K.,** Cleavage of structural proteins during the assembly of the head of bacteriophage T4, *Nature*, 227, 680, 1970.

11. **Voller, A., Birdwell, D., and Bartlett, A.,** Microplate enzyme immunoassays for the immunodiagnosis of virus infections, in *Manual of Clinical Immunology*, Rose, N. and Fishman, H., Eds. American Society for Microbiology, Washington, D.C., 506, 1976.
12. **Timpl, R.,** Immunological studies on collagen, in *Biochemistry of Collagen*, Ramachandran, G. N. and Reddi, A. H., Eds., Plenum Publishing, New York, 319, 1976.
13. **Walls, K. W. and Palmer, D. F.,** Procedural guide for ELISA microtitration test, Public Health Service, Center for Disease Control, U.S. Department of Health, Education and Welfare, Atlanta, 1977.
14. **Yolken, R. H., Kim, H. W., Clem, T., Wyatt, R. G., Kalica, A. R., Chanock, R. M., and Kapikian, A. Z.,** Enzyme-linked immuno-assay (ELISA) for detection of human reovirus-like agent of infantile gastroenteritis, *Lancet*, 2, 263, 1977.
15. **Feldman, H. A. and Rodbard, D.,** Mathematical theory of radioimmunoassay, in *Competitive Protein Binding Assays*, Daughaday, W. D. and Odell, W. A., Eds., Lippincott and Co., Philadelphia, 158, 1971.
16. **Rodbard, D.,** Statistical aspects of radioimmunoassay, in *Competitive Protein Binding Assays*, Daughaday, W. D. and Odell, W. A., Eds., Lippincott and Co., Philadelphia, 204, 1971.
17. **Harris, C., Yolken, R. H., and Hsu, I. C.,** Ultrasensitive enzymatic radioimmunoassay (USERIA) for the detection of human rotavirus and cholera toxin, *Proc. Natl. Acad. Sci. U.S.A.*, 76, 5336, 1979.
18. **Yolken, R. H., and Stopa, P. J.,** Enzyme-linked fluorescence assay: ultrasensitive solid-phase assay for detection of human rotavirus, *J. Clin. Microbiol.*, 10, 317, 1979.
19. **Rennard, S. I., Berg, R., Martin, G. R., Foidart, J. M., Gehron-Robey, P.,** Enzyme Linked Immunoassay (ELISA) for Connective Tissue Components, *Anal. Biochem.*, 104, 205, 1980.
20. **Avrameas, S. and Ternynck, T.,** Peroxidase labelled antibody and Fab conjugates with enhanced intracellular penetration, *Immunochemistry*, 8, 1175, 1971.

INDEX

A

B

C

D

E

F

G

H

I

K

L

M

N

O

P

S

T

U

V

Z